Symmetry through the Eyes of a Chemist

VCH Publishers

© 1987 VCH Publishers, Inc., New York, NY 10010-4606

Distribution:
Softcover edition of ISBN 0-89573-520-2 (hardcover).
Available in USA and Canada only: VCH Publishers, Suite 909, 220 East 23rd Street,
New York, NY 10010-4606, USA

ISBN 0-89573-681-0

István Hargittai
Magdolna Hargittai

Symmetry through the Eyes of a Chemist

1987

VCH
Publishers

New York

Prof. Dr. István Hargittai
Dr. Magdolna Hargittai
Hungarian Academy of Sciences
Structural Chemistry Research Group
H-1088 Budapest, VIII, Puskin utca 11–13
P.O.B. Budapest, Pf. 117, H-1431
Hungary

Library of Congress Cataloging-in-Publication Data

Hargittai, István
  Symmetry through the eyes of a chemist.

  Includes bibliographies and indexes.
  1. Molecular theory. 2. Symmetry (Physics)
I. Hargittai, Magdolna. II. Title.
QD461.H275 1986   541   86–24617
ISBN 0-89573-681-0

Preface

This relatively short book surveys the entire field of chemistry from the point of view of symmetry. We present many examples from chemistry as well as from other fields, in order to emphasize the unifying nature of the concepts of symmetry.

We hope that all those chemists, both academic and industrial, who take broader perspectives, will benefit from our work.

Although this book did not have to be written, we felt the need to write it. We hope that readers will share some of the excitement, aesthetic pleasure and learning that we have experienced during its preparation. In the course of our work we have become ever more conscious of the diverse manifestations of symmetry in chemistry, and in the world at large. We believe that this consciousness will also develop in the reader.

Despite its breadth, our book was not intended to be comprehensive or to be a specialized treatise in any specific area. Rather, after acquiring a broad perspective, the reader may then refer to the excellent monographs which provide detailed treatments of specialized topics. We ourselves have relied heavily on these monographs and have listed them among the references given at the end of each chapter.

We would like especially to note here two classics in the literature of symmetry which have strongly influenced us: Weyl's "Symmetry" and Shubnikov and Koptsik's "Symmetry in Science and Art".

Our book has a simple structure. After the introduction (Chapter 1), the simplest symmetries are presented using chemical and non-chemical examples (Ch. 2). Molecular geometry is then discussed in qualitative terms (Ch. 3). Group theoretical methods (Ch. 4) are applied in an introductory manner to the symmetries of molecular vibrations (Ch. 5), electronic structure (Ch. 6), and chemical reactions (Ch. 7). These chapters are followed by a descriptive discussion of space-group symmetries (Ch. 8) including the symmetries of crystals (Ch. 9).

The general perception of symmetry that most people have is sufficient for reading Chapters 1, 2, 3, 8, and 9. However, in order to appreciate Chapters 5, 6, and 7, the introduction to group theory given in Chapter 4 is necessary. Chapter 4 also deals with antisymmetry.

The progenitor of the present book was published in Hungarian two years ago by Akadémiai Kiadó, Budapest, and we appreciate the permission of the publishers to use some of its materials here. Many friends and colleagues were thanked in that version for their contributions, some of which have been incorporated into the present book.

We express our thanks to those distinguished colleagues who have read one or more chapters and helped us with their criticisms and suggestions. They include James M. Bobbit (University of Connecticut), Russel A. Bonham (Indiana University), Arthur Greenberg (New Jersey Institute of Technology), Joel F. Liebman (University of Maryland), Alan L. Mackay (University of London), Alan P. Marchand (North Texas State University), Kurt Mislow (Princeton University), Ian C. Paul (University of Illinois), Péter Pulay (University of Arkansas), Robert Schor (University of Connecticut), and György Varsányi (Budapest Technical University). Of course, all remaining shortcomings are our responsibility and we shall be grateful to our readers for pointing them out to us.

Professor Claus Bliefert (Fachhochschule Münster) acted as catalyst between the publishers and ourselves at the initial stages of this project. Dipl.-Chem. Rudolf Mutter (Universität Ulm) provided valuable assistance in collecting the literature. We appreciate the support, enthusiasm, and hard work of all VCH staff (Weinheim), including Priv.-Doz. H. F. Ebel and production editor P. J. Biel, in bringing out our book.

We thank those authors and copyright owners who gave us permission to use their illustrations in our book. We made all efforts to identify the sources of all illustrative materials, and regret if, inadvertently, we missed anything in doing so.

Most of the final version was compiled during our stay at the University of Connecticut, 1983/85, and we greatly benefited from the School's creative and inspiring atmosphere. We express our gratitude to Dean Julius A. Elias, IMS Director Leonid V. Azaroff, and to our colleagues of the Departments of Chemistry and Physics.

We dedicate this book to the memory of József Pollák (1901–1973) who was the stepfather of one of us (IH). He was an early and decisive influence in stimulating the interests which eventually led to the creation of this book.

Storrs, CT, March 1985   István and Magdolna Hargittai*
and Budapest, May 1986

---

* Permanent address: Hungarian Academy of Sciences, Budapest, P. O. Box 117, H-1431 Hungary.

Contents Overview

Contents

1 Introduction

Fundamental phenomena and laws of nature are related to symmetry and, accordingly, symmetry is one of science's basic concepts. Perhaps it is so important in human creations because it is omnipresent in the natural world. Symmetry is beautiful although alone it may not be enough for beauty, and absolute perfection may even be irritating. Usefulness and function and aesthetic appeal are the origins of symmetry in the worlds of technology and the arts.

Much has been written about symmetry, for example, in Béla Bartók's music [1-1]. It is not known, and perhaps never will be, however, whether he consciously applied symmetry or was simply led intuitively to the Fibonacci numbers and the golden ratio so often present in his music. Another unanswerable question is how these symmetries contribute to the appeal of Bartók's music, and how much of this appeal originates from our innate sensitivity to symmetry. Bartók himself always refused to discuss the technicalities of his composing and liked merely to state, "We create after Nature".

The above example illustrates how we like to consider symmetry in a broader sense than it appears in geometry. The symmetry concept provides a good opportunity to widen our horizons and to bring chemistry closer to other fields of human activities. An interesting aspect of the relationship of chemistry with other fields was expressed by Vladimir Prelog in his Nobel lecture [1-2]: "Chemistry takes a unique position among the natural sciences for it deals not only with material from natural sources but creates the major parts of its objects by synthesis. In this respect, as stated many years ago by Marcelin Berthelot, chemistry resembles the arts; the potential of creativity is terrifying."

Of course, even the arts are not just for arts' sake and chemistry is certainly not done just for chemistry's sake. But in addition to creating new healing medicines, heat-resistant materials, pesticides, and explosives, chemistry is also a playground for the organic chemist to synthesize exotica including propellane and cubane, for the inorganic chemist to prepare compounds with multiple metal-metal bonds, for the stereochemist to model chemical reactions after a French parlor trick (cf. Sect. 2.7), and for the computational chemist to create undreamed-of molecules and to write exquisitely detailed scenarios of as yet unknown reactions using the computer.

Symmetry considerations play no small role in all these activities. The importance of blending fact and fantasy was succinctly expressed by Arthur Koestler [1-3]: "Artists treat facts as stimuli for the imagination, while scientists use their imagination to coordinate facts." An early illustration of an imaginative use of the concept of shape is furnished by Coulson [1-4] citing Lucretius from the first century B.C. who wrote that "atoms with smooth surfaces would correspond to pleasant tastes, such as honey; but those with rough surfaces would be unpleasant."

Chemical symmetry has been noted and investigated for centuries in crystallography, which is at the border between chemistry and physics. It was probably more physics when crystal morphology and other properties of the crystal were described, and more chemistry when the inner structure of the crystal and the interactions between the building units were considered. Later, discussion of molecular vibrations, the selection rules and other basic principles in all kinds of spectroscopy also led to a uniquely important place for the concept in chemistry with equally important practical implications.

The discovery of the handedness or chirality of crystals and then of molecules led the symmetry concept nearer to the real chemical laboratory. It was still, however, not the chemist, in the classical sense of the profession, who was most concerned about symmetry, but the stereochemist, the structural chemist, the crystallographer, and the spectroscopist. It is not yet two decades from the days when, in discussing the importance of symmetry in chemistry, some sort of an apology was felt necessary for giving so much attention to this concept. Symmetry was then considered to lose its significance as soon as the molecules, the main category of chemical objects, entered the most usual chemical change, the chemical reaction. Orbital theory and the discovery of the conservation of orbital symmetry have removed this last blindfold. The 1981 Nobel prize in chemistry awarded to Fukui [1-5] and Hoffmann [1-6] signifies these achievements.

The question may also be asked as to whether "chemical symmetry" differs from any other kind of symmetry? Symmetries in the various branches of the sciences are perhaps characteristically different, and could be hierarchically related? The symmetry in the great conservation laws of physics (see, for example, [1-7]) is of course present in any chemical system. The symmetry of molecules and their reactions is part of the fabric of biological structure. Left-and-right symmetry is so important for living matter that it may be matched only by the importance of "left-and-right" symmetry (?) in the world of the elementary particles – as if a circle is closed, but that is, of course, a gross oversimplification. What may be safely asserted, however, is that the symmetry concept has some bridging and unifying role not only between, say, science and art, but also among the various branches of the sciences themselves.

But what is symmetry? We may not be able to answer this question satisfactorily, at least not in all its possible aspects. According to Russian crystallographer (and symmetrologist) E. S. Fedorov, "symmetry is the property of geometrical figures to repeat their parts, or more precisely, their property of coinciding with their original position when in different positions". According to the geometer H. S. M. Coxeter [1-8], "When we say that a figure is 'symmetrical' we mean that there is a congruent transformation which leaves it unchanged as a whole, merely permuting its component elements." Fedorov's definition is cited here after another noted symmetrologist (and crystallographer), A. V. Shubnikov [1-9], who adds that while symmetry is a property of geometrical figures, obviously "material figures" may also have symmetry. Shubnikov further says that only parts which are in some sense equal among themselves can be repeated, and notes the existence of two kinds of equality, to wit congruent equality and mirror equality. These two equalities are the subsets of the metric equality concept of Möbius, according to whom "figures are equal if the distances between any given points on one figure are equal to the distances between the corresponding points on another figure" [1-9].

Symmetry also connotes harmony of proportions, however, – a rather vague notion according to Weyl [1-10]. This very vagueness, at the same time, often comes in handy when relating symmetry and chemistry, or generally speaking, whenever the symmetry concept is applied to real systems. Mislow and Bickart [1-11] communicated an epistemological note on chirality in which much of what they have to say about chirality, as this concept is being applied to geometrical figures versus real molecules, solvents, and crystals, is true about the symmetry concept as well. Mislow and Bickart argue "that it is unreasonable to draw a sharp line between chiral and achiral molecular ensembles: in contrast to the crisp classification of geometric figures, one is dealing here with a fuzzy borderline distinction, and the qualifying 'operationally' should be implicitly or explicitly attached to 'achiral' or 'racemic' whenever one uses these terms with reference to observable properties of a macroscopic sample." Further, Mislow and Bickart state that "when one deals with natural phenomena, one enters 'a stage in logic in which we recognize the utility of imprecision' [1-12]." The human ability to geometrize non-geometrical phenomena greatly helps to recognize symmetry even in its "vague" and "fuzzy" variations. In accordance with this, Weyl [1-10] referred to Dürer who "considered his canon of the human figure more as a standard from which to deviate than as a standard toward which to strive."

Symmetry in its rigorous sense helps us to decide problems quickly and qualitatively. The answers lack detail, however [1-13]. On the other hand, the vagueness and fuzziness of the broader interpretation of the symmetry concept allow us to talk about degrees of symmetry, to say that something is more symmetrical than something else. An absolutist geometrical approach would allow us to distinguish only between symmetrical and asymmetrical possibly with dissymmetrical thrown in for good measure. So there must be a range of criteria according to which one can decide whether something is symmetrical, and to what degree. These criteria may very well change with time. A case in point is the question as to whether or not molecules preserve their symmetry upon entering a crystal structure or upon the crystal undergoing phase transition. Our notion about structures and symmetries may evolve as more accurate data become available (though their structures and symmetries are unchanged, of course, by our notions).

There is the remarkable phenomenon of statistical symmetry noted by Loeb [1-14]. There are some apparently totally asymmetrical structures in which characteristic parameters are, however, subject to certain well-defined constrained patterns when averaged according to some system. Recognizing structural and other kinds of regularities has always been considered important in chemistry.

The history of periodic tables, following Mendeleev's seminal discovery, also demonstrates chemists' never-ending quest for beauty and harmony [1-15]. Approximately 700 periodic tables were published during the first one hundred years after the original discovery in 1869. Mazurs [1-15] has collected, systematized and analyzed them in a unique study. Classification of all the tables reduced their number to 146 different types and subtypes which are described by such terms as "helices, space lemniscates, space concentric circles, space squares, spirals, series tables, zigzags, parallel lines, step tables, tables symmetrical about a vertical line, mirror image tables, tables of one revolution and of one row, tables of planes, revolutions, cycles, right side as well as left side electronic configuration tables, tables of concentric circles and parallel lines, right side as well as left side shell and subshell tables." Fig. 1-1 and Fig. 1-2 reproduce two of Mazurs' own tables, one with concentric circles in space, representing subshells and period cones stretched vertically; and the other with parallel lines in the plane with bilateral symmetry.

The quest for symmetry and harmony has, of course, contributed more than mere aesthetics in establishing the Periodic Table of the elements. Beauty and usefulness are blended in it in a natural fashion. C. A. Coulson, theoretical chemist and professor of mathematics, concluded his Faraday lecture on symmetry [1-4] with the words: "Man's sense of shape – his feeling

4

Figure 1-1.
Periodic system by Mazurs [1-15] with concentric circles in space representing the subshells with 2, 6, 10, and 14 elements on them, and with period cones stretched vertically. From *Graphic Representations of the Periodic System During One Hundred Years,* Revised (2nd) edition, © 1974 by The University of Alabama Press. Used by permission.

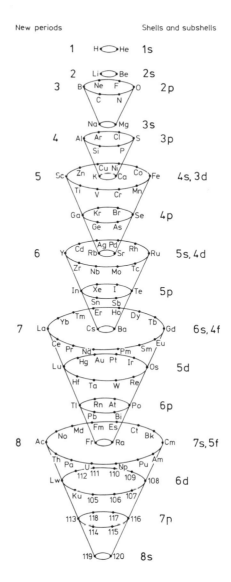

for form – the fact that he exists in three dimensions – these must have conditioned his mind to thinking of structure, and sometimes encouraged him to dream dreams about it. I recall that it was Kekulé himself who said: 'Let us learn to dream, gentlemen, and then we shall learn the truth'. Yet we must not carry this policy too far. Symmetry is important, but it is not everything. To quote Michael Faraday writing of his childhood: 'Do not suppose that I was a very deep thinker, and was marked as a precocious person. I was a lively imaginative person, and could believe in the *Arabian Nights* as easily as in the Encyclopedia. But facts were important to me, and saved me.' It is when symmetry interprets facts that it serves its purpose: and then it delights us because it links our study of chemistry with another

5

Figure 1-2.
Periodic system by Mazurs
[1-15] with parallel lines and
bilateral symmetry. The subshell lines are arranged in the
order in which they are filled
by electrons. The subshell
lines of each period are
connected to form inverted
trapezoids. From *Graphic
Representations of the
Periodic System During One
Hundred Years,* Revised (2nd)
edition, © 1974 by The
University of Alabama Press.
Used by permission.

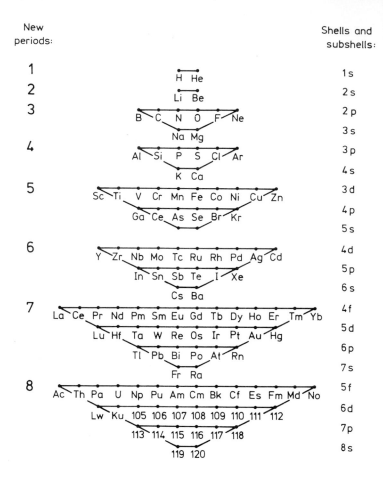

world of the human spirit – the world of order, pattern, beauty,
satisfaction. But facts come first. Symmetry encompasses much
– but not quite all!"

References

[1-1]   E. Lendvai, in *Module, Proportion, Symmetry, Rhythm,* G. Kepes, ed., George Braziller, New York, 1966.

[1-2]   V. Prelog, Science **193**, 17 (1976).

[1-3]   A. Koestler, *Insight and Outlook,* Macmillan, London and New York, 1949.

[1-4]   C. A. Coulson, Chem. Br. **4**, 113 (1968).

[1-5]   K. Fukui, Science **218**, 747 (1982).

[1-6]   R. Hoffmann, Science **211**, 995 (1981).

[1-7]   R. Feynman, *The Character of the Physical Law,* The MIT Press, Cambridge, MA, 1967.

[1-8]   H. S. M. Coxeter, *Regular Polytopes,* Third Edition, Dover Publications, New York, 1973.

[1-9]   A. V. Shubnikov, *Simmetriya i antisimmetriya konechnykh figur,* Izd. Akad. Nauk S.S.S.R., Moskva, 1951. English translation: A. V. Shubnikov, N. V. Belov and others, *Colored Symmetry.* A Series of Publications from the Institute of Crystallography, Academy of Sciences of the U.S.S.R., Moscow, 1951–1958, W. T. Holser, ed., Pergamon Press, New York, 1964.

[1-10]   H. Weyl, *Symmetry,* Princeton University Press, Princeton, NJ, 1952.

[1-11]   K. Mislow and P. Bickart, Israel J. Chem. **15**, 1 (1976/77).

[1-12]   M. Scriven, J. Philos. **56**, 857 (1959).

[1-13]   R. G. Pearson, *Symmetry Rules for Chemical Reactions, Orbital Topology and Elementary Processes,* Wiley-Interscience, New York, 1976.

[1-14]   A. L. Loeb, *Space Structures. Their Harmony and Counterpoint,* Addison-Wesley Publ. Co., Reading, MA, 1976.

[1-15]   E. G. Mazurs, *Graphic Representations of the Periodic System During One Hundred Years,* The University of Alabama Press, University, AL, 1974.

2 Simple and Combined Symmetries

2.1 Bilateral Symmetry

The simplest and most common of all symmetries is bilateral symmetry. Yet at first sight it does not appear so overwhelmingly important in chemistry as in every-day life. The human body has bilateral symmetry, except for the asymmetric location of some internal organs. A unique description of the symmetry of the human body is given by Thomas Mann in *The Magic Mountain* [2-1] as Hans Castorp is telling about his love to Clawdia Chauchat: "How bewitching the beauty of a human body, composed not of paint or stone, but of living, corruptible matter, charged with the secret fevers of life and decay! Consider the wonderful symmetry of this structure: shoulders and hips and nipples swelling on either side of the breast, and ribs arranged in pairs, and the navel centered in the belly's softness, and the dark sex between the thighs. Consider the shoulder blades moving beneath the silky skin of the back, and the backbone in its descent to the paired richness of the cool buttocks, and the great branching of vessels and nerves that passes from the torso to the arms by way of the arm pits, and how the structure of the arms corresponds to that of the legs!"

The bilateral symmetry of the human body is emphasized by the static character of many Egyptian sculptures (Fig. 2-1). Mobility and dynamism in sculptures, however, would not diminish the impression of bilateralness of the human body (Fig. 2-2).

Already Kepler [2-2] noted in connection with the shape of the animals that the "upper and lower depends on their habitat, which is the surface of the earth ... The second distinction of front and back is conferred on animals to put in practice motions that tend from one place to another in a straight line over the surface of the earth ... bodily existence entailed the third diameter, of right and left, should be added, whereby an animal becomes so to speak doubled." The "three diameters" of Kepler suggest a Cartesian coordinate system [2-3].

Bilateral symmetry is indeed very common in the animal kingdom. It always appears when *up* and *down* as well as *forward* and *backward* are different, whereas leftbound and rightbound motion have the same probability. As translational motion along a straight line is the most characteristic for the vast ma-

Figure 2-1.
Egyptian sculpture emphasizing bilateral symmetry from B.C. 2720. Photo of Lehnert & Landrock Art Publishers, Cairo. Used by permission.

Figure 2-2.
Mobility docs not diminish the perception of bilaterality of the human body. Photograph by the authors in Sanssouci, Potsdam.

jority of the animals on Earth, their bilateral symmetry is trivial. This symmetry is characterized by a reflection plane or mirror plane, hence its usual label *m*. The relationship between translational motion and symmetry plane is especially clearly felt in M.C. Escher's "Curl-up" shown in Fig. 2-3.

Figure 2-3.
M. C. Escher: "Curl-up".
Collection Haags Gemeente-
museum – The Hague. Repro-
duced with permission.
© M. C. Escher Heirs c/o
Cordon Art – Baarn – Holland.

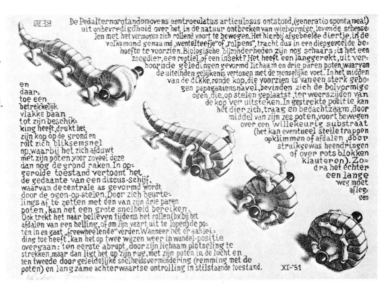

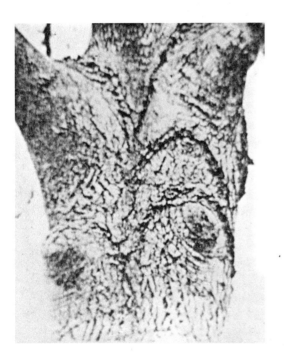

Figure 2-4.
Accidental bilateral symmetry
of trees.
(a) "Tree near Aveley, Essex",
photograph used by kind
permission of C. T. Ballard.
(b) Photograph by the
authors.

In contrast to the widespread occurrence of bilateral symmetry in the animal world, for a tree it may only be accidental
(Fig. 2-4) or man-made (Fig. 2-5). Generally, however, trees as
well as other plants have radial, or cylindrical, or conical symmetries with respect to the trunk or stem. Although these symmetries are very approximate in reality, they are clearly seen in
several examples in Fig. 2-6.

11

Figure 2-5.
Man-made bilateral symmetry
of a tree. Photograph by the
authors.

Figure 2-6.
Approximate radial and
conical symmetries in trees.
Photographs by the authors.

Bilateral symmetry, and symmetry in general has been related to the creation itself (see e.g. Fig. 2-7) and often appears in expressions of religion (see e.g. Fig. 2-8).

Figure 2-7.
(a) William Blake: "The Ancient of Days", 1794. Reproduction by permission of the Syndics of the Fitzwilliam Museum, Cambridge, England.

(b) "Here Creates God Sky and Earth, Sun and Moon and All Elements", from Bible moralisée, c. 1220–1250. Reproduced by permission of the Austrian National Library, Vienna.

Figure 2-8.
(a) From Zagorsk. Photograph by the authors.

(b) Trinity by Rublev.

The symmetry plane of the human face is sometimes emphasized by artists. Some examples are cited in Fig. 2-9. Of course, there are minute variations, or even considerable ones, between the left and right sides of the human face, as is seen, for example, in Fig. 2-10. Faces are often idealized, however, by perfect left-and-right symmetry as in the example of a sculpture from the Charles bridge in Prague (Fig. 2-11).

Figure 2-9.
The bilateral symmetry of the human face as indicated by artists

(a) Henri Matisse: "Woman's portrait". Reproduction by permission of The Hermitage, Leningrad.

(b) George Buday: "Miklós Radnóti", wood-cut, 1969. Reproduced by kind permission of George Buday, R. E.

(c) Jenő Barcsay: "Woman's head", 1961. Reproduced by permission.

Figure 2-10.
Eszter Hargittai in front of a shop-window (1980). Photograph by the authors.

Figure 2-11.
Sculpture from the Charles bridge, Prague. Photograph by the authors.

The origin and meaning of the deviations from bilateral symmetry of the human face are still being intensively studied. Fig. 2-12 shows a face expressing distaste. This photograph is accompanied by two composite pictures according to Sackeim et al. [2-4]. The composite pictures lend added emphasis to the differences between the two sides of the face. There have been speculations that the right-hand side of the human face is generally more "public" while the left-hand side is more "private." It has also been argued that the right-hand side is more characteristic for the whole face than the left-hand side. These views have been a matter of controversy for some time. Recently, a consensus seems to have been reached. The accepted view is that the two sides of the face indeed differ in expressing emotions. The left-hand side probably expresses them more strongly than the right-hand side (cf. Sackeim et al. [2-4] and references therein). An early detection of differences in the two sides of the face can be seen on the triple portrait of the English King Charles I painted by van Dyck about 1637 (Fig. 2-13). An interesting discussion of the differences between the left and right hemispheres of the brain was given by Springer and Deutsch [2-5].

Figure 2-12.
Face expressing distaste [2-4]. Reproduced by permission. Copyright 1978 by the American Association for the Advancement of Science.
(a) Left-side composite.
(b) Original face.
(c) Right-side composite.

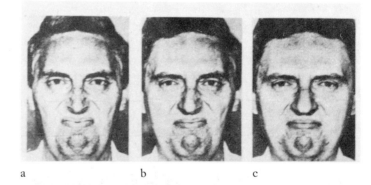

a   b   c

Figure 2-13.
Sir Anthony van Dyck: "Charles I in Three Positions". The Queen's Gallery, London. Reproduced by permission. Copyright reserved.

Bilateral symmetry has outstanding importance in man-made objects. This symmetry, of course, has a functional purpose. The bilateral symmetry of various vehicles, for example, is determined by their translational motion. On the other hand, the cylindrical symmetry of the Lunar Module is consistent with its function of vertical motion with respect to the moon's surface. Shubnikov and Koptsik [2-6] noted that the motorcycle with a side car may be disappearing because its shape suggests circular rather than translational motion.

Fast moving clouds tend to assume shapes with bilateral symmetry as they are moving along the earth's surface. On the other hand, clouds moving primarily in the vertical direction will have cylindrical symmetry, similar to, say, mushrooms or vulcanos, and for similar reasons.

Other examples for cylindrical symmetry, related to the preferential importance of the vertical direction are the salt columns in the Dead Sea (Fig. 2-14a) as well as the stalactites and the stalagmites in caves (Fig. 2-14b), both formed of calcium carbonate.

Figure 2-14.
(a) Salt columns in the Dead Sea. Drawing by Ferenc Lantos after a color slide of Palphot, Ltd., Herzlia, Israel.

(b) Calcium carbonate stalactites and stalagmites in a cave. Photographs by the authors.

Figure 2-15.
(a) Electrolytically deposited copper, magnification x 1000. Courtesy of Dr. Maria Kazinets, University of Beersheva, 1983.

(b) Directionally solidified iron dendrites from an iron-copper alloy after dissolving away the copper, magnification x 2600. Courtesy of Dr. J. Morral, The University of Connecticut, 1984.

The occurrence of radial type symmetries rather than more restricted ones necessitates a spatial freedom in all relevant directions. Thus, for example, the copper formation in Fig. 2-15a has a tendency to form cylindrically symmetric structures. On the other hand, the solidified iron dendrites obtained from iron-copper alloys, after dissolving away the copper, display bilateral symmetry in Fig. 2-15b.

Both folk music and music by master composers are rich in various symmetries. Fig. 2-16 shows two examples with bilateral symmetry. The first example (Fig. 2-16a) is from a Hungarian folk song entitled "Crunchy Cherries Are Ripening". The sequence is A, A^5/A_v^5, A where the upper index indicates a 5 sound shift to higher frequencies and the lower index v indicates some minute variation. Another example is from Bartók's Microcosmos series written specifically for children. Fig. 2-16b of Unisolo No. 6 illustrates a mirror plane which includes a sound.

Figure 2-16.
(a) Hungarian folk-song "Crunchy Cherries Are Ripening".

(b) Bartók: "Microcosmos, Unisono No. 6."

The introductory piece of the Microcosmos is depicted in Fig. 2-17a. It has only approximate bilateral symmetry though the two halves are markedly present. The sound distribution in both halves corresponds to the golden mean as the time

placement of the highest pitch divides the two periods into two unequal parts. The ratio of the durations a/b is equal to the ratio of the larger part to the whole, b/c. That is exactly the definition of the golden mean.

When some school children in their early teens were asked to express their impressions in drawing while listening to this piece of music for the first time, they invariably produced patterns with bilateral symmetry [2-7]. Two of the drawings are reproduced in Fig. 2-17b, and one drawing is singled out in Fig. 2-17c in which the sounds were expressed by (colored) areas with good proportion.

Figure 2-17.
(a) Bartók: "Microcosmos, Unisono No. 1."

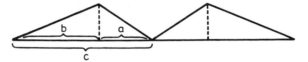

(b) Drawings inspired by the "Unisono No. 1" by early teenagers, Komló Music School. Courtesy of Mária Apagyi, 1983.

(c) Drawing inspired by the "Unisono No. 1" by 12-year old child, Komló Music School. Courtesy of Mária Apagyi, 1983.

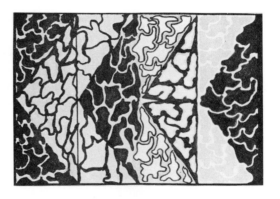

Figure 2-18.
Edge-on view of a typical galaxy and other galaxies in gradually more tilted positions [2-9]. Reproduced by kind permission of R. Jastrow.

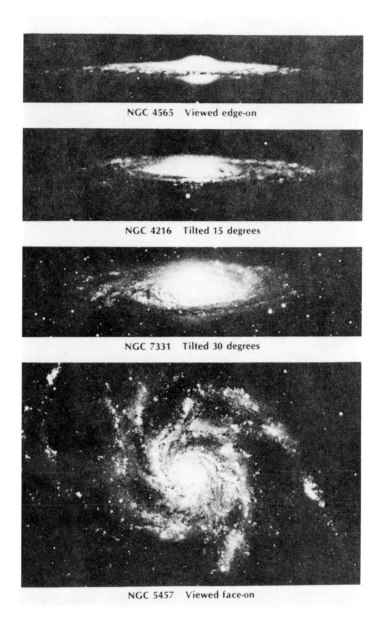

NGC 4565   Viewed edge-on

NGC 4216   Tilted 15 degrees

NGC 7331   Tilted 30 degrees

NGC 5457   Viewed face-on

Changing from Microcosmos to the "macro"cosmos, a typical galaxy of the universe would display bilateral symmetry if viewed edge-on as shown in Fig. 2-18. The more highly tilted pictures reveal the spiral structure of galaxies [2-8, 2-9].

Weyl [2-10] calls bilateral symmetry also heraldic symmetry as it is so common in coats of arms. Characteristically, the Habsburg and the Romanov eagles were double headed (Fig. 2-19) and the Albanian eagle still is.

19

Figure 2-19.
Double-headed eagles.
Photographs by the authors.

(b) The Romanov eagle (from Irkutsk).

(a) The Habsburg eagle (from Prague).

2.2 Rotational Symmetry

Staying with heraldry, the simple and powerful oriental sign yin yang appears with its contour in the South Korean coat of arms in Fig. 2-20a. It has two-fold rotational symmetry in that a half rotation about the axis perpendicular to the midpoint of the drawing brings back the original figure. This rotation axis is called a symmetry axis. The geometrical motif of the trademark in Fig. 2-20b has also this two-fold rotational symmetry with a slight inconsistency in the middle wave line. The Taiwanese stamp of Fig. 2-20c displays a two-fold symmetrical figure of the two fish reminiscent of yin yang.

Figure 2-20.
(a) The contour of yin yang – the principal motif of the South Korean coat of arms.

(b) The trademark of a Hungarian washing machine company HAJDUSÁGI IPARMÜVEK.

(c) Taiwanese stamp.

The *order* of a rotation symmetry axis tells us how many times the original figure is reproduced during a complete rotation. The *elemental angle* is the smallest angle of rotation by which the original figure can be reproduced. Thus for two-fold rotational symmetry the order of the rotation axis is obviously two and the elemental angle is 180°. The corresponding numbers for three-fold rotational symmetry are three and 120°. Fig. 2-21 illustrates two-fold and three-fold rotational symmetries by sculptures of interweaving fish. The one with a trio of fish in Prague is quite analogous to the two fish sculpture in Washington, D.C.

Figure 2-21.
(a) Two-fold rotational symmetry displayed by a Washington, D.C., sculpture of two interweaving fish. Photograph by the authors.

(b) Three-fold rotational symmetry displayed by a Prague sculpture of three interweaving fish. Photograph by the authors.

Fig. 2-22 further illustrates three-fold rotational symmetry. Four-fold rotational symmetry characterizes the popular children's toy the pinwheel shown in Fig. 2-23a. This is also the symmetry of the swastika, an ornament since prehistoric times but also associated with the shameful period of nazism and the Third Reich. Fig. 2-23b illustrates this symmetry in John Heartfield's antinazi poster [2-11] from 1934. A rosette on the regalia of the first Hungarian king (Fig. 2-23c) and an American Indian decoration (Fig. 2-23d) also display four-fold symmetry. American quilts provide a wealth of symmetries. The exclusively rotational symmetries are generally rare but they can often be found in the so-called friendship quilts which were made by exchanging patterns among a circle of friends. The oak leaf wreath quilt cited in Fig. 2-23e has four-fold rotational symmetry and is said to have been used primarily for decorating a man's room as it was believed to have strength and dignity as well as simplicity [2-13]. A four-fold rotationally symmetric jellyfish and a flower are illustrative examples from the living world in Fig. 2-24.

Figure 2-22.
(a) Three-blade propeller.
(b) Window decoration with three-fold rotational symmetry in an ancient Italian town. Photograph by the authors.

Figure 2-23.
(a) Children's toy pinwheel.
(b) John Heartfield: "Blut und Eisen", 1934 [2-11]. Reproduced by permission from Verlag der Kunst, Dresden. All rights reserved.

Der alte Wahlspruch im „neuen" Reich:
BLUT UND EISEN

(c) Rosette on the regalia of the first Hungarian king.

(d) American Indian (Southern Appalachian) decoration [2-12].

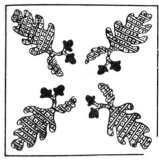

(e) Friendship quilt displaying oak leaf wreath [2-13].

Figure 2-24.
(a) Jellyfish with four-fold rotational symmetry; Aurelia insulinda, after [2-6].
(b) Flower with four-fold rotational symmetry; Vinca minor, photograph by the authors.

It is very rare in the living world to find creatures which would have *only* rotational symmetry [2-6]. An example is the above mentioned jellyfish from the animal world. Such exclusively rotational symmetry may be a consequence of preferential rotational motion in capturing food. Petals folded like a fan lead to a rotational axis and this is not so rare among flowers. Some of these arrangements, however, may have originated from flowers with higher symmetry by genetic mutation.

Fig. 2-25 shows some old ruins along Via Appia Antica in Rome depicting two flowers, one with only rotational symmetry and another with higher symmetry.

Figure 2-25.
Flowers in a stone-carving on old ruins along Via Appia Antica in Rome. The left one has only (four-fold) rotational symmetry while the right one has a combination of a rotation axis and its intersecting symmetry planes. Photograph by the authors.

Five-fold rotational symmetry is displayed by the flowers in Fig. 2-26 while six-fold rotation characterizes the Star of David and the six-blade windmill in Fig. 2-27.

Figure 2-26.
Flowers with exclusively five-fold rotational symmetry.
(a) Vinca minor from Hungary. Photograph by the authors.

(b) Frangipani (Plumeria apocynacea) from Hawaii. Photograph courtesy of John Tucker, Willimantic Photo Club, 1984.

Figure 2-27.
(a) Star of David. New York, NY, 5 Avenue and W 65 Street. Photograph by the authors.
(b) Six-blade windmill.

Figure 2-28.
The rotating circular plate has an infinite-fold rotational symmetry axis. If the rotation stops, an infinite number of symmetry planes intersecting the rotation axis will become apparent.

The order of rotation axes (n) may be 1, 2, 3, ... up to infinity, ∞, thus it may be any integer. The order 1 means that a complete rotation is needed to bring back the original figure, thus there is an absence of symmetry which means asymmetry. A one-fold rotation axis is an identity operator. The other extreme is the infinite order. This means that any, even infinitesimally small, rotation leads to congruency. A circular plate would not be a satisfactory example as it has symmetry planes in addition to the infinite-fold rotation axis. However, a rotating circular plate lacks those symmetry planes and has a mere ∞-fold rotational symmetry axis (Fig. 2-28).

A series of rosettes with only rotational symmetry is shown in Fig. 2-29. Decorations displaying exclusively rotational symmetry often occur in certain folk arts. The otherwise widespread symmetry plane in decorations is easily eliminated by interweaving the motifs. A detailed symmetry analysis of the Pueblo Indian pottery decorations has been presented [2-14]. There is a wealth of design with rotational symmetry only (Fig. 2-30).

Figure 2-29.
Rosettes displaying rotational symmetry. They were prepared with the children's toy Spirograph by Balázs Hargittai, 1980.

25

Figure 2-30.
Decorations with rotational
symmetry only, from Pueblo
Indian pottery design [2-14].
Courtesy and copyright 1977
by the President and Fellows
of Harvard College. The origi-
nal designs were drawn by
Sarah Whitney Powell and
Barbara Westman.

2.3 Combined Symmetries

The symmetry plane and the rotation axis are symmetry elements. If a figure has a symmetry element, it is symmetrical. If it
has no symmetry element, it is asymmetrical. Even an asymmetrical figure has a one-fold rotation axis; or actually, an infinity of
one-fold rotation axes.

The application of a symmetry element is a symmetry
operation. The symmetry elements are also called symmetry
operators. The consequence of a symmetry operation is a symmetry transformation. Strict definitions refer to geometrical
symmetry, and will serve us as guidelines only. They will be
followed qualitatively in our discussion of primarily nongeome-
tric symmetries, according to the ideas of the Introduction.

So far symmetries with *either* a symmetry plane *or* a rotation axis have been discussed. These symmetry elements may also be combined. The simplest case occurs when a symmetry plane includes a rotation axis.

2.3.1 A Rotation Axis with Intersecting Symmetry Planes

A dot between *n* and *m* in the label *n · m* indicates that the axis is in the plane. This combination of a rotation axis and a symmetry plane produces further symmetry planes. Their total number will be *n* as a consequence of the application of the *n*-fold rotational symmetry to the symmetry plane.

The complete set of symmetry operations of a figure is its symmetry group. Fig. 2-31 shows an example of a three-fold rotation axis in a symmetry plane. The rotation axis will, of course, rotate not only the flower but any other symmetry element, in this case the symmetry plane, as well. The 120° rotations will generate altogether three symmetry planes, and these planes will make an angle of 60° with each other. This very symmetry has already been represented by the flower on the right-hand half of the stone carving shown in Fig. 2-25. Some primitive organisms are shown in Fig. 2-32 after Häckel [2-15]. They all have five-fold rotation axes and some of them have intersecting (vertical) symmetry planes as well. The symmetry class of the starfish in the middle is, for example, 5 · *m*. This starfish consists of ten congruent parts, with each pair related by a symmetry plane. The whole starfish is unchanged either by 360°/5 = 72° rotation around the rotation axis, or by mirror reflection through the symmetry planes which intersect at an

Figure 2-31.
Norwegian tulip: example of a three-fold rotational axis at the intersection of the symmetry planes. Note the similarity to the right-hand flower on the Via Appia Antica ruins of Fig. 2-25. Photograph by the authors.

27

Figure 2-32.
Starfish and other primitive organisms possessing a five-fold rotational symmetry axis. The axis may or may not have symmetry planes intersecting it; Häckel [2-15].

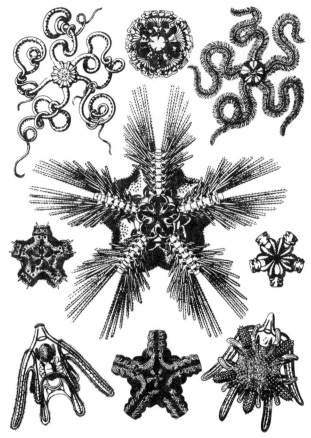

angle of 36°. Five-fold symmetry with five-fold rotation and coinciding mirror reflection is quite common among fruits and flowers. An example is shown in Fig. 2-33. On the other hand, this symmetry is conspicuously absent in the world of crystals as will be discussed in more detail later.

Figure 2-33.
Flower displaying $5 \cdot m$ symmetry. Photograph by the authors.

2.3.2 A Rotation Axis with a Perpendicular Symmetry Plane

Figure 2-34.

Rotating bicone and cylinder possessing $\infty : m$ symmetry.

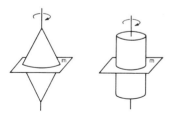

The general label of this symmetry combination is $n : m$, where the colon indicates the orthogonality of the n-fold rotation axis and the symmetry plane. The simplest case is $n = 1$ corresponding to mere bilateral symmetry. The other extreme case is when the rotation axis has infinite order, and there is a symmetry plane perpendicular to it, $\infty : m$. Such is the symmetry of the rotating bicone and rotating cylinder as shown in Fig. 2-34. The rotation excludes the symmetry planes coinciding with the rotation axis. Such symmetry planes would not permit the bicone and the cylinder to have only rotational symmetry.

2.3.3 A Rotation Axis with Intersecting Symmetry Planes and a Perpendicular Symmetry Plane

This combination of symmetries is labeled $m \cdot n : m$ and it is characteristic of highly symmetrical objects. Accordingly, their shapes are relatively simple. As seen in Fig. 2-35, some

Figure 2-35.

Prisms, bipyramids, bicone, cylinder, ellipsoid; examples of $m \cdot n : m$ symmetry

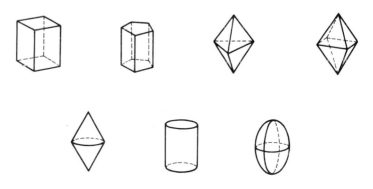

fundamental shapes have $m \cdot n : m$ symmetries. Examples include the square prism, $m \cdot 4 : m$, the pentagonal prism, $m \cdot 5 : m$, the trigonal bipyramid, $m \cdot 3 : m$, the square bipyramid, $m \cdot 4 : m$, and the bicone, cylinder, and the ellipsoid, all having $m \cdot \infty : m$ symmetry. One of the most beautiful and most common examples of this symmetry is the $m \cdot 6 : m$ symmetry of snowflakes.

Figure 2-36.
Snowflake showing virtually perfect symmetry of design after Nakaya [2-16]. Reprinted by permission.
(a) Photomicrograph.
(b) Sketch of part of the crystal.

2.3.3.1 The Shape and Symmetry of Snowflakes

The magnificent hexagonal symmetry of snow crystals, the virtually endless variety of their shapes and their natural beauty make them outstanding examples of symmetry. The fascination in the shape and symmetry of snowflakes goes far beyond the scientific interest in their formation, variety and properties. The morphology of the snowflakes is determined by their internal structures and the external conditions of their formation. It is remarkable, however, how relatively little has been firmly established about the mechanism of snowflake formation. Of course, it is well known that the internal hexagonal arrangement of the water molecules produced by the hydrogen bonds is responsible for the hexagonal symmetry of snowflakes. However, it remains a puzzle why there is a countless number of different shapes of snowflakes, and furthermore why even the smallest variations from the basic underlying shape of a snowflake are repeated in all six directions.

The practically perfect symmetry of snowflake design is illustrated by a photomicrograph and a sketch in Fig. 2-36 after Nakaya [2-16]. The artist Jean Effel's presentation of the origin of the great variety of snowflakes is shown in Fig. 2-37.

As the real puzzling questions concerning snowflakes are related to their morphology rather than to their internal struc-

Figure 2-37.
The story of the creation of the great variety of snowflake shapes according to Jean Effel ("La Création du Monde"). Reproduced by permission. © Mme Jean Effel and Agence Hoffman, Paris.

< CONCOURS D'HEXAGONES DÉCORÉS >

Histoire des cristaux de neige

— Tous premiers *ex æquo* ! Impossible de choisir : on réalisera tous les projets...

tures, these questions will be discussed at some length in the present section. Recently, the process of solidification of fluids into crystals has been simulated using mathematical models. The investigation of the relative stability of various shapes is especially rewarding [2-17]. These simulations showed that crystals with sharp tips grew rapidly and had high stability, while crystals with fat shapes grew slowly and were less stable. However, when these slowly growing shapes were slightly perturbed, they tended to split into sharp, rapidly growing tips. This observation led to the hypothesis of the so-called *points of marginal stability.*

According to the marginal stability model, the snow crystal may start with a relatively stable shape. The crystal may, however, be easily destabilized by a small perturbation. A rapid process of crystallization from the surrounding water vapor ensues. The rapid growth gradually transforms the crystal into another semi-stable shape. A subsequent perturbation may then occur resulting again in a new direction of growth with a different rate. The marginal stability of the snowflake makes the growing crystal very sensitive to even slight changes in its microenvironment. The hypothesis has been worked out by theoretical physicist Langer according to a recent publication [2-17].

The uniqueness of snowflakes may be related to the marginal stability. The ice starts crystallizing in a flat six-fold pattern of water crystals so it is growing in six equivalent directions. As the ice is quickly solidifying, latent heat is released which flows between the growing six bulges. The released latent heat retards the growth in the areas between these bulges. This model accounts for the dendritic or tree-like growth. Both the minute differences in the conditions of two growing crystals and their marginal stability make them develop differently. "Something that is almost unstable, will be very susceptible to changes, and will respond in a large way to a small force" [2-17]. At each step of growth slightly new microenvironmental conditions are encountered, causing new and new variations in the branches. However, it is assumed that each of the six branches will encounter exactly the same microenvironmental conditions, hence their almost exact likeness.

The marginal stability model is attractive in its explanation of the great variety of snowflake shapes. It is somewhat less convincing in explaining the repetitiveness of the minute variations in all six directions since the microenvironmental changes may occur also across the snowflakes themselves and not only between the spaces assigned to different snowflakes.

In order to explain the morphological symmetry of the dendritic snow crystals, McLachlan [2-18] suggested a mechanism about three decades ago which has not yet been seriously challenged. This author posed the very question already mentioned above: "How does one branch of the crystal know what the other branches are doing during growth?" [2-18]. McLachlan noted that the kind of regularity encountered among the snowflakes is not uncommon among flowers and blossoms or among sea animals in which hormones and nerves coordinate the development of the *living* organisms.

McLachlan's explanation for the coordination of the growth among the six branches of a snow crystal is based on the existence of thermal and acoustical standing waves in the crystal. As the snowflake grows by deposition of water molecules upon a small nucleus, it undergoes thermal vibrations at temperatures between 250 and 273 K. The water molecules strike and bounce off the nucleus and those which stay add to the growth. Branching occurs at points with high concentration of water molecules. If the starting ice nucleus has the hexagonal shape shown in Fig. 2-38a and the conditions favor dendritic growth, then the six corners would be receiving more molecules and would be releasing more heat of crystallization than the flat portions. The dendritic development evolving from this situation is shown in Fig. 2-38b. The next stage in the development of a snowflake is the production of a new set of equally spaced dendritic branches determined by the modes of vibration along the spines of the flake. The long spines of Fig. 2-38c are thought to be particular molecular arrays which correspond to the ice structure. The molecules are vibrating and the energy distribution between the modes of vibration is influenced by the boundary conditions. When one of the spines becomes "heavily loaded" at some point, then nodes are induced along this spine. These nodes will eject dendritic branches that are equally spaced as indicated in Fig. 2-38d through f. The question of how the standing waves in one of the six branches are coupled with those in the other branches is answered by considering the torque about an axis through the intersection point. This torque transmits the same frequencies and induces the same nodes in all the branches. Thus, McLachlan asserts that the dendritic development is identical in all six branches and is independent of the particular branch in which the change in the conditions occurred.

An eloquent description of the beauty and symmetry of snowflakes is given by Thomas Mann in *The Magic Mountain* [2-1]: "... Indeed, the little soundless flakes were coming down more quickly as he stood. Hans Castorp put out his arm and let some of them to rest on his sleeve; he viewed them with the knowing eye of the nature-lover. They looked mere shapeless morsels; but he had more than once had their like under his good lens, and was aware of the exquisite precision of form displayed by these little jewels, insignia, orders, agraffes – no jeweller, however skilled, could do finer, more minute work. Yes, he thought, there was a difference, after all, between this light,

Figure 2-38.
(a–f) McLachlan's [2-18] representation of the coordinated growth of the six branches of the snowflake based on his standing wave theory. The original photographs were from Bentley's collection [2-19].

soft, white powder he trod with his skis, that weighed down the trees, and covered the open spaces, a difference between it and the sand on the beaches at home, to which he had likened it. For this powder was not made of tiny grains of stone; but of myriads of tiniest drops of water, which in freezing had darted together in symmetrical variation – parts, then, of the same inorganic substance which was the source of protoplasm, of plant life, of the human body. And among these myriads of enchanting little stars, in their hidden splendour that was too small for man's naked eye to see, there was not one like unto another, and endless inventiveness governed the development and unthinkable differentiation of one and the same basic scheme, the equilateral, equiangular hexagon. Yet each, in itself – this was the

33

uncanny, the anti-organic, the life-denying character of them all – each of them was absolutely symmetrical, icily regular in form. They were too regular, as substance adapted to life never was to this degree – the living principle shuddered at this perfect precision, found it deathly, the very marrow of death – Hans Castorp felt he understood now the reason why the builders of antiquity purposely and secretely introduced minute variation from absolute symmetry in their columnar structures."

The coldness and lifelessness of too much symmetry is as beautifully expressed by Thomas Mann as the beauty of the hexagonal symmetry of the snow crystal. Michael Polányi [2-20] remarked that an environment that was perfectly ordered was not a suitable human habitat. Crystallographers Fedorov and Bernal simply stated "Crystallization is death" [2-21].

Human interest in snowflakes has a long history. The oldest known recorded statement on snowflake forms dates back to the second century B.C. and comes from China according to Needham and Lu Gwei-Djen [2-22]. "Flowers of plants and trees are generally five-pointed, but those of snow are always six-pointed"... was stated as early as in 135 B.C. [2-23]. Six was a symbolic number for water in many classical Chinese writings [2-22]. The contrast between five-pointed plant structures and six-pointed snowflakes has become a literary commonplace in subsequent centuries. Of several other relevant citations collected by Needham and Lu Gwei-Djen [2-22], another is reproduced here, from a statement by a physician from 1189, "... the reason why double-kernelled peaches and apricots are harmful to people is that the flowers of these trees are properly speaking fivepetalled yet if they develop with sixfold (symmetry) twinning will occur. Plants and trees all have the fivefold pattern; only the yellow-berry and snowflake crystals are hexagonal. This is one of the principles of Yin and Yang. So if double-kernelled peaches and apricots with an (aberrant) sixfold (symmetry) are harmful, it is because these trees have lost their standard rule".

The examination of snowflake shapes and their comparison with other shapes has apparently been a great achievement in East Asia. The involvement of Yin and Yang amply demonstrates how much importance could be given to these studies.

As a forerunner of the modern investigations of the correlation between snowflake shapes and environmental, i.e., meteorological conditions, the following passage from the thirteenth century is cited [2-22]: "The Yin embracing the Yang gives hail, the Yang embracing the Yin gives sleet. When snow gets six-pointedness, it becomes snow crystals. When hail gets three-pointedness, it becomes solid. This is the sort of difference that arises from Yin and Yang."

The first known sketches of snowflakes from Europe in the sixteenth century did not reflect their hexagonal shape. Johannes Kepler was the first in Europe, who recognized the hexagonal symmetry of the snowflakes as he described it in his Latin tractate entitled The Six-cornered Snowflake [2-2] published in 1611. By this time Kepler had already discovered the first two laws of planetary motion and thus found the true celestial geometry when he turned his attention to the snowflakes. He considered their perfect form and, for the first time, sought the origin of shape and symmetry in the internal structure. The relationship between crystal habit and the internal structure will be discussed in the chapter on crystals (Chap. 9).

Figure 2-39.
Snow crystals by Descartes from 1635 after Nakaya [2-16]. Reprinted by permission.

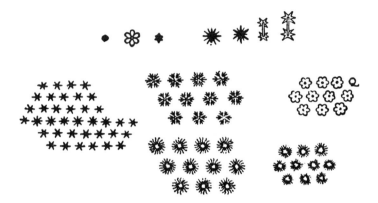

Descartes observed and recorded the shapes of snow crystals. Some of his sketches from 1635 are reproduced in Fig. 2-39 after Nakaya [2-16]. As these were the first drawings of hexagonal snowflakes recorded, it was quite an achievement that even rare versions such as those composed of a hexagonal column with plane crystals developed at both ends could be found among them. More of such important contributions in this field (cf. [2-2, 2-16, 2-24]), among them Hooke's observations using his microscope, occured in the seventeenth century. Branching in snow crystals has also been recorded by several investigators. Among the later works, Scoresby's observations and sketches are especially important [2-25]. Fig. 2-40 reproduces some of them. Scoresby who later became an arctic scientist made these drawings in his log book in 1806 at the age of 16 while he was on a voyage with his father to the Greenland whale fisheries. A few years after the publication of Scoresby's work (1820), the Japanese Doi communicated a series of excellent sketches some of which are reproduced in Fig. 2-41.

There are two fundamental books – collections of snowflake pictures available today as a result of photomicrography. Bentley [2-19] devoted his lifetime to taking photomicrographs of snow crystals and collected at least 6000 of them. About half

Figure 2-40.
Scoresby's sketches of snowflakes from his log book (1806),
after Stamp and Stamp [2-25].
Reproduced by permission.

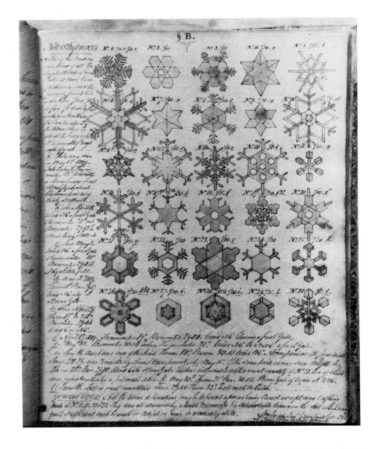

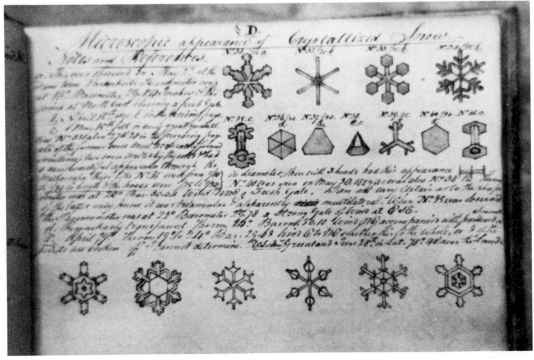

of them appeared in his book co-authored with Humphreys [2-19]. This most well-known book on snowflakes is probably unsurpassable. Bentley's photomicrographs have been reproduced innumerable times in various places – sometimes without indicating the source. Some characteristic examples of snowflakes from this collection are shown in Fig. 2-42.

Figure 2-41.
Snow crystals from Sekka Zusetsu of Doi (from 1832), after Nakaya [2-16]. Reprinted by permission.

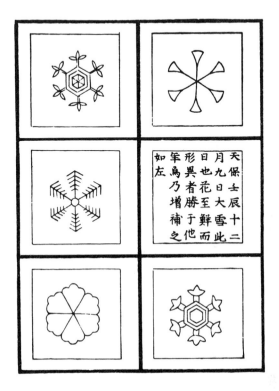

Figure 2-42.
Snowflake photomicrographs by Bentley, after Bentley and Humphreys [2-19].

The other outstanding contribution is Nakaya's [2-16]. He recorded the naturally occurring snow crystals, classified them, and investigated their mass, speed of fall, electrical properties, frequency of occurrence, and so on. In addition Nakaya and co-workers developed methods of producing snow crystals artificially. They succeeded in determining the conditions of formation of all different types of snowflakes. A few examples of Nakaya's studies are cited here [2-16].

The general classification of snow crystals by Nakaya is given in Table 2-1 and Fig. 2-43. The hexagonal plane crystals are the most common and most well-known. They will be described in more detail, following Nakaya's classification.

Simple plate (P1a). Such a shape develops when the degree of supersaturation is low and the temperature goes beyond the limit of dendritic development.

Branches in sector form (P1b) and *broad branches* (P1d) are both intermediate types between the simple plate and the dendritic type.

Simple stellar form (P1e). This is the simplest dendritic type with six straight branches extended from the center.

Dendritic form (P1f). When the simple stellar form further develops there are twigs emanating from the six main branches. This form is also called the regular dendritic form.

Fernlike crystal (P1g). It has well-developed twigs arranged parallel with each other.

Figure 2-43.
Nakaya's general classification of snow crystals [2-16]. Reprinted by permission.

Plate with twigs (P1c) or *with dendritic extensions* (P1i). These are combinations of the plate type and dendritic branches. The plate form usually develops in the upper atmosphere. The dendrites are formed when the snow crystals approach the ground.

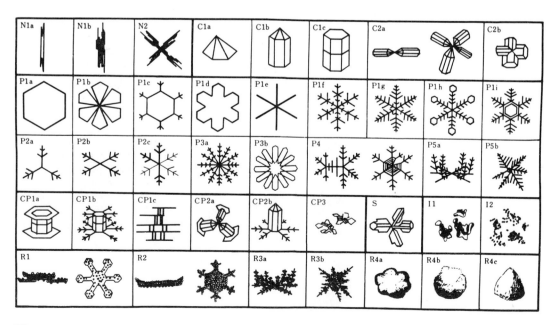

Whether simple extensions appear at the corners or larger dendrites develop, will depend on the thickness of the atmospheric layer and the weather conditions.

Dendritic form with plates (P1h). This is the opposite type to the plate with dendritic extensions (P1i) as the plates develop at the points of the six dendritic branches.

To emphasize the $m \cdot 6 : m$ symmetry of the snowflakes, Fig. 2-44 presents a fernlike crystal. Its central part is somewhat thick and the crystal is becoming thinner nearing the tips.

Nakaya made important contributions not only to observing the perfect or near perfect symmetries of the snow crystals but also distortions from hexagonal symmetry. Of course, the atomic arrangement is always hexagonal, but the morphology or crystal habit may be less than perfectly regular hexagonal. Nakaya calls such crystals *malformed* and states that these asymmetric crystals may be more common than the symmetric ones. Of course, the question of symmetry is a matter of degree. Even the snowflakes which are considered to be most symmetrical may reveal slight differences in their branches when examined closely. Fig. 2-45 shows, however, an obviously asymmetric snowflake whose development was probably influenced by the nonuniformity of the density of the surrounding water vapor.

Figure 2-44.
Fernlike snowflake after Nakaya [2-16]. Reprinted by permission.

Figure 2-45.
Malformed, asymmetric snowflake after Nakaya [2-16]. Reprinted by permission.

Table 2-1. General Classification of Snow Crystals by Nakaya [2-16]

Main Groups	Subgroups	Types
Needle (N)	1. Simple	a. Elementary needle b. Bundle of needles
	2. Combination	
Columnar (C)	1. Simple	a. Pyramid b. Bullet c. Hexagonal
	2. Combination	a. Bullets b. Columns
Plane (P)	1. Regular, developed in plane	a. Simple plate b. Branches in sector form c. Plate with simple extensions d. Broad branches e. Simple stellar form f. Ordinary dendritic form g. Fernlike h. Stellar form with plates at ends i. Plate with dendritic extensions
	2. With irregular number of branches	a. Three-branched b. Four-branched c. Others
	3. With twelve branches	a. Fernlike b. Broad branches
	4. Malformed	many varieties
	5. Spatial assemblage of plane branches	a. Spatial hexagonal b. Radiating

Main Groups	Subgroups	Types
Column/plane combinations (CP)	1. Column with plane at both ends	a. Column with plates b. Column with dendrites c. Capped columns
	2. Bullets with plates	a. Bullets with plates b. Bullets with dendrites
	3. Irregular	
Columnar with extended side planes (S)		
Rimed (R) crystals with cloud particles attached	1. Rimed	
	2. Thick plate	
	3. Graupellike	a. Hexagonal b. Lump
	4. Graupel	a. Hexagonal b. Lump c. Conelike
Irregular snow particle (I)	1. Ice	
	2. Rimed	
	3. Miscellaneous	

2.4 Inversion

What is the symmetry of the 1,2-dibromo-1,2-dichloro-ethane molecule as shown in Fig. 2-46? There is obviously no symmetry plane and no rotation axis. However, any two atoms of the same kind are related by a line connecting them and going through the midpoint of the central bond. This midpoint is the only symmetry element of this molecule and it is called the symmetry center or inversion point. The application of this symmetry element interchanges the atoms, or more generally, any two points located at the same distance from the center along the line going through the center. This interchange is called inversion.

An inversion may also be represented as the consecutive application of two simple symmetry elements, namely a two-fold rotation and mirror-reflection, or *vice versa*. For the molecule of Fig. 2-46 this could be described for example in the following way: (a) rotate the molecule by 180° about the C-C bond as the rotation axis and (b) apply a symmetry plane perpendicular to and bisecting the C-C bond; or (a) apply a two-fold rotation axis perpendicular to the ClCCCl plane and going through the midpoint of the C-C bond and then (b) apply a mirror plane coinciding with the ClCCCl plane. These operations are indicated in Fig. 2-46 and in both examples the results are invariant to the order in which the two operations are performed.

The parallelepiped of Fig. 2-47 is a typical example of an object possessing a center of symmetry. Each apex, edge, and face has its corresponding one through the inversion center. If

Figure 2-46.

The 1,2-dibromo-1,2-dichloro-ethane molecule. Its center of symmetry is the midpoint of the C-C bond. An inversion is equivalent to the consecutive application of a two-fold rotation axis and a reflection plane.

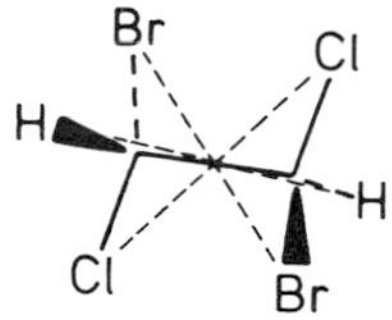

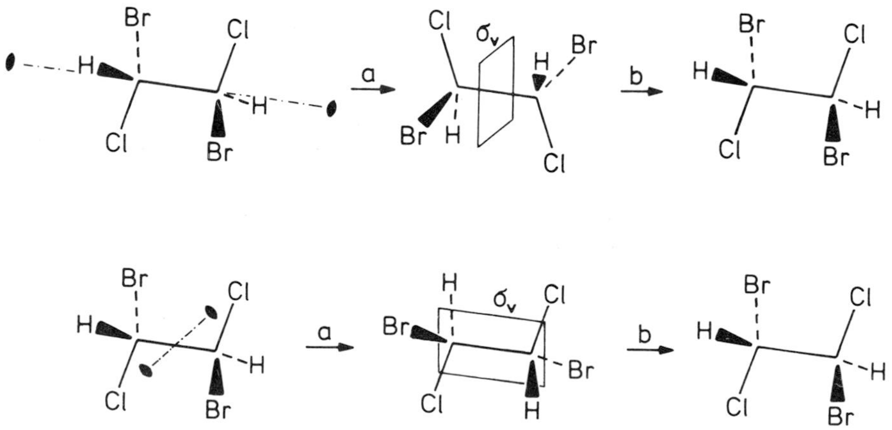

Figure 2-47.
Parallelepiped: Illustration
for an object with center of
symmetry.

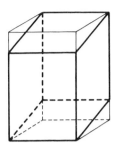

there is any direction of a line or a segment of a face, the symmetry center will invert that direction and the counterpart line or face are obtained.

The sphere is a highly symmetrical object which possesses a center of symmetry. Conjugate locations on the surface of a sphere are related by an inversion through the center of symmetry. The geographical consequences of such an inversion are emphasized in a newspaper article on New Zealand by James Reston in his *Letter from Wellington. Search for End of the Rainbow* [2-26]: "... Nothing is quite the same here. Summer is from December to March. It is warmer in the North Island and colder in the South Island. The people drive on the left rather than on the right. Even the sky is different – dark blue velvet with stars of the Southern Cross – and the fish love the hooks ...". Madrid, Spain corresponds approximately to Wellington, New Zealand, by inversion.

The notation of the symmetry center or inversion center is $\bar{1}$ while the corresponding combined application of two-fold rotation and mirror-reflection may also be considered to be just one symmetry transformation. The symmetry element is called a mirror-rotation symmetry axis of the second order, or two-fold mirror-rotation symmetry axis and it is labeled $\tilde{2}$. Thus $\bar{1} \equiv \tilde{2}$.

A literary example of inversion is taken from a short story entitled *The same in man* by the Hungarian author Frigyes Karinthy [2-27]. It is, in fact, represented by some edited fragments. There are three characters: Bella the beloved lady, Fox the office employee, and Bella's suitor Sándor who is also Fox's director. The editing means that the two meetings of Bella with Sándor and of the director with Fox are presented in a parallel way rather than consecutively.

Bella = B, Fox = F, Sándor/Director = S/D

B ... sometimes I just gaze
before me without thinking
of anything

F ... sometimes I just gaze
before me without thinking
of anything

S/D

Bella! If only you knew
how beautifully you
expressed yourself...

... On my money? Then
you'd better go to a lunatic
asylum, that's where cases
like you are treated...

B Sometimes I have the
feeling that I'd like to be
somewhere other than
where I am. I can't say
where, somewhere I
haven't been before.

F ... I often have the feeling that I'd like to be
somewhere other than
where I am. I don't know
where, anywhere, some-
where, I haven't been
before.

S/D

Bella, how true, how won-
derful... How did you put
it? Let me engrave it in the
records of my mind...

The nuthouse, man, the
nuthouse. That's where
you belong.

B I think people are not
born to what they later
become.

F Sir, I think people are
not born to what they later
become.

S/D

Bella! How very true! How
exquisitely said.

You don't say so? Have
you got any more of that
rubbish to give me? Aren't
you ashamed of yourself, a
grown-up man, to talk all
that rot instead of trying to
apologize for your stupi-
dity?

B (manages an overacted
sigh)

F (sighs)

S/D

Bella ... you must tell me
why you sigh!

What are you puffing for?
Eh? Do you want tu puff
me away?

B Who knows ... I don't
know myself.

F Who knows? I don't
know myself.

S/D

I do, Bella ... it was an answer to my question. After that answer I'm perfectly clear about ourselves. I won't ask you any further questions. I feel that I understand you and admire you. I do not want anything of you. (Moves closer and closer to her)

(getting up) Who knows? You don't know yourself? You've got the cheek to say that to my face? You'll know soon enough, I tell you! (He brandishes the file as he approaches)

B   (brushes off one of the figurines while backing out) Oh dear! What a fright it gave me! How clumsy I am!

F   (backs nervously away and knocks down another figurine)

S/D

(passionately) Clumsy? Oh, no, Bella ... believe me there was so much of yourself in that movement, of your admirable dream-like self. I admire you for that movement. (Kisses her hand).

Idiot! Clumsy idiot! Can't you see? Isn't it enough that you don't do a stroke of work, that you have to break things as well? Let me inform you you're sacked from the first of next month. Take yourself off! (Throws the file after Fox) Sending fools like him here to pester me!

The two-fold mirror-rotation axis is the simplest among the mirror-rotation axes. The object shown in Fig. 2-48a has a four-fold mirror-rotation axis. It was prepared from a square shape with an obliquely inscribed square. The emerging corners are bent alternately up and down. The interesting object obtained in this way has a two-fold rotation axis perpendicular to the square plane and intersecting its midpoint. Moreover, a 90° rotation about the rotation axis plus a reflection through the square plane also brings the object into coincidence with itself. This combined operation is determined by a four-fold mirror-rotation axis labeled $\widetilde{4}$. Generally speaking, a $2n$-fold mirror-rotation axis consists of the following operations: a rotation by $(360/2n)°$ and a reflection by the plane perpendicular to the rotation axis. Another example, a six-fold mirror-rotation axis, $\widetilde{6}$, is shown in Fig. 2-48b. It should be noted that in the objects occuring in Fig. 2-48, only mirror-rotation axes with an even order ($2n$) can be present.

Figure 2-48.

(a) An example of four-fold
mirror-rotation symmetry.

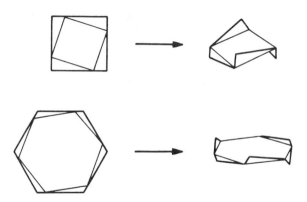

The symmetry of the snowflake involves this type of mirror-rotation axis. The snowflake obviously has a center of symmetry. The symmetry class $m \cdot 6 : m$ contains a center of symmetry at the intersection of the six-fold rotation axis and the perpendicular symmetry plane. In general, for all $m \cdot n : m$ symmetry classes with n even, the point of intersection of the n-fold rotation axis and the perpendicular symmetry plane is also a center of symmetry. When n is odd in an $m \cdot n : m$ symmetry class, however, there is no center of symmetry present.

2.5 Singular Point
and Translational Symmetry

Figure 2-49.

The midpoint of the cylinder
is a singular point.

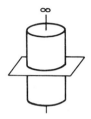

The midpoint of a square is unique, there is no other point equivalent to it. It is called a singular point. A corner of the same square is not singular, the symmetry transformations of the square reproduce it, and there are altogether four equivalent corner points of the square.

Fig. 2-49 shows a cylinder. Its midpoint is singular whereas all the other points on its infinite-order rotation axis are not unique. The symmetry plane perpendicular to this rotation axis doubles all the points of the axis except the midpoint.

An arbitrarily chosen point in a square will have 7 other equivalent points because of the symmetry transformations of the square as shown in Fig. 2-50. Altogether there will therefore be eight equivalent points. However, if the chosen point coincides with one of the corners of the square, there will only be four equivalent points. The same argument applies if the point happens to be on one of the symmetry axes of the square. The

Figure 2-50.
The singular point and the multiplicity of points of a square.

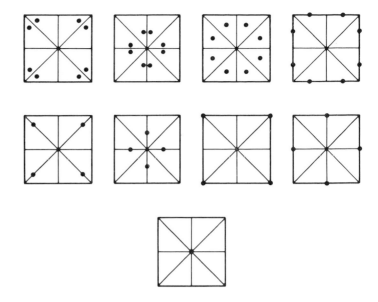

multiplicity of a corner point of the square or any point on a symmetry axis is two. The product of the number of equivalent points and multiplicity is constant (viz. eight for the square). Finally, if the chosen point coincides with the midpoint of the square, the number of equivalent points will be one, and the multiplicity will be eight.

In an asymmetric figure each point is singular and the multiplicity of each point is one.

The symmetry classes characterizing figures or objects which have at least one singular point are called point groups. The point group of the Escher drawing of Fig. 2-51a is expressed by the symmetry class $3 \cdot m$. The drawing presents angels and bats with gradually changing sizes. There is one singular point at the midpoint. Another Escher drawing is shown in Fig. 2-51b. It also presents angels and bats with unchanging sizes. If it is supposed that the drawing is a fragment of an infinitely large one, there is no singular point on it. Assuming an infinite extent for this drawing is natural because of its periodicity. The other drawing, on the other hand, has a circle limit. The absence of a singular point leads to regularity expressed in infinite repetition which characterizes translational symmetry. This kind of symmetry precludes the presence of singular points though does not preclude the presence of a singular line or plane. The symmetry classes characterizing entities with translational symmetry are called space groups. One-dimensional space groups describe the symmetries involving infinite repetition or periodicity in one direction, two-dimensional space groups those involving periodicity in two directions and three-dimen-

47

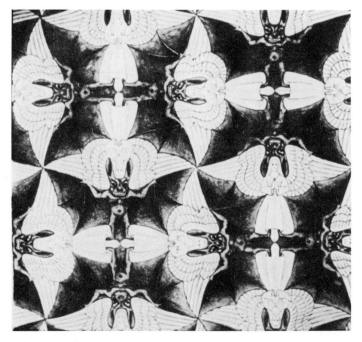

sional space groups describe the symmetry classes when periodicity is present in all three directions. Fig. 2-52 and Table 2-2 summarize the possible cases considering the dimensionality and periodicity. The nomenclature is somewhat inconsistent but has some relationship to Abbott's classic "Flatland" [2-29].

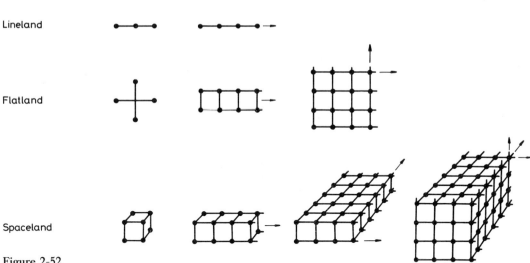

Pointland

Lineland

Flatland

Spaceland

Figure 2-52.
Dimensionality and periodicity
in point groups and space
groups. This figure is consistent with Table 2-2.

Table 2-2. Dimensionality (m) and Periodicity (n) of Symmetry Groups G_n^m after Engelhardt [2-30]

Periodicity → Dimensionality ↓	$n = 0$ No Periodicity	$n = 1$ Periodicity in One Direction	$n = 2$ Periodicity in Two Directions	$n = 3$ Periodicity in Three Directions
$m = 0$ Dimensionless	G_0^0			
$m = 1$ One-dimensional	G_0^1	G_1^1		
$m = 2$ Two-dimensional	G_0^2	G_1^2	G_2^2	
$m = 3$ Three-dimensional	G_0^3	G_1^3	G_2^3	G_3^3

2.6 Polarity

Figure 2-53.

(a) Centrosymmetric rhombic bipyramidal acetanilide crystals from Groth's Chemische Kristallographie [2-32].

(b) Noncentrosymmetric rhombic pyramidal p-chloro-acetanilide crystal from Groth's Chemische Kristallo-graphie [2-32].

A line is polar if its two directions can be distinguished and a plane is polar if its two surfaces are not equivalent. This definition of polarity has, of course, nothing to do with charge separation. A polar line has a "head" and a "tail" and a polar plane has a "front" and a "back". A vertical line on the surface of the Earth is polar with respect to gravity and a sheet of paper with one of its sides painted is polar with respect to its color.

An axis is polar if its two ends are not brought into coincidence by the symmetry transformations of the symmetry group of its figure. An analogous definition applies to the two sides of a polar plane.

If a symmetry group includes a center of symmetry, polarity is excluded. It has already been seen (cf. e.g. Fig. 2-47) that in a centrosymmetric figure a directed line or segment of a face change direction by the inversion. In the case of the absence of a center of symmetry, there will be at least one directed line or face which is not accompanied by parallel counterparts reversed in direction.

The significance of polar axes can be demonstrated, for example, in crystal morphology. Recently, Curtin and Paul [2-31] have summarized the chemical consequences of the polar axis in organic crystal chemistry. A few examples will be mentioned here following Curtin and Paul. Fig. 2-53a shows two centrosymmetric acetanilide crystals. The faces occur in parallel pairs in both habits. On the other hand, the p-chloro-acetanilide crystal shown in Fig. 2-53b is noncentrosymmetric and some of the faces occur without parallel ones at the opposite end of the crystal. This crystal has a polar axis parallel to its long direction.

The morphological symmetry differences between the acetanilide and p-chloroacetanilide crystals originate from their internal structures and are reflected in the differences of the molecular arrangements in Figs. 2-54a and b after Curtin and Paul [2-31]. The acetanilide molecules appear in pairs and the two molecules in each pair are related by an inversion center.

On the other hand, the p-chloroacetanilide molecules are all aligned in one direction.

Even very simple structures may form polar crystals. For example, in a polar crystal composed of diatomic molecules AB, the molecular axis will be oriented more along the polar direc-

Figure 2-54.
(a) The centrosymmetric arrangement of acetanilide molecules of the crystal resulting in centrosymmetric crystal habit. Reprinted with permission from [2-31]. Copyright (1981) American Chemical Society.

(b) The p-chloroacetanilide molecules are aligned in a head-tail orientation resulting in the occurrence of a polar axis of the crystal habit. Reprinted with permission from [2-31]. Copyright (1981) American Chemical Society.

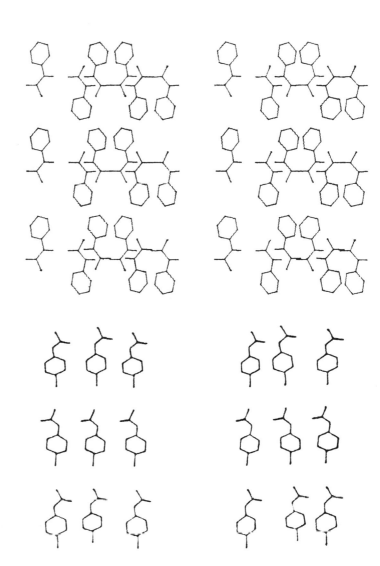

tion of the crystal than perpendicular to it. Furthermore, as there is an ABAB... array in the crystal, it is required that the spacings between the atom A and the two adjacent atoms B be unequal in order to have a polar axis present,

A B   A B   A B ....

Curtin and Paul characterize this situation from the point of view of a submicroscopic traveler proceeding along this array of atoms. The observer is able to determine the direction of travel thanks to the difference in spacings. The distance is always longer from atom B to atom A and shorter from atom A to the next atom B in one direction whereas the reverse is true in the opposite direction.

It is not required that a molecule possess a large dipole moment in order to be suitable for building polar crystal habits.

51

Curtin and Paul cite the nearly "nonpolar" 1-tert-butyl-4-methylbenzene molecule

$$H_3C-\bigcirc-C(CH_3)_3$$

which crystallizes in a polar habit with one end of the crystal being formed by the methyl groups and the other end by the tert-butyl groups. It is not fully understood why some classes of substances prefer to form polar crystals while others with similar potentials do not. Aromatic compounds with certain functional groups (e.g. amino group) more often form polar crystals than do such compounds with other groups (such as carboxyl group). *Meta*-disubstituted benzene derivatives crystallize more often in a polar habit than do *ortho* and *para* derivatives. Sometimes the molecular polar axis is oriented almost perpendicular to the crystal polar axis and only a small component of the molecular polarity contributes to the crystal polarity.

Crystal polarity may have important consequences in the chemical behavior. In solid/gas reactions, for example, crystal polarity may be a source of considerable anisotropy. In the following reactions,

$$C_6H_5COOH + NH_3 \rightarrow C_6H_5COO^- NH_4^+$$
crystalline   gaseous
benzoic acid   ammonia

$$C_6H_5COOCOC_6H_5 + NH_3 \rightarrow C_6H_5CONH_2$$
crystalline   gaseous $+ C_6H_5COO^- NH_4^+$
benzoic anhydride   ammonia

Paul and Curtin [2-33] observed that the attack of the gas occurred at the crystal sides and not at the best developed top crystal faces. These top faces mainly consist of aromatic rings whereas the acid and anhydride groups are exposed at the sides.

Curtin and Paul [2-31] also reported a particularly interesting comparison of two solid/gas reactions. Again the gas was ammonia. In one case the solid was a centrosymmetric p-chlorobenzoic anhydride crystal, and in the other a noncentrosymmetric p-bromobenzoic anhydride crystal:

$$Cl-\bigcirc-C\overset{O}{\diagup}O-C\overset{O}{\diagup}\bigcirc-Cl$$

p-chlorobenzoic anhydride

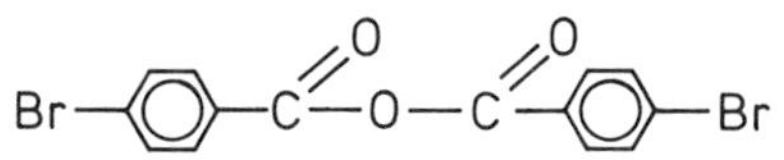

p-bromobenzoic anhydride

The molecular packings of the two crystals are shown in Fig. 2-55 after Curtin and Paul [2-31]. The reaction for the centro-symmetric chloro derivative proceeded uniformly on all sides but not on the top face at which the p-chlorophenyl groups pointed. The situation is rather different in the case of the polar crystals of p-bromobenzoic anhydride, which crystallizes in

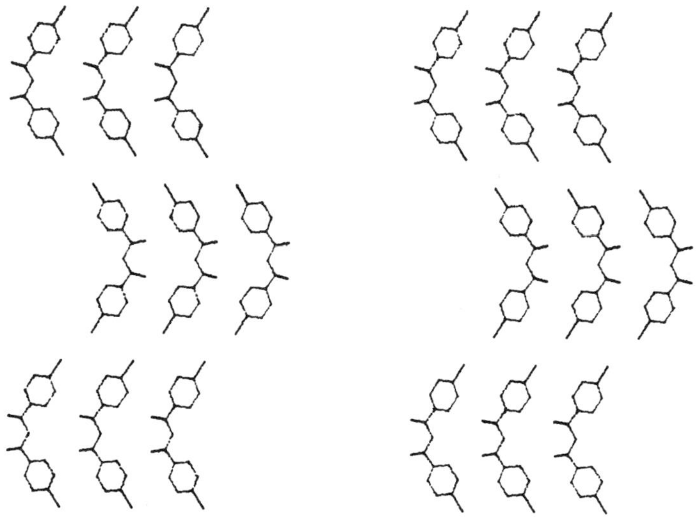

Figure 2-55.
(a) The molecular arrangement in the centrosymmetric p-chlorobenzoic anhydride crystal. Reprinted with permission from [2-31]. Copyright (1981) American Chemical Society.

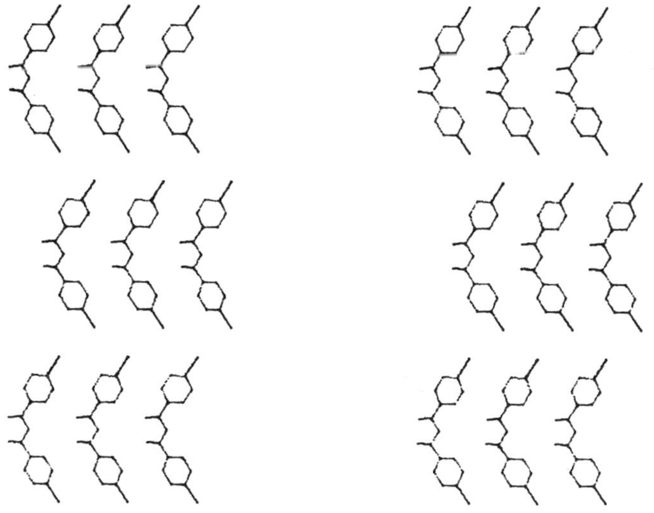

(b) The molecular arrangement in the noncentrosymmetric p-bromobenzoic anhydride crystal. Reprinted with permission from [2-31]. Copyright (1981) American Chemical Society.

different habits (Fig. 2-56) although the internal crystal structure is the same. In the photographs of reacting crystals shown in Fig. 2-56, the polar axis is vertical. The most dramatic effect

53

Figure 2-56.
The reaction of polar p-bromobenzoic anhydride crystals with gaseous ammonia. The crystal plates with the polar axis are along the long direction of the crystals. Reprinted with permission from [2-34]. Copyright (1981) American Chemical Society.

of the reaction sequence is a very clear preference for reaction at one end of the polar direction compared to the other, whereas no such effect was noted for the p-chloroderivative. In addition, there is less reaction at the side faces in the p-bromo crystals than in the p-chloro crystals, an effect not directly attributable to polarity [2-34].

It has been suggested [2-35] that even substances usually crystallizing in a centrosymmetric crystal may be induced to form polar crystals if subjected to an electric field during the crystallization process. Such attempts have not yet been successful for the p-chlorobenzoic anhydride although they have been reported to work for some other compounds [2-31].

2.7 Chirality

There are many objects both animate and inanimate which have no symmetry planes but which occur in pairs related by a symmetry plane and whose mirror images cannot be superposed. Fig. 2-57 shows a pair of rosettes, a detail from Bach's *The Art of the Fugue,* a pair of molecules and a pair of crystals.

Figure 2-57.

(a) Two five-fold rotationally symmetrical rosettes. The individual rosettes have no symmetry plane but the pair is related by mirror symmetry. Prepared with the children's toy Spirograph by Balázs Hargittai, 1980.

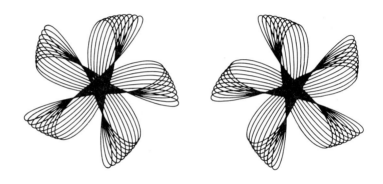

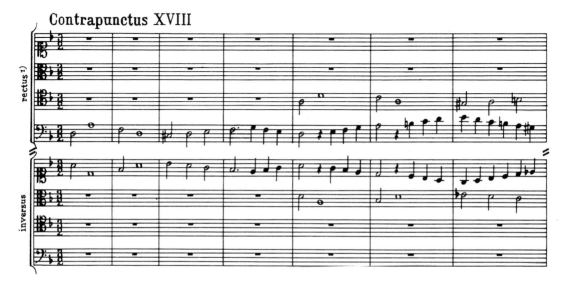

(b) J. S. Bach: "Die Kunst der Fuge, Contrapunctus XVIII", detail.

(c) Glyceraldehyde molecules.

(d) Tartaric acid crystals.

W. H. Thompson - Lord Kelvin wrote [2-36]: "I call any geometrical figure or group of points 'chiral', and say it has 'chirality', if its image in a plane mirror, ideally realized, cannot be brought into coincidence with itself." He called forms of the same sense "homochiral" and forms of the opposite sense "heterochiral". The most common example of a heterochiral form is hands. Indeed, the word chirality itself comes from the greek χειρ, hand. Figs. 2-58 and 2-59 show some heterochiral and homochiral pairs of hands.

a

b

c

Figure 2-58.
Heterochiral pairs of hands.
(a) Albrecht Dürer: "Praying hands". Albertina Graphic Collection, Vienna. Reproduced with permission.
(b) Tomb in the Jewish cemetery, Prague. Photograph by the authors.
(c) Vera Székely: "I would like to be loved". Petőfi Literary Museum, Budapest. Reproduced with permission.
(d) An illustration from a book discussing the possibilities of life in outer space [2-37]. Reproduced by permission from R. N. Bracewell.

d

(e) United Nations stamp for the "World Food Programme".

e

56

Figure 2-59.
A homochiral pair of hands:
Two right hands on a U. S.
stamp.

A chiral object and its mirror image are enantiomorphous, and they are each other's enantiomorphs. Louis Pasteur first suggested that molecules can be chiral. In his famous experiment in 1848, he recrystallized a salt of tartaric acid and obtained two kinds of small crystals which were mirror images of each other. They had the same chemical composition, but differed in their optical activity. One was laevo-active (L) and the other was dextro-active (D). Since the true absolute configuration of molecules could not be determined at the time, an arbitrary convention was applied which, luckily proved to coincide with reality. If a molecule or a crystal is chiral, it is necessarily optically active. The converse is, however, not true. There are, in fact nonenantiomorphous symmetry classes of crystals which may exhibit optical activity.

Whyte [2-38] extended the definition of chirality: "Three-dimensional forms (point arrangements, structures, displacements, and other processes) which possess non-superposable mirror images are called 'chiral'." A chiral process consists of successive states all of which are chiral. The two main classes of chiral forms are screws and skews. Screws may be conical or cylindrical and are ordered with respect to a line. Examples for the latter are the left-handed and right-handed helices in Fig. 2-60. The skews, on the other hand, are ordered around their center. Examples are chiral molecules having a point-group symmetry.

Figure 2-60.
Examples of left-handed and right-handed helices [2-40]. Reproduced with permission, © 1981 by the Benjamin/ Cummings Publishing Co., Menlo Park, CA.

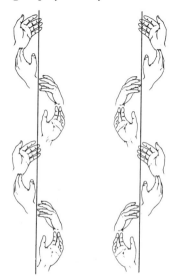

From the point of view of molecules, or crystals, left and right are intrinsically equivalent. An interesting overview of the left/right problem in science is given by Gardner [2-39]. Distinguishing between left and right has also considerable social, political, psychological connotations (see e.g. Weyl [2-10]). For example, left-handedness in children is viewed with varying degrees of tolerance in different parts of the world. Fig. 2-61 shows a classroom at The University of Connecticut with different (homochiral and heterochiral) chairs to accomodate both the right-handed and the left-handed students. Older classrooms at the same University have chairs for the right-handed only (Fig. 2-62).

Figure 2-61.
Classroom at The University of Connecticut with homochiral and heterochiral chairs for both the right-handed and the left-handed students. Photograph by the authors.

Figure 2-62.
Older classrooms have chairs for the right-handed students only. Photograph by the authors.

2.7.1 Asymmetry and Dissymmetry

The simplest chiral molecules are those in which a carbon atom is surrounded by four different ligands – atoms or groups of atoms at the vertices of a tetrahedron. All the naturally occurring amino acids are chiral, except glycine.

Symmetry operations of the first kind and of the second kind are sometimes distinguished in the literature (cf. [2-41]). Operations of the first kind are sometimes also called even numbered operations. For example, the identity operation is equivalent to two consecutive reflections from a symmetry plane. It is an even numbered operation, an operation of the first kind. Simple rotations are also operations of the first kind. Mirror-rotation leads to figures consisting of right-handed and left-handed components, and therefore is an operation of the second kind. Simple reflection is also an operation of the second kind as it may be considered as a mirror-rotation about a one-fold axis. A simple reflection is related to the existence of two enantiomorphic components in a figure. Fig. 2-63 illustrates these distinctions by a series of simple sketches after Shubnikov [2-41]. In accordance with the above description, chirality is sometimes defined as the absence of symmetry elements of the second kind.

Figure 2-63.
Examples of symmetry operations of the first kind (a) and of the second kind (b) after Shubnikov [2-41].

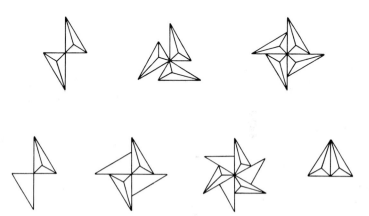

Shubnikov [2-41] noted more than a quarter of a century ago that the terms asymmetry and dissymmetry are often confused in the literature. Unfortunately this confusion has not yet disappeared. Shubnikov pointed out that the scientific meaning of these terms is in complete conformity with the grammar of these words: asymmetry means the absence of symmetry, and dissymmetry means the derangement of symmetry. Pasteur used "dissymmetry" for the first time as he designated the absence of elements of symmetry of the second kind in a figure. Accordingly, dissymmetry did not exclude elements of symmetry of the first kind. Pierre Curie suggested an even broader application of this term. He called a crystal dissymmetric in case of the absence of those elements of symmetry upon which depends the existence of one or another physical property in that crystal. In Pierre Curie's original words [2-42]: "Dissymmetry creates the phenomenon." Namely, a phenomenon exists and is observable due to dissymmetry, i.e., due to the absence of some symmetry elements from the system. Finally, Shubnikov called dissymmetry the *falling out* of one or another element of symmetry from a *given* group. He argued that to speak of the absence of elements of symmetry makes sense only when these symmetry elements are present in some other structures.

Figure 2-64.
Quartz crystals.

Thus from the point of view of chirality, any asymmetric figure is chiral, but asymmetry is not a necessary condition for chirality. All dissymmetric figures are also chiral if dissymmetry means the absence of symmetry elements of the second kind. In this sense, dissymmetry is synonymous with chirality.

An assembly of molecules may be achiral for one of two reasons. Either all the molecules present are achiral, or the two kinds of enantiomorphs are present in equal amounts. Chemical reactions between achiral molecules lead to achiral products. Either all product molecules will be achiral or the two kinds of chiral molecules will be produced in equal amounts. Chiral crystals may sometimes be obtained from achiral solutions. When this happens, the two enantiomorphs will be obtained in (roughly) equal numbers, as was observed by Pasteur. Quartz crystals are an inorganic example of chirality (Fig. 2-64). Roughly equal numbers of left-handed and right-handed crystals are obtained from the achiral silica melt.

2.7.2 Relevance to Origin of Life

The situation with respect to living organisms is unique. Living organisms contain a large number of chiral constituents, but only *L*-amino acids are present in proteins and only *D*-nucleotides in nucleic acids. This happens in spite of the fact that the energy of both enantiomorphs is equal and their formation has equal probability in an achiral environment. However, only one of the two occurs in nature, and the particular enantiomorphs involved in the life processes are the same in humans, animals, plants, and microorganisms. The origin of this phenomenon is a great puzzle which according to Prelog [2-44] may be regarded as a problem of molecular theology.

This problem has long fascinated those interested in the molecular basis of the origin of life, e.g., [2-8, 2-43]. There are in fact two questions. One is why do all the amino acids in a protein have the same *L*-configuration or why do all the components of a nucleic acid, that is, all its nucleotides, have the same *D*-configuration? The other question, the more intriguing one, is why that particular configuration happens to be the *L* for the amino acids and why it happens to be the *D* for nucleotides in all living organisms? This second question seems to be impossible to answer satisfactorily at the present time.

According to Prelog [2-44], a possible explanation is that the creation of living matter was an extremely improbable event, which occurred only once. We may then suppose that if there are living forms similar to ours on a distant planet, their molecular structures may be the mirror image of the corresponding molecular structures on the earth. We know of no structural reason at the molecular level for living organisms to prefer one type of chirality. (There may be reasons at the atomic nuclear level. The violation of parity at the nuclear level has a huge literature, cf., e.g., [2-39]). Of course, once the selection is made, the consequences of this selection must be examined in relation to the first question. The fact remains, however, that chirality is intimately associated with life. This means that at least dissymmetry and possibly asymmetry are basic characteristics of living matter.

Although Pasteur believed that there is a sharp gap between vital and nonliving processes, he attributed the asymmetry of living matter to the asymmetry of the structure of the universe and not to a vital force. Pasteur himself wrote that he was inclined to think that life, as it appears to us, must be a product of the dissymmetry of the universe [2-45].

Concerning the first question, Orgel [2-8] suggests that we compare the DNA structure to a spiral staircase. The regular DNA right-handed double-helix is composed of D-nucleotides. On the other hand, if a DNA double-helix were synthesized from L-nucleotides, it would be left-handed. These two helices can be visualized as right-handed and left-handed spiral staircases, respectively. Both structures can perform useful functions. A DNA double-helix containing both D- and L-nucleotides, however, could not form a truly helical structure at all since its handedness would be changing. Just consider the analogous spiral staircase that Orgel suggested as shown in Fig. 2-65.

If each component of a complex system is replaced by its mirror image, the mirror image of the original system is obtained. However, if only *some* components of the complex system are replaced by their mirror images, a chaotic system emerges. Chemical systems that are perfect mirror images of each other behave identically, whereas systems in which some but not all components are mirror images have quite different chemical properties. If, for example, a naturally occurring enzyme made up of L-amino acids synthesizes a D-nucleotide, then the corresponding artificial enzyme obtained from D-amino acids would synthesize the L-nucleotide. On the other hand, a corresponding polypeptide containing both D- and L-amino acids would probably lack the enzymic activity.

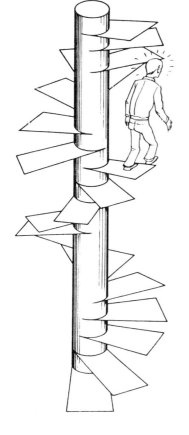

Figure 2-65.
A spiral staircase which changes its chirality [2-8]. Reproduced by permission from L. E. Orgel.

2.7.3 "La coupe du roi"

One of the many chemical processes where chirality/achirality relationships may be important is the fragmentation of some molecules and the reverse process of the association of molecular fragments. An exceptionally interesting case has been described recently. Mislow et al. [2-46] have reported the synthesis of 4-(bromomethyl)-6-(mercaptomethyl)[2.2]metacyclophane

in chiral forms

The stereochemistry of this compound has unique implications for the discussion of chirality. From the chemical point of view, the use of this compound in the synthesis of the dimer represented the first demonstration that an achiral object can be bisected into homochiral halves, and, conversely, that two homochiral objects can be combined to give an achiral whole. The usual cases are those in which an achiral object is bisected into achiral or heterochiral halves. On the other hand, if an achiral object can be bisected into two homochiral halves, it cannot be bisected into two heterochiral ones. A relatively simple case is the tessellation of planar achiral figures into achiral, heterochiral and homochiral segments. Some examples are shown in Fig. 2-66. For a detailed discussion see [2-6, 2-46].

Mislow et al. [2-46] have cited a French parlor trick called "la coupe du roi" – or the royal section in which an apple is bisected into two homochiral halves, as shown in Fig. 2-67. An apple can be easily bisected into two achiral halves. On the other hand, it is impossible to bisect an apple into two heterochiral halves. Two heterochiral halves, however, can be obtained from two apples, both cut into two homochiral halves in the opposite sense, see Fig. 2-67.

According to "la coupe du roi" [2-46] two vertical half cuts are made through the apple. One from the top to the equator, and another, perpendicularly, from the bottom to the equator. In

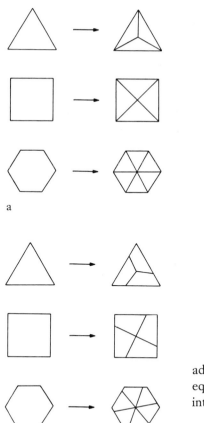

a

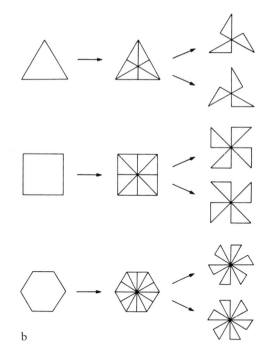

b

c

Figure 2-66.
Dissection of planar achiral figures into (a) achiral, (b) heterochiral and (c) homochiral segments: some examples.

Figure 2-67.
The French parlor trick "La coupe du roi" after Mislow et al. [2-46]. An apple can be cut into two homochiral halves in two ways which are enantiomorphous to each other. An apple cannot be cut into two heterochiral halves. Two heterochiral halves originating from two different apples cannot be combined into one apple. Used with permission. Copyright (1983) American Chemical Society.

addition, two nonadjacent quarter cuts are made along the equator. If all this is properly done, the apple should separate into two homochiral halves as seen in Fig. 2-67.

The work of Mislow et al. [2-46] was preceded by a synthetic study in which Misumi et al. [2-47] prepared the "cis" and "trans" dimers of metacyclophanes:

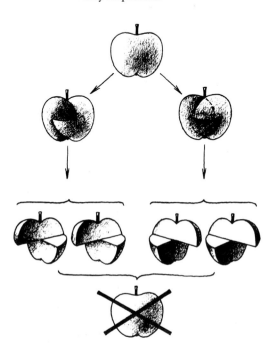

"cis" dimer of metacyclophanes

"trans" dimer of metacyclophanes

The synthesis was achieved by reacting the dibromide and dithiol derivatives of metacyclophane:

Such a reaction yielded an approximately equimolar mixture of the "cis" and "trans" dimers.

Mislow et al. [2-46] have shown that two homochiral 4-(bromomethyl)-6-(mercaptomethyl)[2.2]metacyclophane molecules can only give the "cis" dimer as shown in Fig. 2-68a. This is, of course, true for either form of the homochiral pairs. On the other hand, the reaction between two heterochiral 4-(bromomethyl)-6-(mercaptomethyl)[2.2]metacyclophane molecules can only give the "trans" dimer (Fig. 2-68b). Finally, a racemic mixture of these molecules will yield a 1:1 mixture of the "cis" and "trans" dimers (Fig. 2-68c), as the statistical probability of the process shown in Fig. 2-68b is twice as high as that of each of the two reactions represented in Fig. 2-68a. The reaction of the two homochiral metacyclophane derivatives (Fig. 2-68a) which yields the achiral "cis" dimer is a stereochemical reverse coupe du roi!

2.7.4 A Rotation Axis with Perpendicular Two-fold Rotation Axes

The symmetry class is $n:2$ when a principal axis of rotation of order n occurs with two-fold rotation axes perpendicular to it. All objects having $n:2$ symmetry are chiral. This symmetry may also be derived by eliminating all the symmetry planes from $m \cdot n : m$ symmetry. Thus, for example, if the polyhedra shown in Fig. 2-35 are twisted along their principal axis of rotation, the $m \cdot n : m$ symmetry reduces to $n : 2$ symmetry. The symmetry planes vanish while the principal axis as well as the axes perpendicular to it remain. The twisting along the principal axis may, of course, take place in the left-handed or in the right-handed direction hence the mirror images.

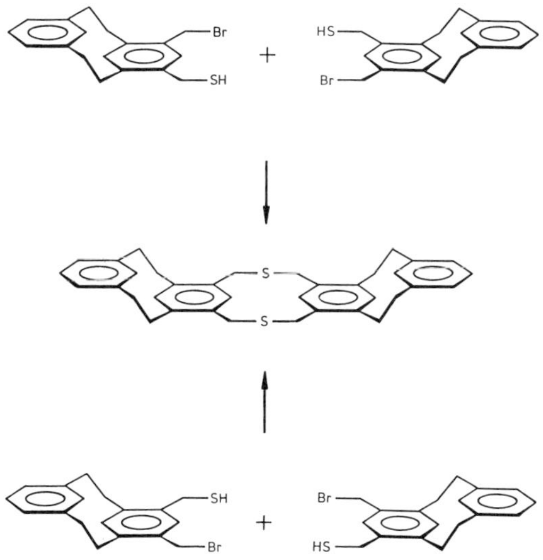

a

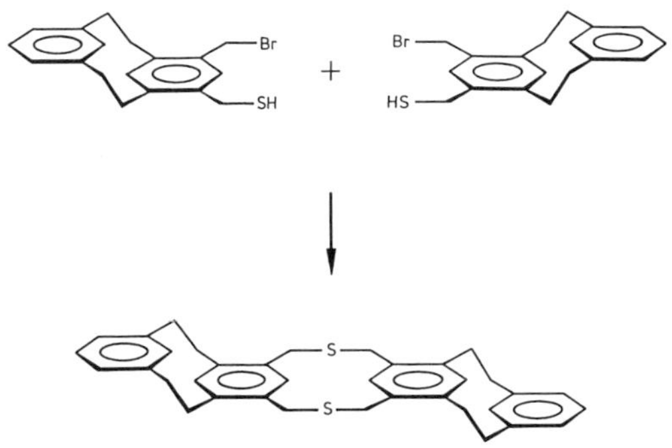

b

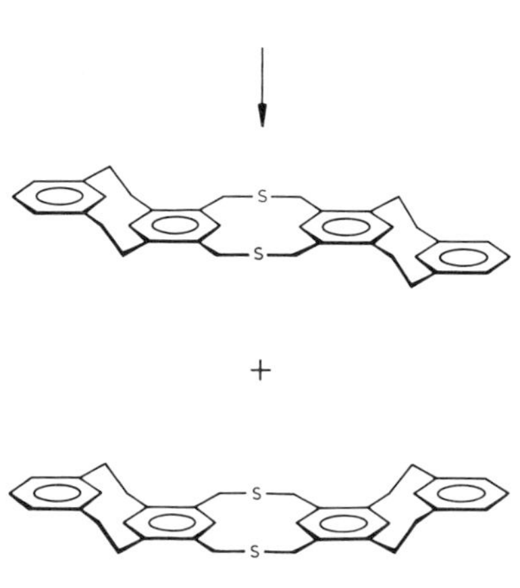

c

racemic mixture

Figure 2-68.
(a) The reaction product is the "cis" dimer of two homochiral 4-(bromomethyl)-6-(mercaptomethyl)[2.2]metacyclophanes.
(b) The reaction product is the "trans" dimer of two heterochiral 4-(bromomethyl)-6-(mercaptomethyl)[2.2]metacyclophane molecules.
(c) The reaction product is a 1:1 mixture of the two dimers when the 4-(bromomethyl)-6-(mercaptomethyl)[2.2]metacyclophane molecules are present in a racemic mixture.

65

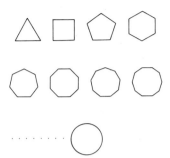

Figure 2-69.
Regular polygons.

2.8 Regular Polyhedra

"A convex polyhedron is said to be *regular* if its faces are regular and equal, while its vertices are all surrounded alike" [2-48]. A polyhedron is convex if every dihedral angle is less than 180°. The dihedral angle is the angle formed by two polygons joined along a common edge.

There are only five regular convex polyhedra, a very small number indeed. The regular convex polyhedra are called Platonic solids because they constituted an important part of Plato's natural philosophy. They are: the tetrahedron, cube (hexahedron), octahedron, dodecahedron, and the icosahedron. The faces are regular polygons, either regular triangles, regular pentagons, or squares.

A regular polygon has equal interior angles and equal sides. Fig. 2-69 presents a regular triangle, a regular quadrangle, i.e., square, a regular pentagon, and so on. The circle is obtained in the limit as the number of sides approaches infinity. The regular polygons have an n-fold rotational symmetry axis perpendicular to their plane and going through their midpoint. Here n is 1, 2, 3, ... up to infinity for the circle.

The five regular polyhedra are shown in Fig. 2-70. Their characteristic parameters are given in Table 2-3. According to Weyl [2-10], the existence of the tetrahedron, cube, and octahedron is a fairly trivial geometric fact. On the other hand, Weyl considered the discovery of the regular dodecahedron and the regular icosahedron "one of the most beautiful and singular discoveries made in the whole history of mathematics". However, to ask the question who first constructed the regular polyhedra is according to Coxeter [2-48] like asking the question who first used fire.

Table 2-3. Characteristics of the Regular Polyhedra

Name	Polygon	Number of Faces	Vertex Figure	Number of Vertices	Number of Edges
Tetrahedron	3	4	3	4	6
Cube	4	6	3	8	12
Octahedron	3	8	4	6	12
Dodecahedron	5	12	3	20	30
Icosahedron	3	20	5	12	30

Figure 2-70.
The five Platonic solids.

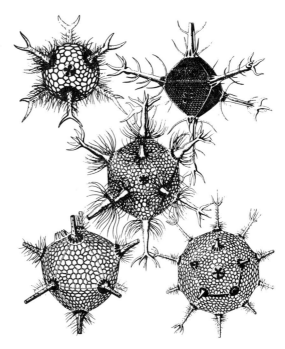

Figure 2-71.
Radiolarians from Häckel's book [2-15].

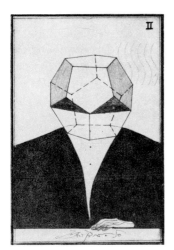

Figure 2-72.
An artist's representation of the pentagonal dodecahedron, Horst Janssen: "ChriStall-Knecht" (crystal slave). Reproduced by permission.

Many primitive organisms have the shape of the pentagonal dodecahedron. As will be seen later, it is not possible to have crystal structures having this symmetry. Belov [2-49] suggested that the pentagonal symmetry of the primitive organisms represents their defense against crystallization. Several radiolarians from Häckel's book [2-15] are shown in Fig. 2-71. An artist's representation of the pentagonal dodecahedron is shown in Fig. 2-72, entitled "ChriStall-Knecht" (crystal slave).

Fig. 2-73 shows Kepler's planetary model based on the regular solids. According to this model the greatest distance of one

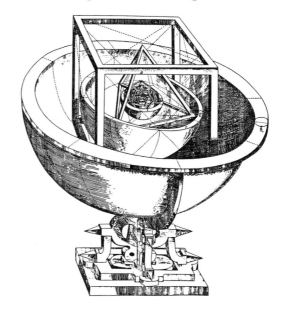

Figure 2-73.
Kepler's planetary model based on the regular solids [2-50].

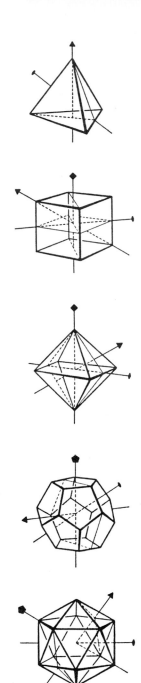

planet from the sun stands in a fixed ratio to the least distance of the next outer planet from the sun. There are five ratios describing the distances of the six planets which were known to Kepler. A regular solid can be interposed between two adjacent planets so that the inner planet, when at its greatest distance from the sun, lays on the inscribed sphere of the solid, while the outer planet, when at its least distance, lays on the circumscribed sphere.

Arthur Koestler in "The Sleepwalkers" [2-51] called this planetary model "Kepler's most spectacular failure". However, the planetary model which is also a densest packing model probably symbolizes Kepler's best attempt at attaining a unified view of his work both in astronomy and in what we would call today crystallography. The ratios of the Copernican distances of the planets to the sun and the ratios of the inscribed spheres to those circumscribing the regular solids are cited in Table 2-4 as given by Schneer [2-52] according to Kepler [2-50].

There are several excellent monographs on regular figures, two of which are especially noteworthy [2-48, 2-53].

The Platonic solids have very high symmetries and one very important common characteristic. None of the rotational symmetry axes of the regular polyhedra is unique, but each axis is associated with several equivalent axes to itself. The five regular solids can be classified into three symmetry classes:

Tetrahedron	$3/2 \cdot m = 3/\widetilde{4}$
Cube and Octahedron	$3/4 \cdot m = \widetilde{6}/4$
Dodecahedron and Icosahedron	$3/5 \cdot m = 3/\widetilde{10}$

Figure 2-74.
Characteristic symmetry elements of the Platonic solids.

It is equivalent to describe the symmetry class of the tetrahedron as $3/2 \cdot m$ or $3/\widetilde{4}$. The skew line relating two axes means that they are not orthogonal. The symbol $3/2 \cdot m$ denotes a three-fold axis, and a two-fold axis which are not perpendicular, and a symmetry plane which includes these axes. These three symmetry elements are indicated in Fig. 2-74. The symmetry class $3/2 \cdot m$ is equivalent to a pair of a three-fold axis and a four-fold mirror-rotation axis. In both cases the three-fold axes connect one of the vertices of the tetrahedron with the midpoint of the opposite face. The four-fold mirror-rotation axes coincide with the two-fold axes. The presence of the four-fold mirror-rotation axis is easily seen if the tetrahedron is rotated by a quarter of rotation about a two-fold axis and is then reflected by a symmetry plane perpendicular to this axis. The symmetry operations chosen as basic will then generate the remaining symmetry elements. Thus the two descriptions are equivalent.

Table 2-4. Kepler's Ratios after Schneer [2-52]

	Ratio of Inscribed to Circumscribing Sphere (x 1000)	Ratio of Inner to Outer Planetary Orbit (x 1000) Using the Copernican Distances	
		1000	Saturn
Cube	577	572	Jupiter/Saturn
Tetrahedron	333	290	Mars/Jupiter
Dodecahedron	795	658	Earth/Mars
Icosahedron	795	719	Venus/Earth
Octahedron	577	500	Mercury/Venus

Characteristic symmetry elements of the cube are shown in Fig. 2-74. Three different kinds of symmetry planes go through the center of the cube parallel to its faces. Furthermore, six symmetry planes connect the opposite edges and also diagonally bisect the faces. The four-fold rotation axes connect the midpoints of opposite faces. The six-fold mirror-rotation axes coincide with three-fold rotation axes. They connect opposite vertices and are located along the body diagonals. The symbol $\tilde{6}/4$ does not directly indicate the symmetry planes connecting the midpoints of opposite edges, the two-fold rotation axes, or the center of symmetry. These latter elements are generated by the others. The presence of a center of symmetry is well seen by the fact that each face and edge of the cube has its parallel counterpart. The tetrahedron, on the other hand, has no center of symmetry.

The octahedron is in the same symmetry class as the cube. The antiparallel character of the octahedron faces is especially conspicuous. As seen in Fig. 2-74, its four-fold symmetry axes go through the vertices, the three-fold axes through the face midpoints, and the two-fold axes go through the edge midpoints.

The pentagonal dodecahedron and the icosahedron are in the same symmetry classes. The five-fold, three-fold and two-fold rotation axes intersect the midpoints of faces, the vertices and the edges of the dodecahedron, respectively (Fig. 2-74). On the other hand, the corresponding axes intersect the vertices, and the midpoints of faces and edges of the icosahedron (Fig. 2-74).

Figure 2-75.
The four regular star polyhedra.

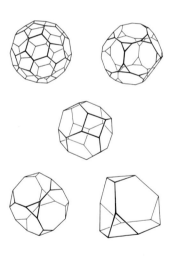

Figure 2-76.
Some semiregular polyhedra.

Consequently, the five regular polyhedra exhibit a dual relationship as regards their faces and vertex figures. The tetrahedron is self dual (Table 2-3).

If the definition of regular polyhedra is not restricted to convex figures, their number rises from five to nine. The additional four are depicted in Fig. 2-75 (for more information see, e.g., [2-48, 2-53]). They are called by the common name of regular star polyhedra.

The *sphere* deserves mention. It is one of the simplest possible figures, and, accordingly one with high and complicated symmetry. It has an infinite number of rotation axes with infinite order. All of them coincide with body diagonals going through the midpoint of the sphere. The midpoint, which is also a singular point, is the center of symmetry of the sphere. The following symmetry elements may be chosen as basic ones: two infinite order rotation axes which are not perpendicular plus one symmetry plane. Therefore, the symmetry class of the sphere is $\infty/\infty \cdot m$. Concerning the symmetry of the sphere Kepes [2-54] quotes Copernicus: "... the spherical is the form of all forms most perfect, having need of no articulation; and the spherical is the form of greatest volumetric capacity, best able to contain and circumscribe all else; and all the separated parts of the world – I mean the sun, the moon, and the stars – are observed to have spherical form; and all things tend to limit themselves under this form – as appears in drops of water and other liquids – whenever of themselves they tend to limit themselves. So no one may doubt that the spherical is the form of the world, the divine body".

In addition to the regular polyhedra, there are various families of polyhedra with decreased degrees of regularity [2-48, 2-53, 2-55]. The so-called *semiregular* or Archimedian polyhedra are similar to the Platonic polyhedra in that all their faces are regular and all their vertices are congruent. However, the polygons of their faces are not all of the same kind. The thirteen semiregular polyhedra are listed in Table 2-5 and some of them are also shown in Fig. 2-76. Table 2-5 also enumerates their rotation axes.

The simplest semiregular polyhedra are obtained by symmetrically shaving off the corners of the regular solids. They are the truncated regular polyhedra and are marked (a) in Table 2-5. Two semiregular polyhedra belong to the so-called quasiregular polyhedra. They have two kinds of faces, and each face of one kind is entirely surrounded by faces of the other kind. They are marked (b) in Table 2-5. The remaining six semiregular polyhedra may be derived from the other semiregular polyhedra.

The prisms and antiprisms are also important polyhedron families. A prism has two congruent and parallel faces and they are joined by a set of parallelograms. An antiprism also has two congruent and parallel faces but they are joined by a set of

Figure 2-77.
Prisms and antiprisms.

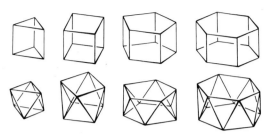

triangles. There is an infinite number of prisms and antiprisms and some of them are shown in Fig. 2-77. A prism or an antiprism is semiregular if all its faces are regular polygons. A cube can be considered a square prism, and an octahedron can be considered a triangular antiprism.

There are additional important polyhedra which are important in discussing molecular geometries and crystal structures.

Table 2-5. The Thirteen Semiregular Polyhedra

No. Name		Number of			Number of Rotation Axes			
		Faces	Vertices	Edges	2-fold	3-fold	4-fold	5-fold
1	Truncated tetrahedron[a]	8	12	18	3	4	0	0
2	Truncated cube[a]	14	24	36	6	4	3	0
3	Truncated octahedron[a]	14	24	36	6	4	3	0
4	Cuboctahedron[b]	14	12	24	6	4	3	0
5	Truncated cuboctahedron	26	48	72	6	4	3	0
6	Rhombicuboctahedron	26	24	48	6	4	3	0
7	Snub cube	38	24	60	6	4	3	0
8	Truncated dodecahedron[a]	32	60	90	15	10	0	6
9	Icosidodecahedron[b]	32	30	60	15	10	0	6
10	Truncated icosahedron[a]	32	60	90	15	10	0	6
11	Truncated icosidodecahedron	62	120	180	15	10	0	6
12	Rhombicosidodecahedron	62	60	120	15	10	0	6
13	Snub dodecahedron	92	60	150	15	10	0	6

[a] Truncated regular polyhedra   [b] Quasiregular polyhedra

71

References

[2-1]   T. Mann, *The Magic Mountain.* The cited passage on p. 9 is in French both in the original German and its English translation. The English translation cited in our text was kindly provided by Dr. Jack M. Davis, Professor of English, The University of Connecticut, 1984.

[2-2]   J. Kepler, *Strena, seu De Nive Sexangula,* 1611. English translation, *The Six-cornered Snowflake,* Clarendon Press, Oxford, 1966.

[2-3]   I. I. Shafranovskii, *Kepler's Crystallographic Ideas and his Tract "The Six-cornered Snowflake",* In *Kepler, Four Hundred Years,* A. Beer and P. Beer (eds.), Proceedings of Conferences held in honor of Johannes Kepler. Vistas Astron. **18**, Pergamon Press, Oxford, etc. (1975).

[2-4]   H. A. Sackeim, R. C. Gur, and M. C. Saucy, Science **202**, 434 (1978).

[2-5]   S. P. Springer and G. Deutsch, *Left Brain, Right Brain,* Freeman & Co., San Francisco, 1981.

[2-6]   A. V. Shubnikov and V. A. Koptsik, *Symmetry in Science and Art,* Plenum Press, New York and London, 1974. Russian original: Nauka, Moscow, 1972.

[2-7]   M. Apagyi, personal communication, 1983.

[2-8]   L. E. Orgel, *The Origins of Life: Molecules and Natural Selection,* John Wiley & Sons, New York, London, Sydney, Toronto, 1973.

[2-9]   R. Jastrow, *Red Giants and White Dwarfs,* W. W. Norton, New York and London, 1979.

[2-10]   H. Weyl, *Symmetry,* Princeton University Press, Princeton, NJ, 1952.

[2-11]   W. Herzfelde, *John Heartfield. Leben und Werk dargestellt von seinem Bruder,* Third revised and expanded edition. VEB Verlag der Kunst, Dresden, 1976.

[2-12]   L. R. H. Appleton, *American Indian Design and Decoration,* Dover Publications, New York, 1971.

[2-13]   M. Ickis, *The Standard Book of Quilt Making and Collecting,* Dover Publications, New York, 1959.

[2-14]   D. K. Washburn, *A Symmetry Analysis of Upper Gila Area Ceramic Design,* Papers of the Peabody Museum of Archeology and Ethnology, Vol. 68, Harvard University, Cambridge, MA, 1977.

[2-15]   E. Häckel, *Kunstformen der Natur,* Vols. 1-10, Verlag des Bibliographischen Instituts, Leipzig, 1899–1904.

[2-16]   U. Nakaya, *Snow Crystals,* Harvard University Press, Cambridge, MA, 1954.

[2-17]   G. Taubes, Discover, p. 75, January, 1984.

[2-18]   D. McLachlan, Proc. Natl. Acad. Sci. **43**, 143 (1957).

[2-19]   W. A. Bentley and W. J. Humphreys, *Snow Crystals,* McGraw-Hill, New York and London, 1931.

[2-20]   Attributed to M. Polányi. Private communication from Professor W. Jim Neidhardt, New Jersey Institute of Technology, 1984.

[2-21]   A. L. Mackay, Jugosl. Cent. Kristalogr. **10**, 5 (1975); A. L. Mackay, personal communication, 1982.

[2-22]   J. Needham and Lu Gwei-Djen, Whether **16,** 319 (1961).

[2-23]   Han Ying, 135 B.C. as cited in [2-22].

[2-24]   G. Hellmann, *Schneekrystalle,* Berlin, 1893.

[2-25]   T. Stamp and C. Stamp, *William Scoresby, Arctic Scientist,* Caedmon of Whitby, 1976.

[2-26]   J. Reston, International Herald Tribune, Thursday, May 7, p. 4, 1981.

[2-27]   F. Karinthy, *Grave and Gray, Selections from his Work,* Corvina Press, Budapest, 1973.

[2-28]   C. H. MacGillavry, *Symmetry Aspects of M. C. Escher's Periodic Drawings,* Bohn, Scheltema and Holkema, Utrecht, 1976.

[2-29]   E. A. Abbott, *Flatland. A Romance of Many Dimensions,* Barnes and Noble Books, Fifth Edition, New York, Cambridge, Philadelphia, etc., 1983.

[2-30]   W. Engelhardt, Mathematischer Unterricht **9** (2), 49 (1963).

[2-31]   D. Y. Curtin and I. C. Paul, Chem. Rev. **81,** 525 (1981).

[2-32]   P. Groth, *Chemische Kristallographie,* 5 Vols., Verlag von Wilhelm Engelmann, Leipzig, 1906–1919.

[2-33]   I. C. Paul and D. Y. Curtin, Science **187,** 19 (1975).

[2-34]   E. N. Duesler, R. B. Kress, C.-T. Lin, W.-I. Shiau, I. C. Paul, and D. Y. Curtin, J. Am. Chem. Soc. **103,** 875 (1981).

[2-35]   J. C. Burfoot and G. W. Taylor, *Polar Dielectrics,* University of California Press, Berkeley and Los Angeles, 1979, as cited by Curtin and Paul [2-31].

[2-36]   Lord Kelvin, *Baltimore Lectures,* C. J. Clay and Sons, London, 1904.

[2-37]   R. N. Bracewell, *The Galactic Club. Intelligent Life in the Outer Space,* W. H. Freeman & Co., San Francisco, 1975.

[2-38]   L. L. Whyte, Leonardo **8,** 245 (1975); Nature **182,** 198 (1958).

[2-39]   M. Gardner, *The Ambidextrous Universe,* The New American Library, New York and Toronto, 1969.

[2-40]   W. B. Wood, J. H. Wilson, R. M. Benbow, and L. E. Hood, *Biochemistry: A Problems Approach,* Benjamin/Cummings Publishing Co., Menlo Park, CA, 1981.

[2-41]   A. V. Shubnikov, N. V. Belov and others, *Colored Symmetry.* A Series of Publications from the Institute of Crystallography, Academy of Sciences of the U.S.S.R., Moscow, 1951–1958, W. T. Holser, ed., Pergamon Press, New York, 1964.

[2-42]   "C'est la dissymétrie qui crée le phénomène", P. Curie, J. Phys. Paris **3,** 393 (1894).

[2-43]   J. D. Bernal, *The Origin of Life,* The World Publ. Co., Cleveland and New York, 1967.

[2-44]   V. Prelog, Science **193,** 17 (1976).

[2-45]   *See* J. B. S. Haldane, Nature **185,** 87 (1960), citing L. Pasteur, C. R. Acad. Sci. Paris, June 1, 1874.

[2-46]   F. A. L. Anet, S. S. Miura, J. Siegel, and K. Mislow, J. Am. Chem. Soc. **105,** 1419 (1983).

[2-47]   T. Umemoto, T. Otsubo, Y. Sakata, and S. Misumi, Tetrahedron Lett. 593 (1973); T. Umemoto, T. Otsubo, and S. Misumi, Tetrahedron Lett. 1573 (1974).

[2-48]   H. S. M. Coxeter, *Regular Polytopes,* Third Edition, Dover Publications, New York, 1973.

[2-49]   N. V. Belov, *Notes on Structural Mineralogy* (in Russian, *"Ocherki po strukturnoi mineralogii"),* Nedra, Moscow, 1976.

[2-50]   J. Kepler, *Mysterium cosmographicum,* 1595.

[2-51]   A. Koestler, *The Sleepwalkers,* The Universal Library, Grosset and Dunlap, New York, 1963.

[2-52]   C. J. Schneer, Am. Sci. **71,** 254 (1983).

[2-53]   L. Fejes Tóth, *Regular Figures,* Pergamon Press, New York, 1964.

[2-54]   N. Copernicus, *De Revolutionibus Orbium Caelestium,* 1543, as cited in G. Kepes, *The New Landscape in Art and Science,* Theobald & Co., Chicago, 1956.

[2-55]   P. Pearce and S. Pearce, *Polyhedra Primer,* Van Nostrand Reinhold Co., New York, 1978.

3 Molecules: Shape and Geometry

A molecule is not simply a collection of its constituent atoms. It is kept together by interactions among those atoms. Thus for some purposes it is better to consider the molecule as consisting of the nuclei of its constituent atoms and its electron density distribution. Generally it is the geometry and symmetry of the arrangement of the atomic nuclei that is considered to be the geometry and symmetry of the molecule itself.

Molecules are finite figures with at least one singular point in their symmetry description. Thus point groups are applicable to them. There is no inherent limitation on the available symmetries for molecules. On the other hand, severe restrictions apply to the symmetry of crystals, as will be seen later. In fact the molecules occupy a lower, more fundamental level in the hierarchy of structures than do crystals. Many crystals themselves are built from molecules.

Molecules in the gas phase are considered to be *free* molecules. They are so far apart that they are unperturbed by interactions from other molecules. On the other hand, intermolecular interactions may occur between the molecules in condensed phases, i. e. in liquids, melts, amorphous solids, or crystals. In the present discussion all molecules will be assumed unperturbed by their environment, regardless of the phase or state of matter in which they exist. The possible influence of intermolecular interactions will be considered later, in the discussion of crystals.

The molecules are never motionless. They are performing vibrations all the time. In addition, the gaseous molecules, and also the molecules in liquid, are performing rotational and translational motion as well. Molecular vibrations constitute relative displacements of the atomic nuclei with respect to their equilibrium positions and occur in all phases, including the crystalline state, and even at the lowest possible temperatures. The magnitude of molecular vibrations is relatively large, amounting to several percent of the internuclear distances. Typically, there are about 10^{12} to 10^{14} vibrations per second.

Symmetry considerations are fundamental in any description of molecular vibrations, as will be seen later in detail. First, however, the molecular symmetries will be discussed, ignoring entirely the motion of the molecules. Various molecular symmetries will be illustrated by numerous examples. A simple model will also be discussed to gain some insight into the origins of the various shapes and symmetries in the world of molecules. Our considerations will be restricted, however, to relatively simple, thus rather symmetrical systems. The importance and consequences of intramolecular motion involving relatively large amplitudes, will be commented upon in the final section of this chapter.

3.1 Formulae, Isomers

The empirical formula of a chemical compound expresses its composition. For example $C_2H_4O_2$ indicates that the molecule consists of two carbon, four hydrogen and two oxygen atoms. This formulation, however, provides no information on the order in which these atoms are linked. This empirical formula may correspond to methyl formate, acetic acid, and glycol aldehyde. Only the structural formulae for these compounds, shown in Fig. 3-1, distinguish among them. This is called structural isomerism. The formulae in Fig. 3-1 may also be written in a simpler way,

$$HCOOCH_3 \qquad CH_3COOH \qquad HCOCH_2OH$$

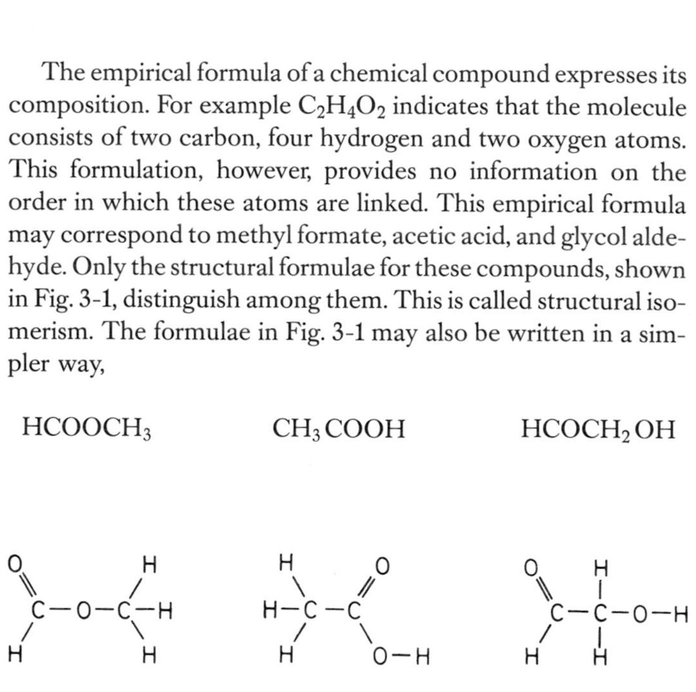

Figure 3-1.
Structural isomerism of the molecules with the same empirical formula $C_2H_4O_2$.

Although these molecules, as a whole, are not symmetric, some of their component parts may be symmetrical. They possess what is called local symmetry. Similar atomic groups in different molecules often have similar geometries, and thus similar local symmetries. The structural formulae reveal considerable information about these local symmetries, or at least their similarities and differences in various molecules. The above simplified structural formulae are especially useful in this respect. This approach is widely applicable in organic chemistry where relatively few kinds of atoms build an enormous number of different molecules. A far greater diversity of structural peculiarities is characteristic for inorganic compounds.

The symbol for the carbon atom occurs twice in all three simplified structural formulae above, a fact that indicates differences in their structural positions. The same argument applies to the oxygen atoms. On the other hand, three hydrogens are equivalent in both methyl formate and acetic acid, with the fourth being different in the two molecules. There are three different types of hydrogen positions in glycol aldehyde.

Structural isomerism is also called constitutional isomerism. The constitution of the molecule is a description of the way in which the atoms are bonded, i. e., their connectivity. Molecules which have the same constitution may differ in the spatial arrangement of their atoms. Two different kinds of isomerism arise from the differences in these spacial arrangements. One is called *enantiomerism* and this is when the isomers are related as an object is to its mirror image. This isomerism is the consequence of chirality and thus has already been discussed. The other is *rotational isomerism*. It may occur when two bonds in a molecule are separated by a third one.

3.2 Rotational Isomerism

The four-atomic chain is the simplest system for which rotational isomerism is possible. It is shown in Fig. 3-2.

Rotational isomers, or conformers, are various forms of the same molecule related by rotation around a bond as axis. The various rotational forms of a molecule are described by the same empirical formula and by the same structural formula. Only the relative positions of the two bonds (or groups of atoms) at the two ends of the rotation axis are changed. The molecular point groups for various rotational isomers may be entirely different.

Figure 3-2.
Rotational isomerism of a four-atomic chain.

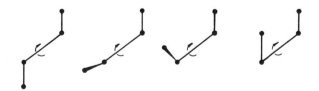

Rotational isomers can be conveniently represented by so-called projection formulae in which the two bonds (or groups of atoms) at the two ends are projected onto a plane which is perpendicular to the central bond. This plane is denoted by a circle whose center coincides with the projection of the rotation axis.

The bonds in front of this plane are drawn as originating from the center. The bonds behind this plane, i. e. the bonds from the other end of the rotation axis, are drawn as originating from the perimeter of the circle. A rotational isomer of 1,2-dibromo-1,2-dichloroethane, BrClHC-CHClBr, is represented in this way in Fig. 3-3. This kind of representation is also called a Newman projection. The form shown in Fig. 3-3 is a so-called *staggered* one as the bonds of one group appear alternately between the bonds of the other group in the projection. When the bonds coincide in the projection, the rotational isomer is said to be *eclipsed.*

Figure 3-3.
Newman projection representation of a rotational isomer of 1,2-dibromo-1,2-dichloroethane.

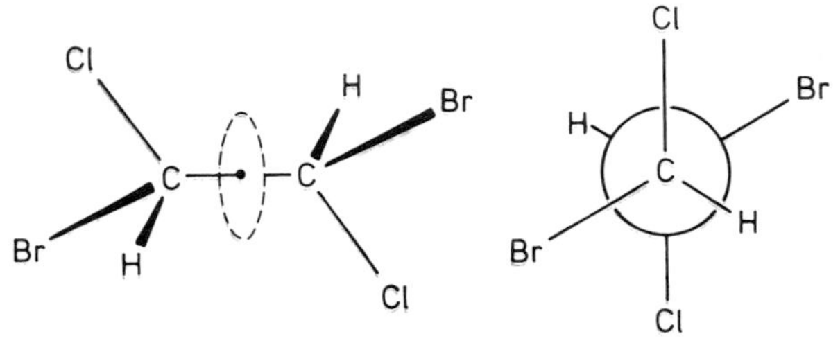

The drawings by Degas "End of the Arabesque" and "Seated Dancer Adjusting Her Shoes" may be looked at as illustrations of the staggered and eclipsed conformations of A_2B-BC_2 molecules. They are shown in Fig. 3-4a. Their projection-like representations are given in Fig. 3-4b, while the conformers of the molecules are depicted in Fig. 3-4c. Degas' drawings are also helpful in understanding the representation for the rotational isomers described above. The projections in Fig. 3-4 represent views along the B-B bond, i. e. the dancer's body. The plane bisecting the B-B bond is shown by the circle and it corresponds to the dancer's skirt. The dancer's arms and legs refer to the bonds B-A and B-C, respectively. Incidentally, the bouquet in the right hand of the dancer in the staggered conformation may be viewed as a different substituent.

Two important cases in rotational isomerism may be distinguished by considering the nature of the central bond. When it is a double bond, rotation of one form into another is hindered by a very high potential barrier. This barrier may be so high that the two rotational isomers will be stable enough to make their physical separation possible. An example is 1,2-dichloroethylene (Fig. 3-5). The symmetry of the *cis* isomer is characterized by two mutually perpendicular mirror planes generating also a two-fold rotational axis. This symmetry class is labeled *mm.* An equivalent notation is C_{2v} as will be seen in the next section. The *trans* isomer has one two-fold rotation axis with a perpendicular symmetry plane, its symmetry class is $2/m$ (C_{2h}).

Figure 3-4.
Illustration for the projectional representation of rotational isomers [3-1].
(a) Left: a drawing after Degas' "End of the Arabesque" by Ferenc Lantos [3-2(a)].
Right: a drawing after Degas' "Seated Dancer Adjusting Her Shoes" by Ferenc Lantos [3-2(b)].

(b) Contour drawings of the dancers.

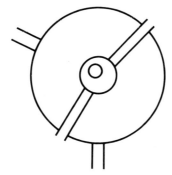 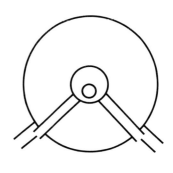

(c) Staggered and eclipsed rotational isomers of the A_2BBC_2 molecule by Newman projections representing view along the B-B bond.

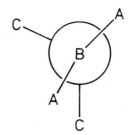 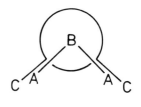

Figure 3-5.
1,2-Dichloroethylene: *cis* and *trans* isomers.

79

Rotational isomerism relative to a single bond is illustrated by ethane and 1,2-dichloroethane in Fig. 3-6. During a complete rotation of one methyl group around the C-C bond relative to the other methyl group, the ethane molecule appears three times in the stable staggered form and three times in the unstable eclipsed form. As all the hydrogen atoms of one methyl group are equivalent, the three energy minima are equivalent, and so are the three energy maxima, as seen in Fig. 3-6a. The situation becomes more complicated when the three ligands bonded to the carbon atoms are not the same. This is seen for 1,2-dichloroethane in Fig. 3-6b. There are three highly symmetrical forms. Of these two are staggered with C_{2h} and C_2 symmetries, respectively. The third is an eclipsed form with C_{2v} symmetry. This form has Cl/Cl and H/H eclipsing. There is no other fully eclipsed form because of insufficient symmetry [3-3].

Figure 3-6.
Potential energy functions for rotation about a single bond, φ is the angle of rotation. Newman projections of the symmetrical rotational isomers are also shown with their point-group symmetries. (a) Ethane, $H_3C\text{-}CH_3$. There are two different symmetrical forms. Both the staggered form with D_{3d} symmetry and the eclipsed form with D_{3h} symmetry occur three times in a complete rotational circuit.

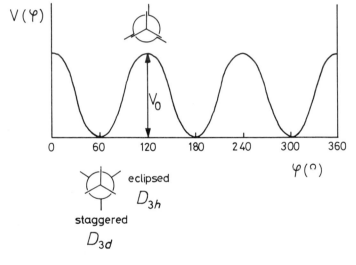

(b) 1,2-Dichloroethane, $ClH_2C\text{-}CH_2Cl$. There is no other symmetrical form in the region between the two symmetrical staggered forms shown. Only partial eclipsing can occur here because of insufficient symmetry (cf. [3-3]). The eclipsed form with C_{2v} symmetry and the staggered form with C_{2h} symmetry occur once, while the staggered form with C_2 symmetry occurs twice in a complete rotational circuit.

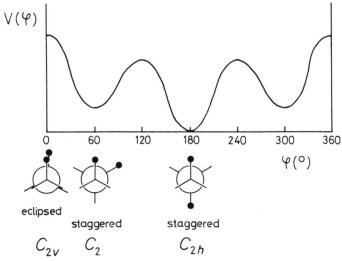

Fig. 3-6 shows only the symmetrical conformers by projection formulae. The symmetrical forms always belong to extreme energies, either minima or maxima. The barriers to internal rotation in the potential energy functions depicted in Fig. 3-6 are about 10 kJ/mol. Typical barriers for systems where the double bonds would be considered to be the "rotational axis" may be as much as 30 times greater than those for systems with single bonds.

3.3 Symmetry Notations

So far the so-called International or Hermann–Mauguin symmetry notations have been used in the descriptions in this text. Another, older system by Schoenflies is generally used, however, to describe the molecular point-group symmetries. In fact, this notation has been given in parenthesis following the International notation in the preceding section. The Schoenflies notation has the advantage of succinct expression for even complicated symmetry classes combining various symmetry elements. The two systems are compiled in Table 3-1 (see e. g. [3-4]) for a selected set of 32 symmetry classes. Point-group symmetries in the world of crystals are, in fact, restricted to these 32 classes. The reasons and significance of these restrictions will be discussed later, in the chapter on crystals. There are no restrictions on the point-group symmetries for individual molecules.

The Schoenflies notation for rotation axes is C_n, and for mirror-rotation axes the notation is S_{2n}, where n is the order of the rotation. The symbol for a center of symmetry is i. Symmetry planes are labeled σ; σ_v is a vertical plane which always coincides with the rotation axis with an order higher than two, and σ_h is a horizontal plane which is always perpendicular to the rotation axis when it has an order higher than two.

Point-group symmetries not listed in Table 3-1 may easily be assigned the appropriate Schoenflies notation by analogy. Thus, e. g., C_{5v}, D_{5h}, C_7, C_8, etc. can be established. Such symmetries may well occur among real molecules.

Table 3-1. Symmetry Notations

Number	Hermann-Mauguin	Schoenflies
1	1	C_1
2	$\bar{1}$	C_i
3	m	C_s
4	2	C_2
5	$2/m$	C_{2h}
6	mm	C_{2v}
7	222	D_2
8	mmm	D_{2h}
9	4	C_4
10	$\bar{4}$	S_4
11	$4/m$	C_{4h}
12	$4mm$	C_{4v}
13	$\bar{4}2m$	D_{2d}
14	422	D_4
15	$4/mmm$	D_{4h}
16	3	C_3
17	$\bar{3}$	S_6
18	$3m$	C_{3v}
19	32	D_3
20	$\bar{3}m$	D_{3d}
21	$\bar{6}$	C_{3h}
22	6	C_6
23	$6/m$	C_{6h}
24	$\bar{6}m2$	D_{3h}
25	$6mm$	C_{6v}
26	622	D_6
27	$6/mmm$	D_{6h}
28	23	T
29	$m\bar{3}$	T_h
30	$43m$	T_d
31	432	O
32	$m\bar{3}m$	O_h

3.4 Establishing the Molecular Point Group

Fig. 3-7 shows a possible scheme for establishing the molecular point group (cf. [3-5, 3-6]). The symmetry of most molecules may be reliably established by this scheme.

First an examination is carried out whether the molecule belongs to some "special" group. If the molecule is linear it may have a perpendicular symmetry plane ($D_{\infty h}$) or it may not have one ($C_{\infty v}$). Very high symmetries are easy to recognize. Each of the groups T, T_h, T_d, O, and O_h, has four three-fold rotation axes. Both icosahedral I and I_h groups require ten three-fold rotation axes and six five-fold rotation axes. The molecules belonging to these groups have a central tetrahedron, octahedron, cube, or icosahedron.

If the molecule does not belong to one of these "special" groups, a systematic approach is followed. Firstly, the possible presence of rotation axes in the molecule is checked. If there is no rotation axis, then it is determined whether there is a symmetry plane (C_s). In the absence of rotational axes and mirror planes, there may only be a center of symmetry (C_i), or there may be no symmetry element at all (C_1). If the molecule has rotation axes, it may have a mirror-rotation axis with even-number order (S_{2n}) coinciding with the rotation axis. For S_4 there will be a coinciding C_2, for S_6 a coinciding C_3, and for S_8, both C_2 and C_4.

Figure 3-7.
Scheme for establishing the molecular point groups (cf. [3-5, 3-6]).

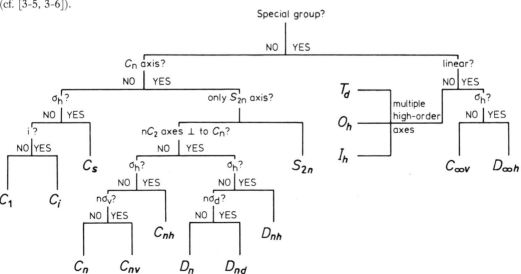

In any case the search is for the highest order C_n axis. Then it is examined whether there are n C_2 axes present perpendicular to the C_n axis. If such C_2 axes are present, then there is D symmetry. If in addition to D symmetry there is a σ_n plane, the point group is D_{nh}, while if there are n symmetry planes (σ_d) bisecting the two-fold axes, the point group is D_{nd}. If there are no symmetry planes in a molecule with D symmetry, the point group is D_n.

Finally, if no C_2 axes perpendicular to C_n are present, then the lowest symmetry will be C_n, when a perpendicular symmetry plane is present, it will be C_{nh}, and when there are n coinciding symmetry planes, the point group will be C_{nv}.

3.5 Examples

In the section that follows, actual molecules will be shown for the various point groups along with some rosettes or other illustrations familiar from every day life. The Schoenflies notation is used and the characteristic symmetry elements are enumerated.

C_1

There are no symmetry elements (except the one-fold rotation axis, or identity, of course). Some examples are shown in Fig. 3-8.

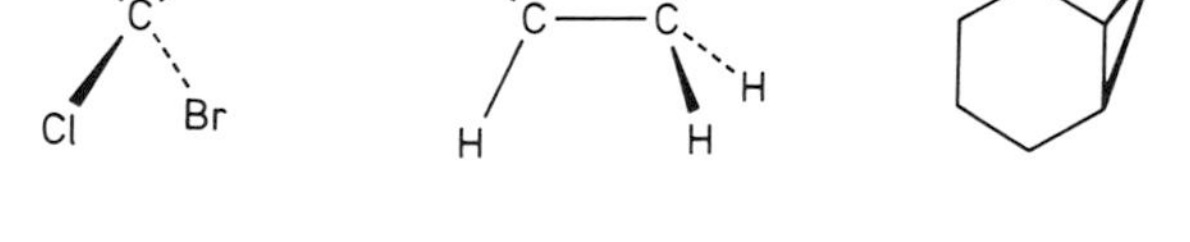

Figure 3-8.
C_1 symmetry: no symmetry elements except the one-fold rotation axis (C_1 symmetry is asymmetry).

C_2

One two-fold rotation axis. Examples: Fig. 3-9a.

C_3, C_4, C_5, C_6

One three-fold, four-fold, five-fold, six-fold rotation axis, respectively. Examples: Fig. 3-9b-e.

C_7, ..., C_n

Can be continued by analogy. C_n has one n-fold rotation axis.

Figure 3-9.
(a) C_2. Logos: Security First National Bank, California (left) and United Banks of Colorado. Source of logos [3-7(a)].

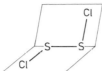

(b) C_3. Logos: Pittsburgh National Bank (left) and Woolmark [3-7(a)].

[3]-rotane

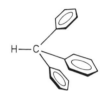

(c) C_4. Logo: Chase Manhattan Bank [3-7(a)].

[4]-rotane

(d) C_5. Logo: First American National Bank, Tennessee [3-7(a)].

[5]-rotane

(e) C_6. Logo: Crocker Bank [3-7(a)].

[6]-rotane

85

C_i

Center of symmetry. Examples: Fig. 3-10.

Figure 3-10.
C_i.

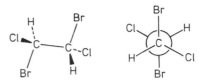

C_s

Figure 3-11.
C_s. The picture shows the tail
of a whale, off Plymouth, MA.
Photograph by the authors.

One symmetry plane. Examples: Fig. 3-11.

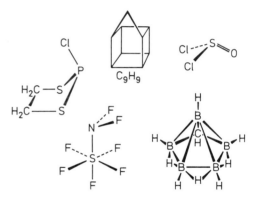

S_4

One four-fold mirror-rotation axis. Example: Fig. 3-12a.

S_6

One six-fold mirror-rotation axis, which is, of course, equivalent to one three-fold rotation axis plus center of symmetry. Example: Fig. 3-12b.

Figure 3-12.

(b) *S_6.*

a) *S_4.*

$[(CH_2)_2CH]_3CC[CH(CH_2)_2]_3$

One two-fold rotation axis with a perpendicular symmetry plane. Examples: Fig. 3-13a.

$C_{3h}, C_{4h}, C_{5h}, ..., C_{nh}$
One n-fold rotation axis plus a symmetry plane perpendicular to it. Examples: Fig. 3-13b-d.

Figure 3-13.
(a) C_{2h}.

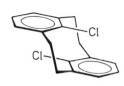

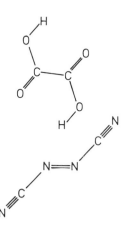

(b) C_{3h}. The molecule bicyclo [3.3.3]undecane is also called "manxane". It has C_{3h} symmetry indeed. The Isle of Man coin shows a one-sided rosette whose symmetry is only C_3.

(c) C_{4h}.

(d) C_{6h}.

C_{2v}
Two perpendicular symmetry planes whose crossing line is a two-fold rotation axis. Examples: Fig. 3-14a.

C_{3v}
One three-fold rotation axis with three symmetry planes which include the rotation axis. The angle is 60° between two symmetry planes. Examples: Fig. 3-14b.

C_{4v}
One four-fold rotation axis with four symmetry planes which include the rotation axis. The four planes are grouped in two non-equivalent pairs. One pair is rotated relative to the other pair by 45°. The angle between the two planes within each pair is 90°. Examples: Fig. 3-14c.

C_{5v}, C_{6v}, ..., C_{nv}

This series can be continued by analogy. When n is even, there are two sets of symmetry planes. One set is rotated relative to the other set by $(180/n)°$. The angle between the planes within each set is $(360/n)°$. When n is odd, the angle between the symmetry planes is $(180/n)°$. Example: Fig. 3-14d.

Figure 3-14.
(a) C_{2v}.

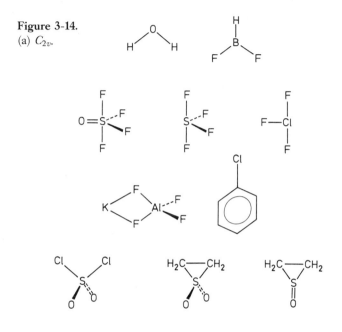

(b) C_{3v}.

(c) C_{4v}. Logo: Sarajevo Winter Olympics, 1984. Indian stamp. "Standing Brahma", by permission of the Metropolitan Museum of Art, New York.

(d) C_{5v}.

$C_{\infty v}$

One infinite-fold rotation axis with infinite number of symmetry planes which include the rotation axis. Examples: Fig. 3-15.

D_2

Three mutually perpendicular two-fold rotation axes. Example: Fig. 3-16a.

D_3

One three-fold rotation axis and three two-fold rotation axes perpendicular to the three-fold axis. The two-fold axes are at

Figure 3-16.
(a) D_2.

(b) D_3.

twistane

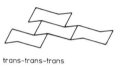

trans-trans-trans
perhydro-triphenylene

Figure 3-15.
$C_{\infty v}$.

H–Cl   H–B–S   H–C≡C–C≡C–Cl

Figure 3-17.
(a) D_{2d}.

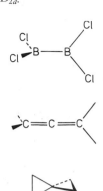

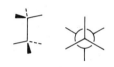

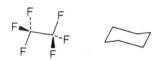

(b) D_{3d}. The drawing of the radiolarum is from [3-7(b)].

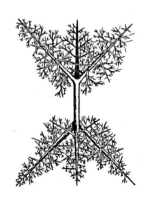

(c) D_{4d}. The drawing of the plant is from [3-7(b)].

120°, so the minimum angle between two such axes is 60°. Examples: Fig. 3-16b.

D_4

One four-fold rotation axis and four two-fold rotation axes which are perpendicular to the four-fold axis. The four axes are grouped in two non-equivalent pairs. One pair is rotated relative to the other pair by 45°. The angle between the two axes within each pair is 90°.

D_5

One five-fold rotation axis with five two-fold rotation axes perpendicular to it. The angle between the two-fold axes is 36°.

D_6, D_7 ..., D_n

This series can be continued by analogy. It is characterized by one n-fold rotation axis and n two-fold rotation axes perpendicular to the n-fold axis.

D_{2d}

Three mutually perpendicular two-fold rotation axes and two symmetry planes. The planes include one of the three rotation axes and bisect the angle between the other two. Examples: Fig. 3-17a.

D_{3d}

One three-fold rotation axis with three two-fold rotation axes perpendicular to it, and three symmetry planes. The angle between the two-fold axes is 60°. The symmetry planes include the three-fold axis and bisect the angles between the two-fold axes. Examples: Fig. 3-17b.

D_{4d}

One four-fold rotation axis with four two-fold rotation axes perpendicular to it, and four symmetry planes. The angle between the two-fold axes is 45°. The symmetry planes include the four-fold axis and bisect the angles between the two-fold axes. Examples: Fig. 3-17c.

(d) D_{5d}.

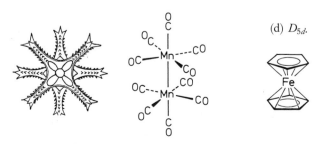

90

Figure 3-18.
(a) D_{2h}.

(b) D_{3h}.

(c) D_{4h}.

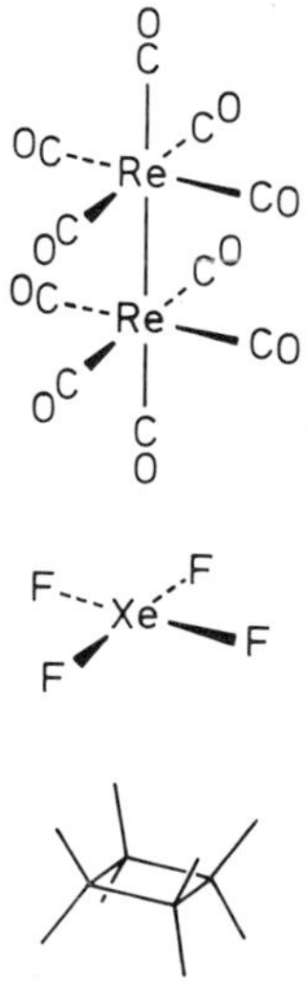

D_{5d}, D_{6d}, D_{7d}, ..., D_{nd}

The series can be continued by analogy. Example: Fig. 3-17d.

D_{2h}

Three mutually perpendicular symmetry planes. Their three crossing lines are three two-fold rotation axes, and their crossing point is a center of symmetry. Examples: Fig. 3-18a.

D_{3h}

One three-fold rotation axis, three symmetry planes (at 60°) which contain the three-fold axis, and another symmetry plane perpendicular to the three-fold axis. Examples: Fig. 3-18b.

D_{4h}

One four-fold axis, one symmetry plane perpendicular to it, and four symmetry planes which include the four-fold axis. The four planes make two pairs. One pair is rotated relative to the other pair by 45°. The two planes in each pair are perpendicular to each other. Examples: Fig. 3-18c.

D_{5h}

One five-fold rotation axis, one symmetry plane perpendicular to it, and five symmetry planes which include the five-fold rotation axis. The angle between the adjacent five planes is 36°. Examples: Fig. 3-18d.

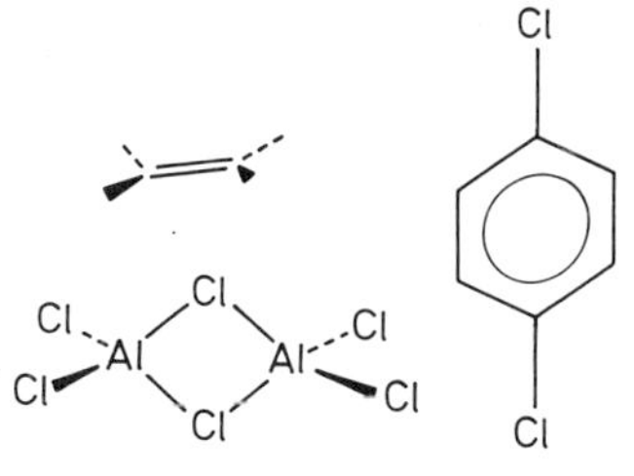

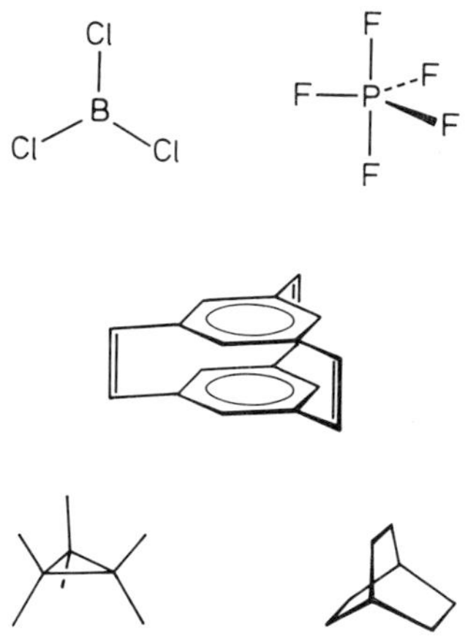

(d) D_{5h}.

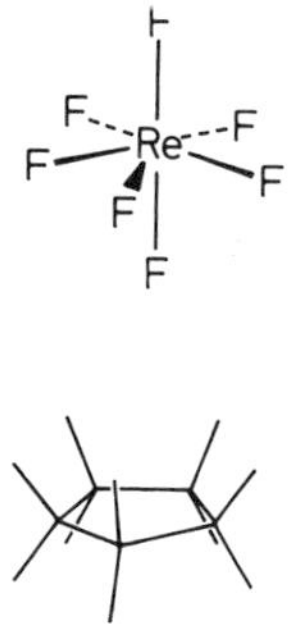

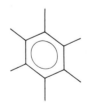

(e) D_{6h}.

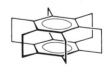

91

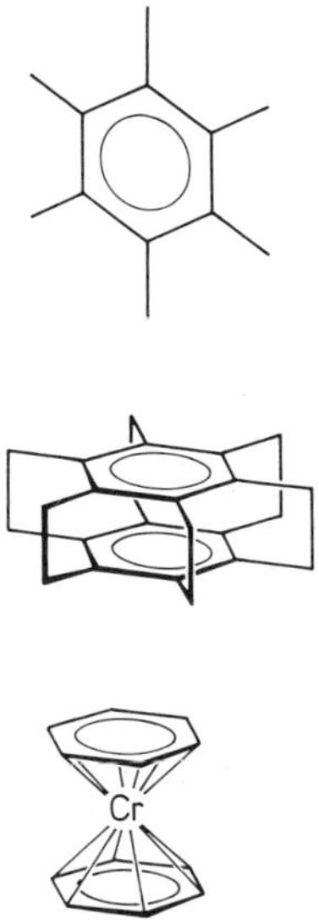

D_{6h}

One six-fold rotation axis, one symmetry plane perpendicular to it, and six symmetry planes which include the six-fold axis. The six planes are grouped in two sets. One set is rotated relative to the other set by 30°. The angle between the planes within each set is 60°. Examples: Fig. 3-18e.

D_{nh}

The series can be continued by analogy. There will be one n-fold rotation axis, one symmetry plane perpendicular to it, and n symmetry planes which include the n-fold axis. When n is even, there are two sets of symmetry planes. One set is rotated relative to the other set by $(180/n)°$. The angle between the planes within each set is $(360/n)°$. When n is odd, the angle between the symmetry planes is $(180/n)°$.

$D_{\infty h}$

One ∞-fold axis and a symmetry plane perpendicular to it. Of course, there are also ∞ number of symmetry planes which include the ∞-fold rotation axis. Examples: Fig. 3-19.

Figure 3-19.
$D_{\infty h}$.

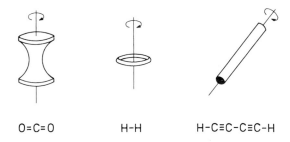

$$O=C=O \qquad H-H \qquad H-C\equiv C-C\equiv C-H$$

T

Three mutually perpendicular two-fold rotation axes and four three-fold rotation axes. The three-fold axes all go through a vertex of a tetrahedron and the midpoint of the opposite face center. The two-fold axes connect the midpoints of opposite edges of this tetrahedron. Examples: Fig. 3-20a.

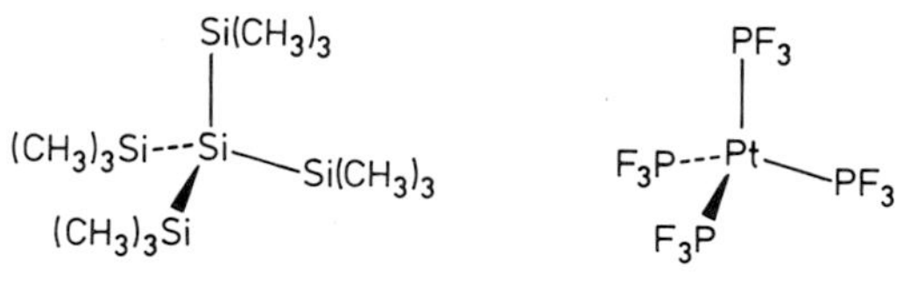

Figure 3-20.
(a) T.

$$\begin{array}{cc} Si(CH_3)_3 & PF_3 \\ | & | \\ (CH_3)_3Si\text{---}Si\diagdown Si(CH_3)_3 & F_3P\text{---}Pt\diagdown PF_3 \\ (CH_3)_3Si & F_3P \end{array}$$

T_d

In addition to the symmetry elements of symmetry T, there are six symmetry planes, each two of them being mutually perpendicular. All of these symmetry planes contain two three-fold axes. Examples: Fig. 3-20b.

(b) T_d.

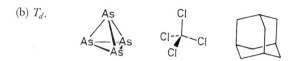

T_h

In addition to the symmetry elements of symmetry T, there is a center of symmetry which introduces also three symmetry planes perpendicular to the two-fold axes. Example: Fig. 3-20c.

(c) T_h.

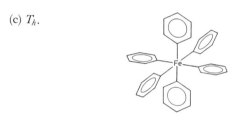

O

Three mutually perpendicular four-fold rotation axes and four three-fold rotation axes which are tilted with respect to the four-fold axes in a uniform manner.

O_h

In addition to the symmetry elements of symmetry O, there is a center of symmetry. Examples: Fig. 3-21.

Figure 3-21.
O_h.

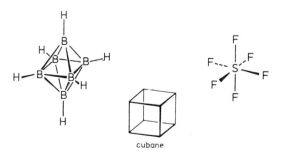

cubane

93

3.6 Consequences of Substitution

A tetrahedral AX_4 molecule, e.g. methane, CH_4, has the point group of the regular tetrahedron T_d. Gradual substitution of the X ligands by B ligands leads to less symmetrical tetrahedral configurations with the following point groups (Fig. 3-22a), until complete substitution is accomplished:

AX_4	AX_3B	AX_2B_2	AXB_3	AB_4
T_d	C_{3v}	C_{2v}	C_{3v}	T_d

As the sites of all X ligands are equivalent in each of these configurations, the symmetry changes accompanying the substitution are determined *a priori*.

Figure 3-22.
Substitution in a tetrahedral AX_4 molecule.
(a) Gradual substitution of the ligands X by ligands B.

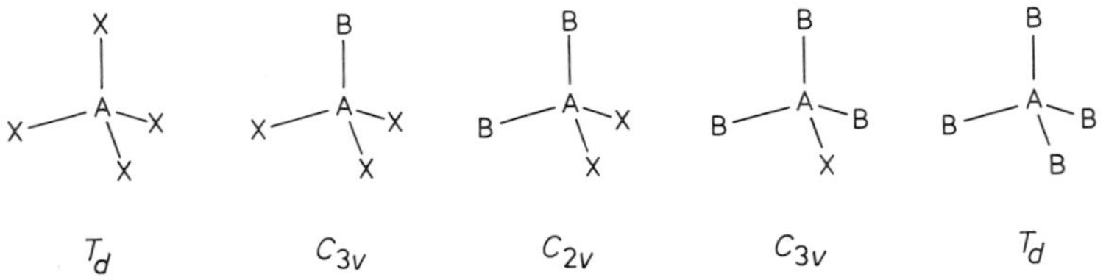

Let us consider now an octahedral AX_6 molecule, e.g. sulfur hexafluoride, SF_6, which has the symmetry of the regular octahedron O_h. Substitution of an X ligand by a B ligand results in an AX_5B molecule whose symmetry is again determined *a priori* to be C_{4v}. The substitution of a second X ligand by another ligand B may lead to alternative structures as the sites of the five X ligands after the first substitution are no longer equivalent. The symmetry variations in this substitution process are illustrated in Fig. 3-23.

If each consecutive substitution introduces a new kind of ligand then the symmetry will continue to decrease. This is shown for the tetrahedral case in Fig. 3-22b. There is, of course, a larger variety of structures obtained in such a substitution sequence for the octahedral configuration.

(b) Substitution of the ligands X by various ligands.

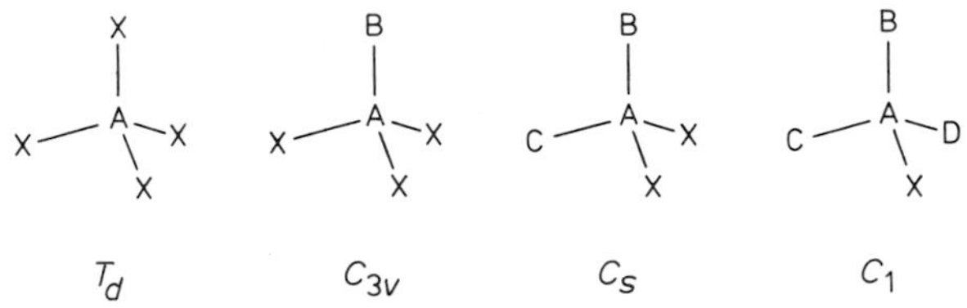

Figure 3-23.
Gradual substitution of the
ligands X in an octahedral
AX_6 molecule by ligands B.

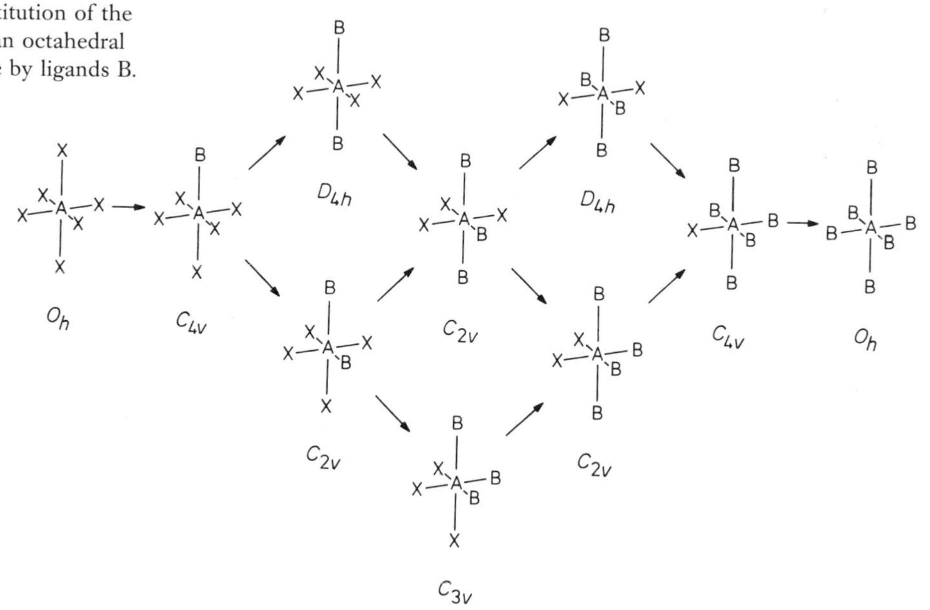

Another example among fundamental structures is the benzene geometry, D_{6h}. Gradual substitution of an increasing number of hydrogens by ligands X results in the symmetry variations illustrated in Fig. 3-24. As regards the molecular point group, the monosubstituted and the pentasubstituted derivatives are equivalent. All derivatives can be grouped in such pairs with each of the trisubstituted benzenes constituting a pair by itself. Again only the simplest case is considered here with one kind of ligand used in all substituted positions. The decrease in the symmetry in the molecular point group for the substituted derivatives occurs because of the presence of the substituent ligands. It does not presuppose a change in the hexagonal symmetry of the benzene ring itself. Modern structure analyses have determined, however, that an appreciable deformation of the ring from regularity may also take place depending on the nature of the substituents. The largest deformation usually occurs at the so-called *ipso* angle adjacent to the substituent. According to the general observation, electronegative substituents tend to compress the ring while electropositive substituents elongate it. Fig. 3-25 presents a correlation [3-8] between the *ipso* angles α and the C1 ... C4 non-bonded ring distances for a series of para-disubstituted derivatives. The dashed line was obtained from the experimental data points. The α/C1 ... C4 interdependence reflects in part purely geometric relations represented by the solid line. This description corresponds to an α/C1 ... C4 correlation with no changes in the ring bond lengths. The difference between the dashed and solid lines seems to be characteristic.

Figure 3-24.
The symmetries of benzene
and its $C_6H_nX_{6-n}$ derivatives.

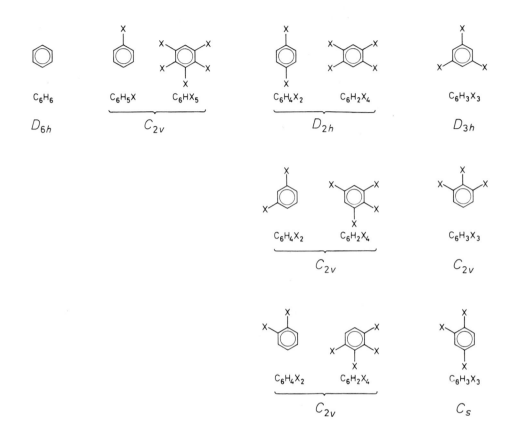

C_6H_6

D_{6h}

C_6H_5X C_6HX_5

C_{2v}

$C_6H_4X_2$ $C_6H_2X_4$

D_{2h}

$C_6H_3X_3$

D_{3h}

$C_6H_4X_2$ $C_6H_2X_4$

C_{2v}

$C_6H_3X_3$

C_{2v}

$C_6H_4X_2$ $C_6H_2X_4$

C_{2v}

$C_6H_3X_3$

C_s

Figure 3-25.
Correlation between the ring
ipso angles and the C1 ... C4
distances for a series of symmetrically para-disubstituted
benzene derivatives [3-8]. For
the origin of experimental
structures, see [3-9].

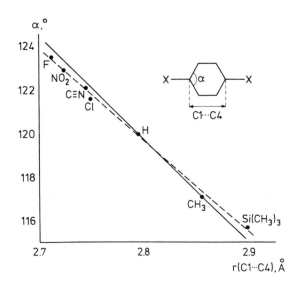

Complex formation usually implies the association of molecules or other species which may also exist separately in chemically non-extreme conditions. Complex formation often has important consequences on the shapes and symmetries of the constituent molecules [3-10]. The $H_3N \cdot AlCl_3$ donor-acceptor complex [3-11], for example, has a triangular antiprismatic shape with C_{3v} symmetry as seen in Fig. 3-26. The component molecules in their free uncomplexed states are, of course, ammonia (C_{3v}) and $AlCl_3$ (D_{3h}). The symmetry of the donor part (NH_3) remains unchanged in the complex and the geometrical changes are relatively small. On the other hand, there are more drastic geometrical changes in the acceptor part ($AlCl_3$) due to loss of coplanarity of the four atoms and this results in a reduction in the point group. However, the structural change in the acceptor part may also be viewed as if the complex formation would complete the tetrahedral configuration around the central atoms in the component molecules. The nitrogen configuration may be considered to be tetrahedral already in ammonia with the lone pair of electrons being the fourth ligand. For aluminum, it is indeed the complexation which makes the tetrahedral configuration complete. Coordination molecules often demonstrate the utility of polyhedra in describing molecular shapes, symmetries, and geometries. Of course, such description may be useful in many other classes of compounds as well.

Figure 3-26.
The triangular antiprismatic shape of the $H_3N \cdot AlCl_3$ donor-acceptor complex and the uncomplexed ammonia and aluminum trichloride molecules.

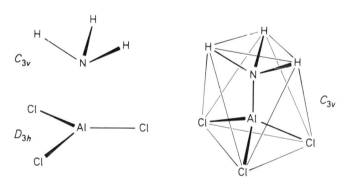

3.7 Polyhedral Molecular Geometries

In the Preface to the Third Edition of his "Regular Polytopes" [3-12], Coxeter calls attention to the icosahedral structure of a boron compound in which twelve boron atoms are arranged like the vertices of an icosahedron. It had been widely believed that there would be no *inanimate* occurrence of an icosahedron, or of a regular dodecahedron either.

In 1982 the synthesis and properties of a new polycyclic $C_{20}H_{20}$ hydrocarbon, dodecahedrane, was reported [3-13]. The twenty carbon atoms of this molecule are arranged like the vertices of a regular dodecahedron. When in the early sixties Schultz [3-14] discussed the topology of the polyhedrane and prismane molecules (*vide infra*), at that time it was in terms of a geometrical diversion rather than truelife chemistry. Since then it has become real chemistry.

It should be reemphasized that the above high-symmetry examples refer to isolated molecules and not to crystal structures. Crystallography has, of course, been one of the main domains where the importance of polyhedra has been long recognized, together with some limitations which forbid the occurrence of regular *pentagonal* figures in crystals. Polyhedra are not less important in the world of molecules, where the limitations existing in crystals do not apply.

In the First Edition of the "Regular Polytopes" [3-12], Coxeter stated, "... the chief reason for studying regular polyhedra is still the same as in the times of the Pythagoreans, namely, that their symmetrical shapes appeal to one's artistic sense." The success of modern molecular chemistry does not diminish the validity of this statement. On the contrary. There is no doubt that the aesthetic appeal has much contributed to the rapid development of what could be termed polyhedral chemistry.

One of the pioneers in the area of polyhedral borane chemistry, Muetterties movingly described [3-15] his attraction to the chemistry of boron hydrides, comparing it to Escher's devotion to periodic drawings [3-16]. Muetterties' words [3-15] are quoted here:

"When I retrace my early attraction to boron hydride chemistry, Escher's poetic introspections strike a familiar note. As a student intrigued by early descriptions of the extraordinary hydrides, I had not the prescience to see the future synthesis developments nor did I have then a scientific appreciation of symmetry, symmetry operations, and group theory. Nevertheless, some inner force also seemed to drive me but in the direction of boron hydride chemistry. In my initial synthesis efforts, I was not the master of these molecules; they seemed to have destinies unperturbed by my then amateurish tactics. Later as the developments in polyhedral borane chemistry were evident on the horizon, I found my general outlook changed in a characteristic fashion. For example, my doodling, an inevitable activity of mine during meetings, changed from characters of nondescript form to polyhedra, fused polyhedra and graphs.

I (and others, my own discoveries were not unique nor were they the first) was profoundly impressed by the ubiquitous character of the three-center relationship in bonding (e.g., the boranes) and nonbonding situations. I found a singular uniformity in geometric relationships throughout organic, inorganic, and organometallic chemistry: The favored geometry in coordination compounds, boron hydrides, and metal clusters is the polyhedron that has all faces equilateral or near equilateral triangles ..."

The polyhedral description of molecular geometries is, of course, generally applicable as these geometries are spatial constructions. To emphasize that even planar or linear molecules are also included, the term polytopal could be used rather than polyhedral. The real utility of the polyhedral description is for molecules possessing a certain amount of symmetry. Because of this and also because of the introductory character of our discussion, only molecules with relatively high symmetries will be mentioned.

The polyhedral description may be useful for widely different systems. Thus, for example, both the tetraarsene, As_4, and the methane, CH_4, molecules have tetrahedral shapes (Fig. 3-27) and T_d symmetry. However, there is an important difference in their structures. In the As_4 molecule all the four constituting nuclei are located at the vertices of a regular tetrahedron, and all the edges of this tetrahedron are chemical bonds between the As atoms. In the methane molecule, there is a central carbon atom, and four chemical bonds are directed from it to the four vertices of a regular tetrahedron where the four protons are located. The edges are not chemical bonds.

Figure 3-27.
The molecular shapes of As_4 and CH_4.

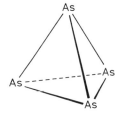

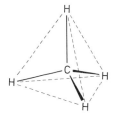

The As$_4$ and CH$_4$ molecules are clear-cut examples of the two distinctly different arrangements. However, these distinctions are not always so unambiguous. An interesting example is the structure of zirconium borohydride, Zr(BH$_4$)$_4$. Two independent studies [3-17, 3-18] described its structure by the same polyhedral configuration, while they differed in the assignment of the chemical bonds (Fig. 3-28). The most important difference in the two interpretations concerns the linkage between the central zirconium atom and the four boron atoms situated in the four vertices of a regular tetrahedron. According to one interpretation [3-17], there are four Zr-B bonds in the tetrahedral arrangement. On the other hand, there is no direct Zr-B bond according to the other interpretation [3-18], but each boron atom is linked to the zirconium atom by three hydrogen bridges. Zirconium borohydride is one of the interesting metal borohydrides whose molecular geometries have presented a challenge to the structural chemist [3-10]. The boron hydrides themselves are one of the most beautiful classes of polyhedral compounds whose representatives range from the simplest to the most complicated systems.

Figure 3-28.
The molecular configuration of zirconium borohydride, Zr(BH$_4$)$_4$, in two interpretations but described by the same polyhedral shape.
(a) According to one interpretation [3-17], the zirconium atom is directly bonded to the four tetrahedrally arranged boron atoms.

(b) According to another interpretation [3-18], the zirconium and the tetrahedrally arranged boron atoms are not bonded directly. Their linkage is established by four times three hydrogen bridges.

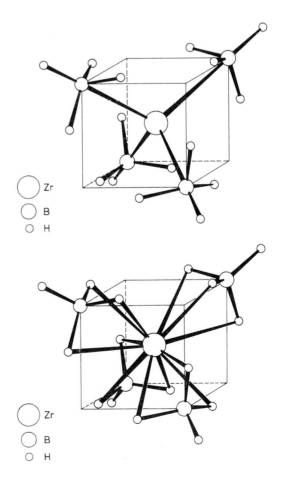

100

3.7.1 Boron Hydride Cages

Figure 3-29.
The regular icosahedral configuration of the $B_{12}H_{12}^{2-}$ ion; the boron skeleton is shown.

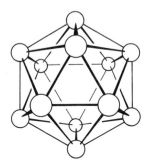

Our description here is purely phenomenological. Only a passing reference is made to the relationship of the characteristic polyhedral cage arrangements of the boron hydrides and the peculiarities of multicenter, and most notably, three-center bonding (see, e.g. [3-19, 3-20]).

The boron hydride polyhedra are characterized by having all faces equilateral or nearly equilateral triangles. Those boron hydrides that have a complete polyhedral shape are called *closo* boranes (the Greek *closo* meaning closed). One of the most symmetrical, and accordingly, most stable polyhedral boranes is the $B_{12}H_{12}^{2-}$ ion. Its regular icosahedral configuration is shown in Fig. 3-29. The structural systematics of $B_nH_n^{2-}$ *closo* boranes and related compounds, notably $C_2B_{n-2}H_n$ *closo* carboranes, are presented in Table 3-2 after Muetterties [3-15]. In carboranes some of the boron sites are taken by carbon atoms.

Another structural class of the boron hydrides is the so-called quasi-*closo* boranes. They are related to the *closo* boranes by removing a framework atom from the latter and adding in its stead a pair of electrons. Thus one of the polyhedron framework sites is taken by an electron pair.

Table 3-2. Structural Systematics of $B_nH_n^{2-}$ *closo* Boranes and $C_2B_{n-2}H_n$ *closo* Carboranes after Muetterties [3-15].

Polyhedron and Point Group	Boranes	Dicarbaboranes
Tetrahedron, T_d	$(B_4Cl_4)^{a)}$	–
Trigonal bipyramid, D_{3h}	–	$C_2B_3H_5$
Octahedron, O_h	$B_6H_6^{2-}$	$C_2B_4H_6$
Pentagonal bipyramid, D_{5h}	$B_7H_7^{2-}$	$C_2B_5H_7$
Dodecahedron (triangulated), D_{2d}	$B_8H_8^{2-}$	$C_2B_6H_8$
Tricapped trigonal prism, D_{3h}	$B_9H_9^{2-}$	$C_2B_7H_9$
Bicapped square antiprism, D_{4d}	$B_{10}H_{10}^{2-}$	$C_2B_8H_{10}$
Octadecahedron, C_{2v}	$B_{11}H_{11}^{2-}$	$C_2B_9H_{11}$
Icosahedron, I_h	$B_{12}H_{12}^{2-}$	$C_2B_{10}H_{12}$

a) No boron hydride

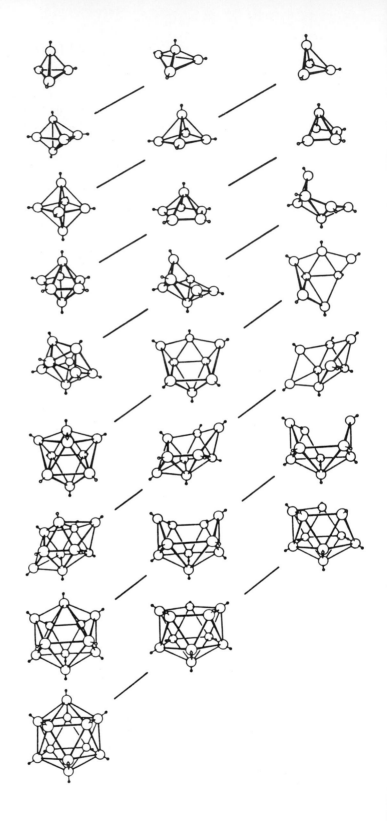

Figure 3-30.
Closo, nido, and *arachno* boranes after Williams [3-21] and Rudolph [3-20]. The genetic relationships are indicated by diagonal lines. Reprinted with permission. Copyright (1976) American Chemical Society.

There are boron hydrides in which one or more of the polyhedral sites are truly removed. Fig. 3-30 shows the systematics of borane polyhedral fragments as obtained from *closo* boranes, after Williams [3-21] and Rudolph [3-20]. As all the faces of the polyhedral skeletons are triangular, they may be termed deltahedra and the derived fragments deltahedral [3-22]. The starting deltahedra are the tetrahedron, the trigonal bipyramid, the octahedron, the pentagonal bipyramid, the bisdisphenoid, the symmetrically tricapped trigonal prism, the bicapped square antiprism, the octadecahedron, and the icosahedron.

A *nido* (nest-like) boron hydride is derived from a *closo* borane by the removal of one skeleton atom. If the starting *closo* borane is not a regular polyhedron, then the atom removed is the one at a vertex with the highest connectivity. An *arachno* (web-like) boron hydride is derived from a *closo* borane by the removal of two adjacent skeleton atoms. If the starting *closo* borane is not a regular polyhedron, then again, one of the two atoms removed is at a vertex with the highest connectivity. Complete *nido* and *arachno* structures are shown in Fig. 3-31 together with the starting boranes [3-15]. The fragmented structures are completed by a number of bridging and terminal hydrogens. The above examples are, of course, from among the simplest boranes and their derivatives.

Figure 3-31.
Examples of *closo/nido* and *closo/arachno* structural relationships after Muetterties [3-15].
(a) *Closo*-$B_6H_6^{2-}$ and *nido*-B_5H_9.

(b) *Closo*-$B_7H_7^{2-}$ and *arachno*-B_5H_{11}.

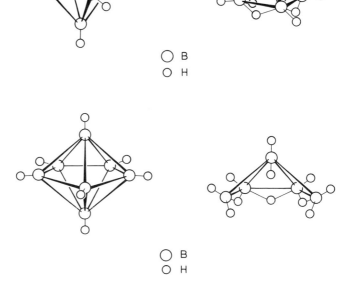

103

3.7.2 Polycyclic Hydrocarbons

Some fundamental polyhedral shapes are realized among polycyclic hydrocarbons where the edges are C-C bonds and there is no central atom. The bond arrangements around the carbon atoms in such configurations may be far from the energetically most advantageous, causing *strain* in these structures [3-23, 3-24]. The strain may be so large as to render particular arrangements too unstable to exist under any reasonable conditions. On the other hand, the fundamental character of these shapes, their high symmetry and aesthetic appeal make them an attractive and challenging "playground" to the organic chemist [3-23, 3-25]. Incidentally, these substances have also great practical importance as they are building blocks for such natural products as steroids, alkaloids, vitamins, carbohydrides, antibiotics, etc.

Tetrahedrane, $(CH)_4$, would be the simplest regular polyhedral polycyclic hydrocarbon (Fig. 3-32). However, since it has such a high strain energy and provides easy access to attacking reagents, its preparation may not be possible. Its derivative, tetra-tert-butyltetrahedrane (Fig. 3-32), however, has been prepared [3-26]. This compound is amazingly stable perhaps because the substituents help "clasp" the molecule together.

Figure 3-32.
(a) Tetrahedrane, $(CH)_4$. It has very high strain energy and has not (yet?) been prepared.

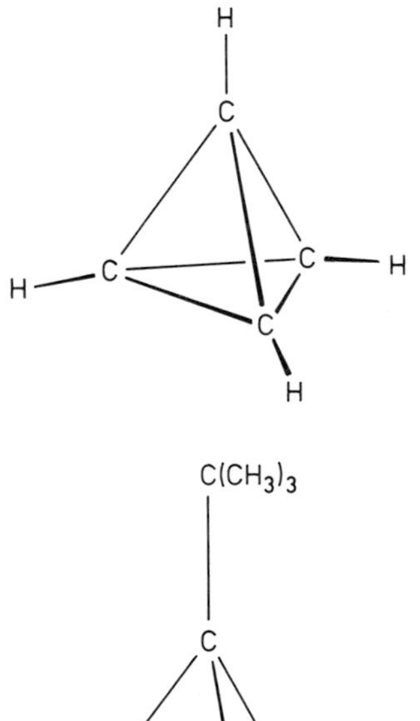

(b) Tetra-tert-butyltetrahedrane, $\{C[C(CH_3)_3]\}_4$, [3-26].

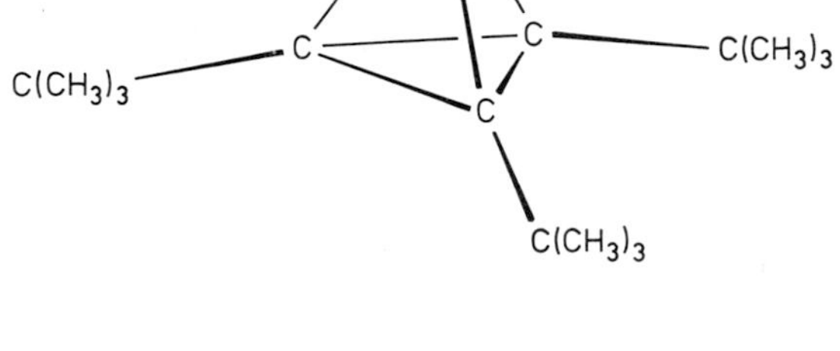

(c) Cubane, $(CH)_8$, [3-27].
(d) Dodecahedrane, $(CH)_{20}$, [3-13].

The next Platonic solid is the cube, and the corresponding polycyclic hydrocarbon cubane, $(CH)_8$, Fig. 3-32, has been known for some time [3-27]. The preparation of dodecahedrane (Fig. 3-32) by Paquette et al. [3-13] is much more recent. Almost two decades prior to this work, Schultz [3-14] made the following prediction for dodecahedrane as he was describing the possible hydrocarbon polyhedranes: "Dodecahedrane is the one substance of the series with almost ideal geometry, physically the molecule is practically a miniature ball bearing! One would expect the substance to have a low viscosity, a high melting point but low boiling point, high thermal stability, a very simple infrared spectrum and perhaps an aromatic-like p.m.r. spectrum. Chemically one might expect a relatively easy (for an aliphatic hydrocarbon) removal of a tertiary proton from the molecule, for the negative charge thus deposited on the molecule could be accommodated on any one of the twenty completely equivalent carbon atoms, the carbanion being stabilized by a 'rolling charge' effect that delocalizes the extra electron."

In the $(CH)_n$ convex polyhedral hydrocarbon series each carbon atom is bonded to three other carbon atoms. The fourth bond is directed externally to a hydrogen atom. Around the all-carbon polyhedron, there is thus a similar polyhedron whose vertices are protons. The edges of the all-carbon polyhedron are carbon-carbon chemical bonds, while the edges of the larger all-proton polyhedron do not correspond to any chemical bonds. This kind of arrangement of the polycyclic hydrocarbons is not possible for the remaining two Platonic solids. There are four bonds meeting at the vertices of the octahedron and five at the vertices of the icosahedron. For similar reasons, only seven of the 13 Archimedian polyhedra can be considered in the $(CH)_n$ polyhedral series. Table 3-3 presents some characteristics of the polyhedranes after Schultz [3-14]. It is also indicated, which of the hydrocarbon polyhedranes have already been synthesized at the time of the present writing.

105

Table 3-3. Characterization of Polyhedrane Molecules after Schultz [3-14]

Name	Formula	Number of Faces (all regular)	Face Angles	Has been prepared?
Tetrahedrane	$(CH)_4$	Triangle, 4	60°	–
Cubane	$(CH)_8$	Square, 6	90°	Yes
Truncated tetrahedrane	$(CH)_{12}$	Triangle, 4	60°	–
		Hexagon, 4		
Dodecahedrane	$(CH)_{20}$	Pentagon, 12	108°	Yes
Truncated octahedrane	$(CH)_{24}$	Square, 6	90°	–
		Hexagon, 8	120°	
Truncated cubane	$(CH)_{24}$	Triangle, 8	60°	–
		Octagon, 6	135°	
Truncated cuboctahedrane	$(CH)_{48}$	Square, 12	90°	–
		Hexagon, 8	120°	
		Octagon, 6	135°	
Truncated icosahedrane	$(CH)_{60}$	Pentagon, 12	108°	–
		Hexagon, 20	120°	
Truncated dodecahedrane	$(CH)_{60}$	Triangle, 20	60°	–
		Decagon, 12	144°	
Truncated icosidodecahedrane	$(CH)_{120}$	Square, 30	90°	–
		Hexagon, 20	120°	
		Decagon, 12	144°	

The cubane molecule may also be considered and called tetraprismane, cf. Figs. 3-32 and 3-33. It may be described as composed of eight identical methine units arranged at the corners of a regular tetragonal prism with O_h symmetry and bound into two parallel four-membered rings cojoined by four four-membered rings. Triprismane, $(CH)_6$, [3-28] has D_{3h} symmetry and pentaprismane, $(CH)_{10}$, [3-29] has D_{5h} symmetry. Both are depicted in Fig. 3-33. The quest for a synthesis of pentaprismane is a long story with a happy ending [3-29]. At the time of this writing, hexaprismane, $(CH)_{12}$, which is the face-to-face dimer of benzene, (Fig. 3-33), has not yet been prepared. Table 3-4 presents some characteristic geometric information on the hydrocarbon prismane molecules after Schultz [3-14]. The description of the general n-prismane is that it is composed of $2n$ identical methine units arranged at the corners of a regular prism with D_{nh} symmetry, and bound into two parallel n-membered rings cojoined by n four-membered rings.

Figure 3-33.
(a) Triprismane, $(CH)_6$, [3-28].
(b) Tetraprismane (cubane), $(CH)_8$ [3-27].
(c) Pentaprismane, $(CH)_{10}$, [3-29].
(d) Hexaprismane, $(CH)_{12}$, not yet prepared.

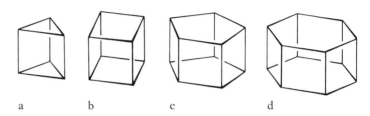

a   b   c   d

Table 3-4. Characterization of Prismane Molecules after Schultz [3-14]

Name	Formula	Number of Faces	Face Angles	Has been prepared?
Triprismane	C_6H_6	Triangle, 2	60°	Yes
		Square, 3	90°	
Tetraprismane (cubane)	C_8H_8	Square, 6	90°	Yes
Pentaprismane	$C_{10}H_{10}$	Pentagon, 2	108°	Yes
		Square, 5	90°	
Hexaprismane	$C_{12}H_{12}$	Hexagon, 2	120°	
		Square, 6	90°	
Heptaprismane	$C_{14}H_{14}$	Heptagon, 2	128°34'	
		Square, 7	90°	
Octaprismane	$C_{16}H_{16}$	Octagon, 2	135°	
		Square, 8	90°	
Nonaprismane	$C_{18}H_{18}$	Nonagon, 2	140°	
		Square, 9	90°	
Decaprismane	$C_{20}H_{20}$	Decagon, 2	144°	
		Square, 10	90°	
Undecaprismane	$C_{22}H_{22}$	Undecagon, 2	147°16'	
		Square, 12	90°	
Dodecaprismane	$C_{24}H_{24}$	Dodecagon, 2	150°	
		Square, 12	90°	
n-Prismane	$C_{2n}H_{2n}$	n-gon, 2	180° (approaches)	
		Square, n	90°	

Incidentally, the *regular* prisms and the *regular* antiprisms are also semiregular, i.e. Archimedian, solids. Moreover, the second prism in its most symmetrical configuration is a regular solid, the cube; and the first antiprism in its most symmetrical configuration is also a regular solid, the octahedron.

Only a few highly symmetrical structures have been mentioned above. The varieties become virtually endless if one reaches beyond the most symmetrical convex polyhedral shapes. For example, the number of possible isomers is 5,291 for the tetracyclic structures of the $C_{12}H_{18}$ hydrocarbons with 12 skeletal carbon atoms [3-30]. Of all these geometric possibilities, however, only a few stable [3-31]. One is *iceane* shown in Fig. 3-34 [3-32]. The molecule may be visualized as two chair cyclohexanes connected to each other by three axial bonds. Alternatively, the molecule may be viewed as consisting of three fused boat cyclohexanes. The trivial name iceane had been proposed for this molecule by Fieser [3-33] almost a decade before its preparation [3-32]. As Fieser was considering the arrangement of the water molecules in the ice crystal (Fig. 3-34), he noticed three *vertical* hexagons with boat conformations. The emerging *horizontal* $(H_2O)_6$ units possess three equatorial hydrogen atoms and three equatorial hydrogen bonds available for horizontal building. Fieser [3-33] further notes that this structure "suggests the possible existence of a hydrocarbon of analogous conformation of the formula $C_{12}H_{18}$, which might be named 'iceane'. The model indicates a stable strain-free structure analogous to adamantane and twistane. 'Iccane' thus presents a challenging target for synthesis." Within a decade the challenge was met [3-32].

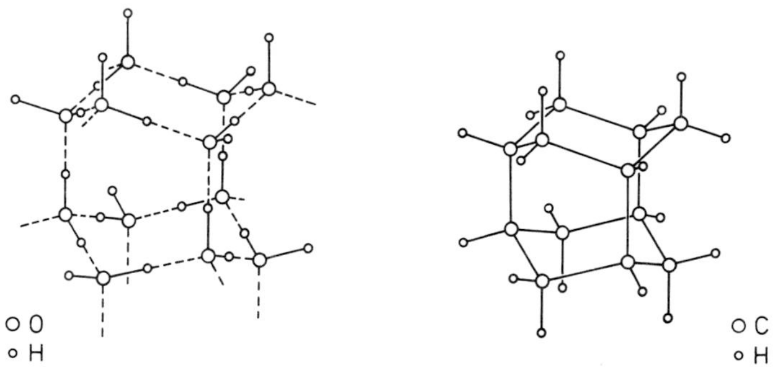

Figure 3-34.
Ice crystal structure and the iceane hydrocarbon molecule after Fieser [3-33] and Cupas and Hodakowski [3-32], respectively.

There is a close relationship between the adamantane, $C_{10}H_{16}$, molecule and the diamond crystal. Diamond has even been termed the "infinite adamantylogue to adamantane" [3-34]. While iceane has D_{3h} symmetry, adamantane has T_d. This high symmetry can be clearly seen when the configuration of adamantane is described by four imaginary cubes packed one inside the other, two of which are shown in Fig. 3-35 [3-35]. The

Figure 3-35.
Adamantane, $C_{10}H_{16}$ or $(CH)_4(CH_2)_6$ in two representations.

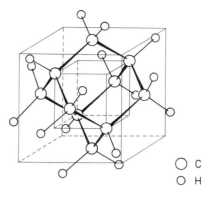

○ C
○ H

Figure 3-36.
Adamantane-analogs among inorganic polymeric oxides, e.g., P_4O_6 and $(PO)_4O_6$.

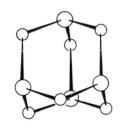

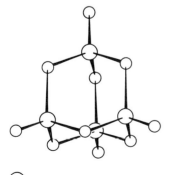

○ P
○ O

adamantane geometry may be described, for example, by the following four parameters: R_1, the C-C bond length, R_2, the mean C-H bond length, R_3, the bond angle C-CH$_2$-C at the secondary carbon, and R_4, the H-C-H bond angle. Then the edge lengths of the four imaginary cubes can be expressed by these parameters. Four of the vertices of the smallest cube are occupied by the four tertiary carbon atoms. This cube is the smaller of the two cubes shown in Fig. 3-35. The length of its edge is

$$TC = [2R_1 \sin(R_3/2)]/\sqrt{2}.$$

The hydrogen atoms adjoining the tertiary carbons occupy four tetrahedrally related vertices of a cube with edge length

$$TH = 2(TC + R_2/\sqrt{3}).$$

This cube is not shown in Fig. 3-35. The six secondary carbon atoms are located at the face centers of a cube with edge length

$$SC = 2[TC + R_1 \cos(R_3/2)].$$

This corresponds to the larger of the two cubes shown in Fig. 3-35. Finally the hydrogens adjoining the secondary carbons lie on the diagonals of the faces of a cube whose edge length is

$$SH = 2[SC + R_2 \cos(R_4/2)].$$

Similar structures are found among the inorganic polyoxides, where by analogy to adamantane, $(CH)_4(CH_2)_6$, the general formula is A_4O_6. Here A may be, e.g., P, As, Sb, or the P=O group, as illustrated in Fig. 3-36.

Adamantane molecules may be imagined to join at vertices, edges, or even at faces. Examples are shown in Fig. 3-37; most of them, however, have not yet been synthesized (for references, see [3-23]).

Figure 3-37.
Joined adamantanes.
(a) At vertices, [1]diadamantane [3-36].

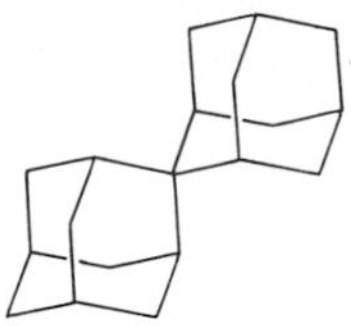

(b) At edges, [2]diadamantane [3-37].

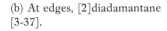

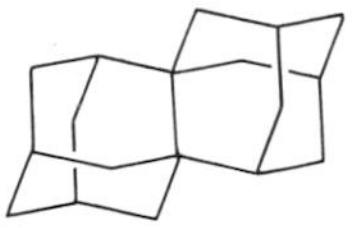

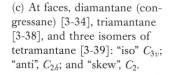

(c) At faces, diamantane (congressane) [3-34], triamantane [3-38], and three isomers of tetramantane [3-39]: "iso" C_{3v}; "anti", C_{2h}; and "skew", C_2.

3.7.3 Structures with Central Atom

Adamantane is sometimes regarded as the cage analog of methane while diamantane and triamantane as the analogs of ethane and propane. Methane has, of course, a tetrahedral structure with the point group of the regular tetrahedron, T_d. Important structures may be derived by joining two tetrahedra, or for example two octahedra, at a common vertex, edge, or face as shown in Fig. 3-38. Ethane, H_3C-CH_3, ethylene, $H_2C=CH_2$, and acetylene, $HC\equiv CH$, may be derived formally from joined tetrahedra in such a way. The analogy with the joining tetrahedra is even more obvious in some metal halide structures with halogen bridges [3-40]. Thus, e.g., the $Al_2Cl_7^-$ ion may be considered as two aluminumtetrachloride tetrahedra joined at a common vertex, or the Al_2Cl_6 molecule may be looked at as two such tetrahedra joined at a common edge. These examples are shown in Fig. 3-39.

In mixed-halogen complexes, as e.g. potassium tetrafluoroaluminate, $KAlF_4$, [3-41] there is also a tetrahedral metal coordination. In fact, the regular or nearly regular tetrahedral tetrafluoroaluminate part of the molecule is an especially well

Figure 3-38.
Joined tetrahedra and octahedra.

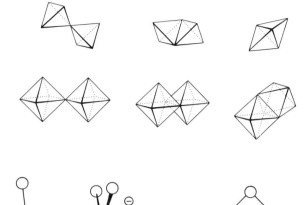

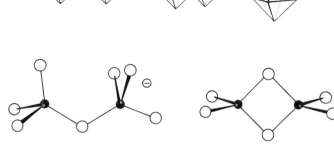

Figure 3-39.
The configurations of the $Al_2Cl_7^-$ ion and Al_2Cl_6 molecule.

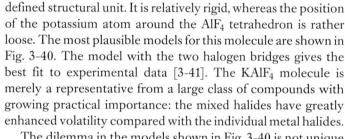

Figure 3-40.
Models of the $KAlF_4$ molecule [3-41].

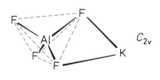

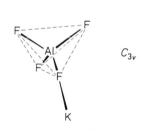

defined structural unit. It is relatively rigid, whereas the position of the potassium atom around the AlF_4 tetrahedron is rather loose. The most plausible models for this molecule are shown in Fig. 3-40. The model with the two halogen bridges gives the best fit to experimental data [3-41]. The $KAlF_4$ molecule is merely a representative from a large class of compounds with growing practical importance: the mixed halides have greatly enhanced volatility compared with the individual metal halides.

The dilemma in the models shown in Fig. 3-40 is not unique for systems with central atoms. For tetralithiotetrahedrane, $(CLi)_4$, the structure with the lithium atoms above the faces of the carbon tetrahedron was found in the calculations to be more stable than with the lithium atoms above the vertices [3-42]. The two models are shown in Fig. 3-41. The stable structure of $(CLi)_4$ is reminiscent of the $(CH_3Li)_4$ molecule [3-43] in which the methyl groups are above the faces of the tetrahedron consisting of the lithium atoms.

Figure 3-41.
Models of the $(CLi)_4$ molecule [3-42].

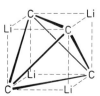

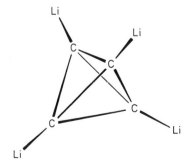

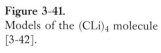

111

Figure 3-42.
(a) Ferrocene: Prismatic (D_{5h}) and antiprismatic (D_{5d}) models.

(b) Dibenzene chromium: prismatic model (D_{6h}).

The prismatic cyclopentadienyl and benzene complexes of transition metals (see e.g. [3-10]) are reminiscent of the polycyclic hydrocarbon prismanes. Fig. 3-42 shows ferrocene, $(C_5H_5)_2Fe$, for which both the barrier to rotation and the free energy difference between the prismatic (eclipsed) and antiprismatic (staggered) conformations are very small [3-44]. Fig. 3-42 presents also a prismatic model with D_{6h} symmetry for dibenzene chromium, $(C_6H_6)_2Cr$.

Molecules with multiple bonds between metal atoms often have structures with beautiful and highly symmetrical polyhedral shapes [3-45]. Only two examples are mentioned here. One is the square prismatic $[Re_2Cl_8]^{2-}$ ion [3-46] shown in Fig. 3-43 which played an important role in the history of the discovery of metal-metal multiple bonds. The other is the paddle-like structure of the anhydrous dimolybdenum tetraacetate, $Mo_2(O_2CCH_3)_4$, in the vapor phase [3-47] shown in Fig. 3-44.

Figure 3-43.
The square prismatic structure of the $[Re_2Cl_8]^{2-}$ ion which played a historic role in the discovery of metal-metal multiple bonds ([3-46], cf. [3-45]).

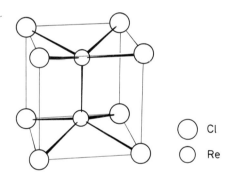

Figure 3-44.
The paddle-like structure of the anhydrous dimolybdenum tetra-acetate, $Mo_2(O_2CCH_3)_4$, from a gas electron diffraction study [3-47].

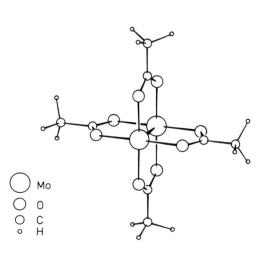

Incidentally, there is a class of organic hydrocarbons which are called paddlanes on account of their similarity to the shape of the paddles that propel riverboats [3-48]. The very symmetrical one called [2.2.2.2]paddlane (Fig. 3-45) has not yet been prepared due to its high strain energy. In fact, the most unusual

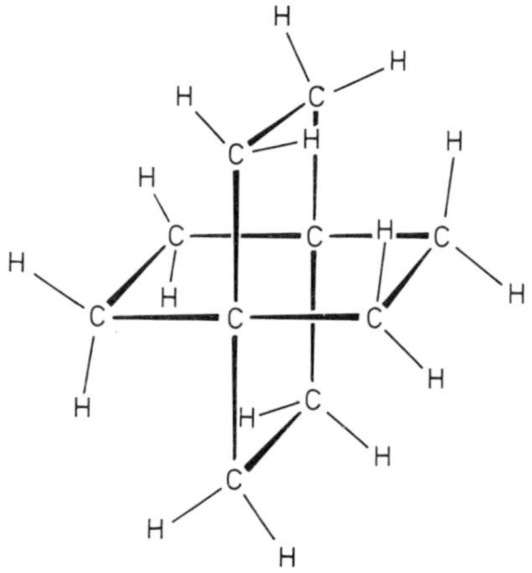

Figure 3-45.
[2.2.2.2]Paddlane, a highly strained polycyclic hydrocarbon. It has not yet been prepared.

Figure 3-45.
[2.2.2.2]Paddlane, a highly strained polycyclic hydrocarbon. It has not yet been prepared.

Figure 3-46.
(a) [1.1.1]Propellane
(b) 1,3-Diborabicyclo[1.1.1]-pentane. Analogous structures and bonding situations were found by a theoretical investigation [3-50].

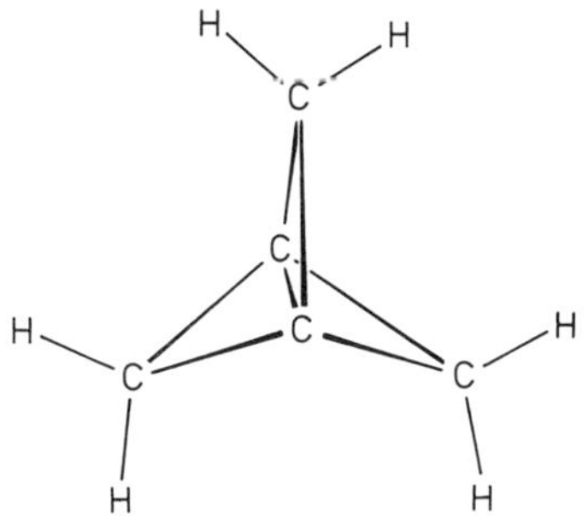

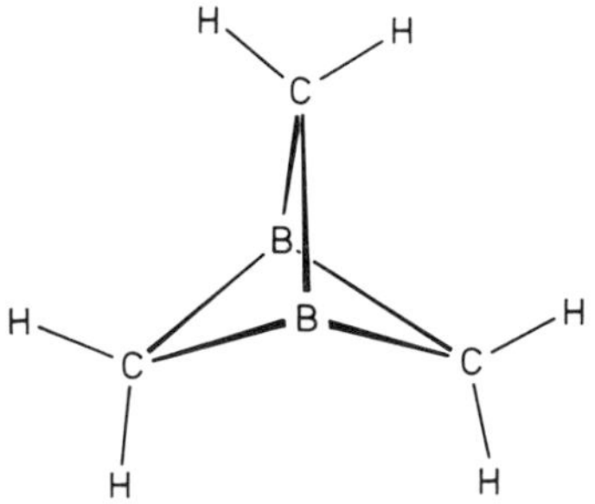

parent polycyclic hydrocarbon known to date is a somewhat related substance, [1.1.1]propellane [3-49] shown in Fig. 3-46a. This highly strained molecule has an interesting structure in that the two bridgehead carbon atoms possess inverted configurations, and there is no formal bond between them. On the other hand, the separation of the two bridgehead carbons corresponds to the distance for a normal chemical bond according to a theoretical investigation by Jackson and Allen [3-50]. They found a similar bonding situation in 1,3-diborabicyclo-[1.1.1]pentane (Fig. 3-46b). In fact, as the interaction between the bridgehead carbons of [1.1.1]propellane has been interpreted by three-center, two-electron orbitals, there is a natural analogy with the boron bonding situations in borohydrides and carboranes. The comparison of these two molecules is a good example of how an understanding of the correlation between chemical bonding and the geometrical aspects of molecular structure facilitates bridging the two domains of organic and inorganic chemistry. Reference should also be made to the fundamental work by Hoffmann [3-51] with respect to the structural concepts bridging organic and inorganic chemistry. The hydrocarbon skeleton of [1.1.1]propellane seems to be "electrondeficient", while the extra electron density which gives the molecule its normal electron complement is on the outside of the skeleton.

3.7.4 Regularities in Non-Bonded Distances

The structure of [1.1.1]propellane is interesting as it shows a sort of "pseudobonding" situation with proper "bonding geometry" of the bridgehead carbon atoms between which, however, there is no chemical bond. In some respects, the situation is reversed in the structure of the ONF_3 molecule (Fig. 3-47) which can be looked at as a regular tetrahedron formed by three fluorines and one oxygen while there is, of course, true bonding from each of these atoms as ligands to the central nitrogen atom. The non-bonded F... F and F... O distances representing the lengths of the edges of a tetrahedron are indeed equal within the experimental errors of their determination [3-52] as shown in Fig. 3-47. The bond lengths and bond angles are also given. The molecule has C_{3v} symmetry, and the central nitrogen atom is obviously not in the center of the essentially regular tetrahedron of its ligands.

Figure 3-47.
The molecular geometry of ONF_3 [3-52].
(a) Bond lengths and bond angles.
(b) Non-bonded distances.

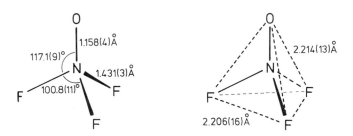

In some molecular geometries, the so-called intramolecular 1,3 separations are remarkably constant. The "1,3" label refers to the interactions between two atoms in the molecule which are separated by a third atom. The near equality of the non-bonded distances in the ONF_3 molecule is a special case. What is usually observed is the constancy of a certain 1,3 non-bonded distance throughout a series of related molecules. Significantly, this constancy of 1,3-distances may be accompanied by considerable changes in the bond lengths and bond angles within the three atom group. The intramolecular 1,3 interactions have also been called intramolecular van-der-Waals interactions, and Bartell [3-53] postulated a set of intramolecular non-bonded 1,3 radii. These 1,3 non-bonded radii are intermediate in value between the corresponding covalent radii and "traditional" van-der-Waals radii, all compiled for some elements in Table 3-5.

Table 3-5. Covalent, 1,3 Intramolecular Non-bonded, and Van-der-Waals Radii of Some Elements (in Å units)

Element	Covalent Radii[a]	1,3 Intra-molecular Non-bonded Radii[b]	Van-der-Waals Radii[a]
B	0.817	1.33	
C	0.772	1.25	
N	0.70	1.14	1.5
O	0.66	1.13	1.40
F	0.64	1.08	1.35
Al	1.202	1.66	
Si	1.17	1.55	
P	1.10	1.45	1.9
S	1.04	1.45	1.85
Cl	0.99	1.44	1.80
Ga	1.26	1.72	
Ge	1.22	1.58	
As	1.21	1.61	2.0
Se	1.17	1.58	2.00
Br	1.14	1.59	1.95

[a] after Pauling [3-55]
[b] after Bartell [3-53] and Glidewell [3-56]

Fig. 3-48 shows some structural peculiarities which originally prompted Bartell [3-54] to recognize the importance of the intramolecular non-bonded interactions. It was an interesting observation that the three outer carbon atoms in $H_2C = C(CH_3)_2$ were arranged as if they were at the corners of an approximately equilateral triangle, as shown in Fig. 3-48a. Since the central carbon atom in this arrangement is obviously not in the center of the triangle, the bond angle between the bulky methyl groups is smaller than the ideal 120°. In the other example, in Fig. 3-48b, the C-C bond lengthening is related to the increasing number of non-bonded interactions.

Figure 3-48.
Geometrical consequences of non-bonded interactions after Bartell [3-54].
(a) The three outer carbon atoms of $H_2C=C(CH_3)_2$ are in the corners of an approximately equilateral triangle, leading to a relaxation of the bond angle between the methyl groups.
(b) Considerations of non-bonded interactions in the interpretation of the C–C single bond length changes in a series of molecules.

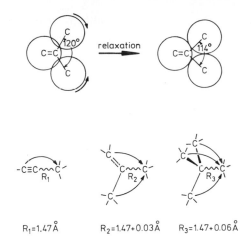

$R_1=1.47 Å$ $\quad R_2=1.47+0.03 Å$ $\quad R_3=1.47+0.06 Å$

Of course, the 1,3 intramolecular non-bonded radii (Table 3-5) are purely empirical, but so are the other kinds of radii. Some of the 1,3 non-bonded radii have been up-dated [3-56] based on more recent experimental data. The 1,3 non-bonded radius for fluorine has proved to be remarkably well established. Some time ago a controversy between two structure determinations of tetrafluoro-1,3-dithietane,

F_2C S CF_2 S

was settled by considering the constancy of the F...F non-bonded distances [3-57]. The mean of the F...F 1,3-distances in 40 molecules containing a CF_3 group was found to be 2.162 Å with a standard deviation of 0.008 Å! The postulated 1,3 non-bonded radius for fluorine was 1.08 Å [3-53].

The O...O non-bonded distances in XSO_2Y sulfones have been observed to be remarkably constant at 2.48 Å [3-58] in a relatively large series of compounds. At the same time the S=O bond lengths varied up to 0.05 Å and the O=S=O bond angles up to 5° depending on the nature of the X and Y ligands. Fig. 3-49 illustrates this structural feature. The geometrical variations in the sulfone series could be visualized (Fig. 3-50a) as if the two oxygen ligands were firmly attached to two of the four vertices of the ligand tetrahedron around the sulfur atom, and this central atom were moving along the bisector of the OSO angle depending on the X and Y ligands [3-59].

116

Figure 3-49.
Two empirical relationships are shown here concerning the geometry of the SO_2 parts of a series of XSO_2Y sulfone molecules [3-58]. One of the curves expresses linear correlation between the $r(S=O)$ bond lengths and $\angle$ (OSO) bond angles:
$\angle = -147.7\ r + 331.7\ (\sigma = 0.6°)$.
The other utilizes the constancy of the O ... O distance observed at 2.484 Å ($\sigma = 0.004$ Å) in this series of substances:
$\angle = 2$ arc sin $(2.484/2\ r)$.

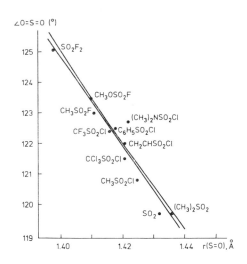

Figure 3-50.
(a) The configuration of XSO_2Y sulfone molecules

The sulfuric acid, H_2SO_4 or $(HO)SO_2(OH)$, molecule has its four oxygens around the sulfur at the vertices of a nearly regular tetrahedron, Fig. 3-50b. Compared with the differences in the various OSO angles (up to 20°) and in the two kinds of SO bonds (up to 0.15 Å), the greatest difference among the six O ... O non-bonded distances is only 0.07 Å [3-60].

(b) The configuration of the sulfuric acid molecule

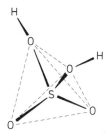

The alkali sulfate molecules used to appear in old textbooks with the following structural formula:

$$Na\!-\!O \diagdown \quad \diagup O$$

It has been established (see e.g. [3 61]), however, that the SO_4 groups in such molecules have nearly regular tetrahedral configuration. The metal atoms are located on axes perpendicular to the edges of the SO_4 tetrahedron. Thus this structure is bicyclic as shown in Fig. 3-50c. There is a marked similarity between the sulfate and the $KAlF_4$ tetrafluoroaluminate structures in that both have a well-defined, rather rigid tetrahedral nucleus around which the alkali atoms occupy relatively loose positions.

(c) The configuration of alkali sulfate molecules, M_2SO_4.

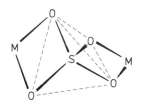

3.7.5 The VSEPR Model

Numerous examples of molecular structures have been introduced in the preceding sections. They are all confirmed by modern experiments and/or calculations. We would like to know, however, not only *what* is the structure of a molecule and its symmetry, but also, *why* a certain structure with a certain symmetry is realized?

It has been a long-standing goal in chemistry to determine the shape and measure the size of molecules, and also to calculate these properties. Today, quantum chemistry is capable of determining the molecular structure, at least for relatively simple molecules, starting from the mere knowledge of the atomic composition, and without using any empirical information. Such calculations are called *ab initio*. The primary results from these calculations are, however, wave functions and energies which may also be considered "raw measurements" similar to some experimental data. At the same time there is a desire to understand molecular structures in simple terms, for example, the localized chemical bond, that have proved so useful to chemists' thinking. There is a need for a bridge between the measurements and calculations on one hand, and simple qualitative ideas, on the other hand. There are several qualitative models for molecular structure that serve this purpose well. These models can explain, for example, why the methane molecule is regular tetrahedral, T_d, why ammonia is pyramidal,

C_{3v}, why water is bent, C_{2v}, and why the xenon tetrafluoride molecule is square planar, D_{4h}, etc. It is also important to understand why seemingly analogous molecules like OPF_3 and $OClF_3$ have so different symmetries, the former C_{3v}, and the latter C_s as seen in Fig. 3-51?

Figure 3-51.
The molecular configuration of OPF_3 and $OClF_3$.

The structure of a series of the simplest AX_n type molecules will be examined in terms of one of these useful and successful qualitative models. A is the central atom, X are the ligands, and not necessarily all n ligands are the same. The qualitative models simplify the picture. They usually consider only a few, if not just one, of the many effects which are obviously present and are interacting in a most complex way. The measure of the success of these qualitative models is in their ability to create consistent patterns for interpreting individual structures and structural variations in a series of molecules, and above all, in their ability to correctly predict the structures of molecules, not yet studied or possibly not yet prepared.

One of the simplest models is based on the following postulate [3-62]: *The geometry of the molecule is determined by the repulsions among the electron pairs in the valence shell of its central atom.* The valence shell is the outermost shell of the electron cloud surrounding the atomic nucleus, and it is the electrons of this shell that participate in the chemical bonding. The atoms in the molecule are usually linked by pairs of electrons, one of the electrons originating from one atom and the other from the other atom. These two electrons are called the bonding pair. The valence shell of an atom may also have other electron pairs that do not participate in bonding and belong to this atom alone. They are called unshared or lone pairs of electrons. The above postulate emphasizes the importance of *both* bonding pairs and lone pairs in establishing the molecular geometry. The model is appropriately called the Valence Shell Electron Pair Repulsion or VSEPR model. The bond configuration around atom A in the molecule AX_n, and accordingly, the geometry of the AX_n molecule, is such that the electron pairs of the valence shell be at maximum distances from each other, as if the electron pairs were mutually repelling each other. Thus the situation may be visualized in such a way that the electron pairs occupy well-defined parts of the space around the central atom corresponding to the concept of localized molecular orbitals.

If it is assumed that the valence shell of the central atom retains its spherical symmetry in the molecule, then the electron pairs will be at equal distances from the nucleus of the central atom. In this case the arrangements at which the distances among the electron pairs are at maximum, will be the following:

number of electron pairs in the valence shell	arrangement
2	linear
3	equilateral triangle
4	tetrahedron
5	trigonal bipyramid
6	octahedron

For spherical symmetry, the electron pairs can be represented by points on the surface of a sphere. Then the shapes shown in Fig. 3-52 are obtained by connecting these points. Of the three polyhedra shown in Fig. 3-52, only two are regular, viz. the tetrahedron and the octahedron. The trigonal bipyramid is not a regular polyhedron; although its six faces are equivalent, its edges and vertices are not. Incidentally, the trigonal bipyramid is not a unique solution to the five-point problem. Another, only slightly less advantageous arrangement is the square pyramidal configuration and there are numerous intermediate ones between the trigonal bipyramid and the square pyramid.

Figure 3-52.
Molecular shapes from a points-on-the-sphere model.

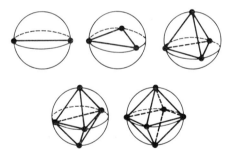

The repulsions considered in the VSEPR model may be expressed by the potential energy terms
$$V_{ij} = K/r_{ij}^n,$$
where K is a constant, r_{ij} is the distance between the points i and j, and the exponent n is large for strong or "hard" repulsion interactions and small for weak or "soft" repulsion interactions. Experience shows [3-63] that the value of n is much larger than it would be for simple electrostatic Coulomb interactions. Indeed, when n is larger than 3, the results become rather insensitive to the value of n. That is very fortunate because n is not really known. This insensitivity to the choice of n is what provides the wide applicability of the VSEPR model.

3.7.5.1 Analogies

It is easy not only to imagine the three-dimensional consequences of the VSEPR model, but it is also easy to demonstrate them in reality. We need only to blow up a few balloons that children play with [3-64]. If groups of two, three, four, five, and six balloons, respectively, are connected at the ends near their openings, the resulting arrangements are shown in Fig. 3-53. Obviously, the space requirements of the various groups of balloons acting as mutual repulsions, determine the shapes and symmetries of these assemblies. There is no difficulty in recognizing the linear arrangement of two balloons, the trigonal planar shape of three, the regular tetrahedron of four, the trigonal bipyramid of five, and the octahedron of six balloons. Thus the balloons here play the role of the electron pairs of the valence shell.

Figure 3-53.
Shapes of groups of balloons.

Another beautiful analogy with the VSEPR model and one found directly in nature, is demonstrated in Fig. 3-54. These are drawings of walnuts growing together (cf. [3-65]). The small clusters of walnuts have exactly the same arrangements for two, three, four, and five walnuts in assemblies as predicted for the electron pairs in the valence shell by the VSEPR model or as those shown by the balloons. The walnuts are required to accommodate themselves to each other's company, and find the arrangements that are most advantageous considering the space requirements of all. Incidentally, the balloons and the walnuts may be considered as "soft" and "hard" objects, with weak and strong interactions, respectively. The complete consistency in their resulting configurations is comforting.

3.7.5.2 Molecular Shapes

Using the VSEPR model it is simple to predict the shape and symmetry of a molecule from the *total* number of bonding pairs, n, and lone pairs, m, of electrons in the valence shell of its central atom. The molecule may then be written as AX_nE_m where E denotes a lone pair of electrons.

Figure 3-54.
Walnut clusters drawn by
Ferenc Lantos.

For the present discussion, only a few examples will be described for illustration purposes. For a comprehensive coverage see, e.g., [3-62].

First, we shall consider the methane molecule, shown in Fig. 3-55 together with ammonia and water. Originally there were four electrons in the carbon valence shell, and these formed four C-H bonds, with the four hydrogens contributing the other four electrons. Thus methane is expressed by AX_4 and its symmetry is, accordingly, regular tetrahedral. In ammonia originally there were five electrons in the nitrogen valence shell, and the formation of the three N-H bonds added three more. With the three bonding pairs and one lone pair in the nitrogen valence shell, ammonia may be written as AX_3E and, accordingly, the arrangement of the molecule is related to a tetrahedron. However, only in three of its four directions do we find bonds, and consequently ligands, while in the fourth there is a lone pair of electrons. Hence a pyramidal geometry is found for the ammonia molecule. The bent configuration of the water molecule can be similarly deduced.

Figure 3-55.
The molecular configuration
of methane, ammonia and
water.

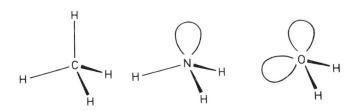

In order to establish the total number of electron pairs in the valence shell, the number of electrons originally present and the number of bonds formed need to be considered. A summary of geometrical arrangements for a series of various types of simple molecules is shown in Fig. 3-56.

Figure 3-56.
Bond configurations with 2, 3, 4, 5, and 6 electron pairs in the valence shell of the central atom after [3-62]. Reproduced with permission from R. J. Gillespie.

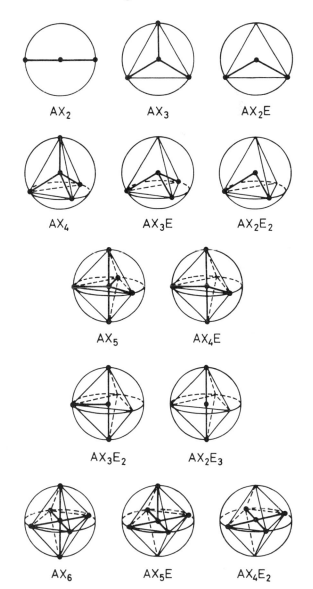

123

The molecular shape to a large extent determines the bond angles. Thus the bond angle X-A-X is 180° in the linear AX_2 molecule, it is 120° in the trigonal planar AX_3 molecule, and 109°28' in the tetrahedral AX_4 molecule. The arrangements shown in Fig. 3-56 correspond to the assumption that the strengths of the repulsions from all electron pairs are equal. In reality, however, the space requirements and accordingly, the strengths of the repulsions from various electron pairs may be different depending on various circumstances as described in the following three subrules [3-62].

1. A lone pair, E, in the valence shell of the central atom has a greater space requirement in the vicinity of the central atom than does a bonding pair. Thus a lone pair exercises a stronger repulsion towards the neighboring electron pairs than does a bonding pair, b. The repulsion strengths weaken in the following order:

    $$E/E > E/b > b/b.$$

    This order is well illustrated by the various angles in sulfur difluoride in Fig. 3-57 as determined by *ab initio* molecular orbital calculations [3-66]. This is also why, for example, the bond angles H-N-H of ammonia, 106.7° [3-67], are smaller than the ideal tetrahedral value, 109.5°. Unless stated otherwise, the parameters in the present discussion are taken from [3-67].

2. Multiple bonds, b_m, have greater space requirements than do single bonds and thus exercise stronger repulsions towards the neighboring electron pairs than do single bonds. The repulsion strengths weaken in the following order:

    $$b_m/b_m > b_m/b > b/b$$

    A consequence of this is that the bond angles will be larger between multiple bonds than between single bonds. The structure of dimethyl sulfate provides a good example as shown in Fig. 3-58. This molecule has three different types of OSO bond angles and they change in the following order:

    $$S=O/S=O > S=O/S-O > S-O/S-O$$

    Another example is the structure of the sulfuric acid molecule, or more generally, the configurations of the XSO_2Y sulfones (cf. Fig. 3-50) for which

    $$S=O/S=O > S=O/S-X \text{ (or } S=O/S-Y) > S-X/S-Y$$

Figure 3-57.
The angles of sulfur difluoride as determined by *ab initio* molecular orbital calculations [3-66].

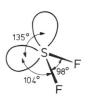

Figure 3-58.
The three different kinds of oxygen-sulfur-oxygen bond angles in the dimethyl sulfate molecule as determined by electron diffraction [3-68].

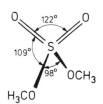

3. A more electronegative ligand decreases the electron density in the vicinity of the central atom as compared with a less electronegative ligand. Accordingly, the bond to a less electronegative ligand, b_X, has a greater space requirement than the bond to a more electronegative ligand, b_Y. The repulsion strengths then weaken in the following order:

$$b_X/b_X > b_X/b_Y > b_Y/b_Y$$

Consequently, the bond angles are smaller for more electronegative ligands than for less electronegative ligands. An example of this effect can be seen in a comparison of sulfur difluoride and sulfur dichloride, for each of which the bond angles are shown in Fig. 3-59. Another example is provided again by the XSO_2Y sulfone series in Fig. 3-49. As the electronegativities of the ligands X and Y increase, the $S=O$ bonds shorten and the $O=S=O$ bond angles open somewhat. With decreasing electronegativities the opposite effect is observed.

Figure 3-59.
Experimentally determined bond angles of sulfur difluoride [3-69] and sulfur dichloride [3-70].

It is interesting to compare the implications expressed by these subrules with the depiction of some localized molecular orbitals shown in Fig. 3-60 after Schmiedekamp et al. [3-66]. It is seen indeed that the lone pair of electrons occupies more space than do the bonding pairs in the vicinity of the central atom. Also, a bond to a more electronegative ligand such as fluorine occupies less space in the vicinity of the central atom than does a bond to a less electronegative ligand such as hydrogen. It is also seen that a double bond occupies more space than a single bond. The angular ranges of the corresponding contours in the

Figure 3-60.
Localized molecular orbitals represented by contour lines denoting electron densities of 0.02, 0.04, 0.06 etc. electron/Bohr^3 from theoretical calculations [3-66].

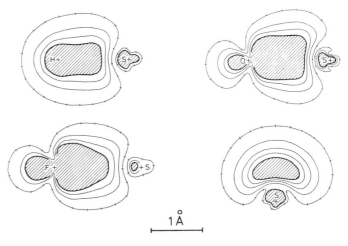

125

electron density plots are all in good qualitative agreement with the postulates of the VSEPR model, the difference between the lone pair and bonding pairs being especially large.

The VSEPR model has a fourth subrule that concerns the relative availability of space in the valence shell:

4. There is less space available in a completely filled valence shell than in a partially filled valence shell. Accordingly, the repulsions are stronger and the possibility for angular changes are smaller in the filled valence shell than in the partially filled one. Thus, for example, the bond angles of ammonia are closer to the ideal tetrahedral value than are those of phosphine. The experimental values are given in Fig. 3-61.

It has been shown that the differences in the electron pair repulsions may account for the bond angle variations in several series of molecules. The question now arises as to whether these differences have any effect in the *symmetry choice* of the molecules. In the four-electron-pair systems the differences in the electron pair repulsions have a decisive role in the sense that the AX_4, EBX_3, E_2CX_2 molecules have T_d, C_{3v}, C_{2v} symmetries, respectively. Within each series, however, the symmetry is preserved regardless of the changes in the ligand electronegativities. For example, only the bond angles change in the molecules EBX_3 and EBY_3, the symmetry remains the same.

Ligand electronegativity changes may have decisive effects, however, on the symmetry choices of various bipyramidal systems, of which the trigonal bipyramidal configuration is the simplest.

When five electron pairs are present in the valence shell of the central atom, the trigonal bipyramidal configuration is usually found, although a tetragonal pyramidal arrangement cannot be excluded in some cases. Even intermediate arrangements between these two may appear to be the most stable in some special structures. The trigonal bipyramidal configuration with an equilateral triangle in the equatorial plane has D_{3h} symmetry while the square pyramidal has C_{4v} symmetry. The intermediate arrangements have C_{2v} symmetry or nearly so. Indeed, rearrangements often occur in trigonal bipyramidal structures performing low-frequency large-amplitude motion. Such rearrangements will be illustrated later.

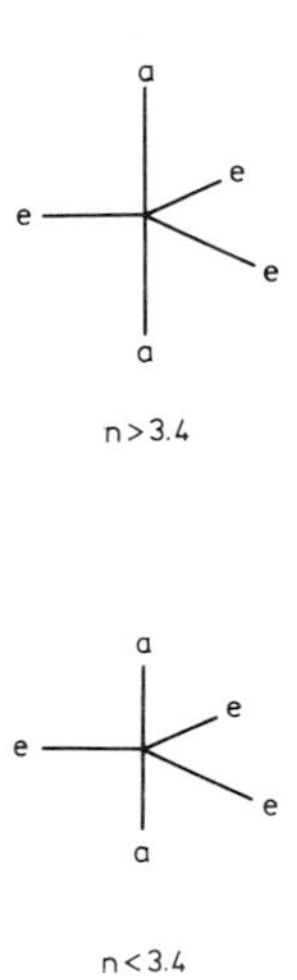

The positions in the D_{3h} trigonal bipyramid are generally not equivalent. For n values larger than 3.4, in the expression $V = K/r^n$, the axial ligand position is further away from the central atom than the equatorial one. The reverse is true for n values smaller than 3.4, as is illustrated in Fig. 3-62. The axial and equatorial positions are at equal distances from the central atom when $n = 3.4$. These variations, however, have no effect on the symmetry of the AX_5 structures and this is comforting from

Figure 3-61.
The bond angles of ammonia and phosphine [3-67].

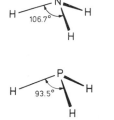

Figure 3-62.
Changing relationship between the axial and equatorial bond lengths depending on the strength of the repulsion interactions.

the point of view of the applicability of the VSEPR model in establishing the point group symmetries of such molecules.

On the other hand, when there is inequality among the electron pairs, the differences in the axial and equatorial positions do have importance for symmetry considerations. The PF_5 molecule as an AX_5 system shows unambiguously D_{3h} symmetry in its trigonal bipyramidal configuration. However, the prediction of the symmetry of the SF_4 molecule that may be written as AX_4E, is less obvious. For SF_4 the problem is where will the lone pair of electrons occur?

An axial position in the trigonal bipyramidal arrangement has three nearest neighbors at 90° away, and one more neighbor at 180°. For an equatorial position there are two nearest neighbors at 90° and two further ones at 120°. As the closest electron pairs exercise by far the strongest repulsion, the axial positions are affected more than the equatorial ones. In agreement with this reasoning, the axial bonds are usually found to be longer than the equatorial ones. If there is a lone pair of electrons with a relatively large space requirement, it should be found in the more advantageous equatorial position. Accordingly, the SF_4 structure has C_{2v} symmetry, as does the ClF_3 molecule, which is of the AX_3E_2 type. Finally, the XeF_2 molecule is AX_2E_3 with all three lone pairs in the equatorial plane, hence its symmetry is $D_{\infty h}$. All these structures are depicted in Fig. 3-63.

By similar reasoning, the VSEPR model predicts that a double bond will also occupy an equatorial position. Thus the point group may easily be established for the molecules $O=SF_4$, $O=ClF_3$, XeO_3F_2, and XeO_2F_2, as is also seen in

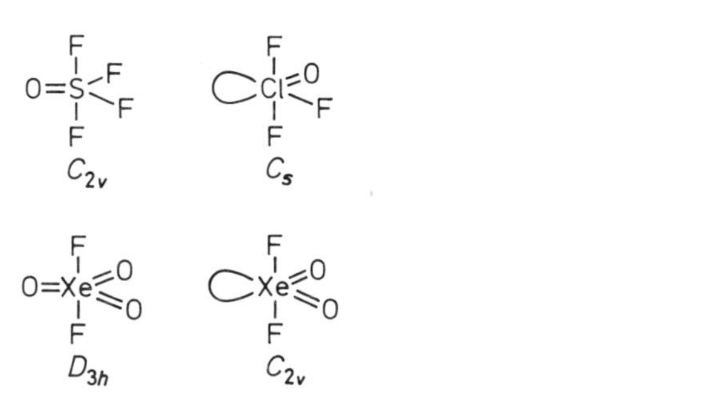

Figure 3-63.
Molecules with trigonal bipyramidal and related configurations.

AX_5 AX_4E AX_3E_2 AX_2E_3

127

Fig. 3-63. We note the C_s symmetry for the OClF$_3$ molecule (cf. Fig. 3-51) as a consequence of the bipyramidal geometry with both the Cl=O double bond and the lone pair in the equatorial plane. The molecule OPF$_3$ (cf. Fig. 3-51) is only seemingly analogous. There is no lone pair in the phosphorus valence shell, and thus the molecule has a distorted tetrahedral bond configuration. The P=O double bond is along the three-fold axis, and the point group is C_{3v}, similar to that for ammonia.

Of the arrangements shown in Fig. 3-63 only those for PF$_5$, XeO$_3$F$_2$, and XeF$_2$ are equilateral trigonal bipyramids. The others do not have trigonal symmetry.

Lone pairs and/or double bonds replaced single bonds in the above examples. Similar considerations are applicable when only ligand electronegativity changes take place. A typical and very simple example can be demonstrated by a comparison of the structures of PF$_2$Cl$_3$ and PF$_3$Cl$_2$ [3-71]. The chlorine atoms are less electronegative ligands than the fluorines, and they will be in equatorial positions in *both* structures as seen in Fig. 3-64. The point groups are C_{2v} for PF$_3$Cl$_2$ and D_{3h} for PF$_2$Cl$_3$. Were the chlorines in the axial positions in PF$_3$Cl$_2$, this molecule would also have the much higher symmetry D_{3h}.

Figure 3-64.
The molecular structures of PF$_3$Cl$_2$ and PF$_2$Cl$_3$ are not analogous: the chlorine ligands occupy equatorial positions in both cases [3-71].

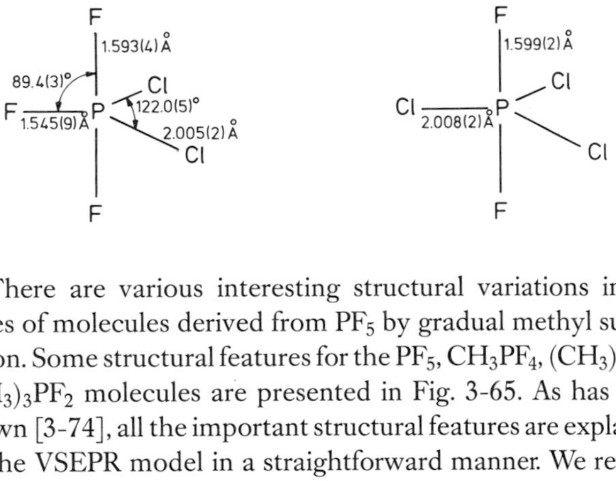

There are various interesting structural variations in the series of molecules derived from PF$_5$ by gradual methyl substitution. Some structural features for the PF$_5$, CH$_3$PF$_4$, (CH$_3$)$_2$PF$_3$, (CH$_3$)$_3$PF$_2$ molecules are presented in Fig. 3-65. As has been shown [3-74], all the important structural features are explained by the VSEPR model in a straightforward manner. We restrict ourselves here to enumerating only the following structural features [3-74]:

1. The molecules are trigonal bipyramids.
2. The methyl groups occupy equatorial positions.
3. The axial bonds are longer than the corresponding equatorial bonds.
4. As the number of methyl substituents increases, all bond lengths increase, but the ratio of the length of the axial bonds to the length of the equatorial bonds also increases.
5. The P-F bonds bend away from the P-C bonds.

There is complete analogy in the following two series:

| PF$_5$ | SF$_4$ | ClF$_3$ | ArF$_2$ |
| PF$_5$ | CH$_3$PF$_4$ | (CH$_3$)$_2$PF$_3$ | (CH$_3$)$_3$PF$_2$ |

Figure 3-65.
Molecular geometries of the series PF$_5$ [3-72], CH$_3$PF$_4$ [3-72], (CH$_3$)$_2$PF$_3$ [3-72], (CH$_3$)$_3$PF$_2$ [3-73].

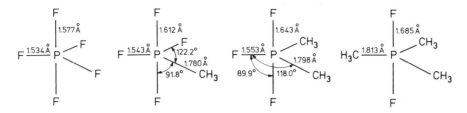

Figure 3-66.
Molecular configuration of phosphorus pentachloride and its CF$_3$-substituted derivatives: PCl$_5$ [3-75], CF$_3$PCl$_4$ [3-76], (CF$_3$)$_2$PCl$_3$ [3-77, 3-78], (CF$_3$)$_3$PCl$_2$ [3-79].

Of these, argon difluoride has not yet been studied. Its linear structure with a 1.76 Å bond length has been suggested, in fact, by this comparison [3-73].

The agreement between CF$_3$-substituted phosphorus pentachloride configurations and the VSEPR model is also excellent. The slightly more electronegative trifluoromethyl ligands occur in the axial positions as seen in Fig. 3-66.

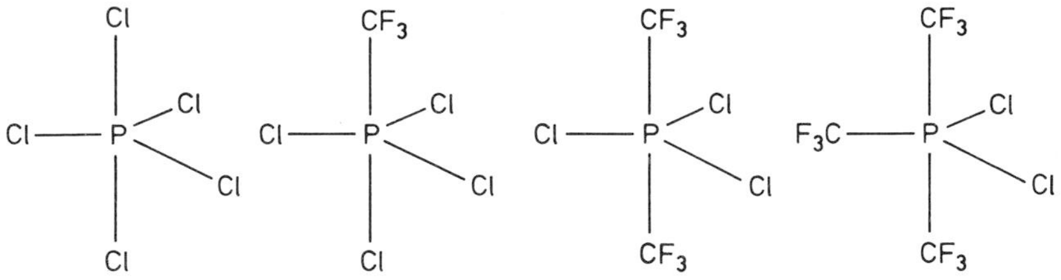

Figure 3-67.
Molecular configurations of phosphorus pentafluoride and its CF$_3$-substituted derivatives: PF$_5$ [3-72], CF$_3$PF$_4$ [3-79 – 3-81], (CF$_3$)$_2$PF$_3$ [3-79], (CF$_3$)$_3$PF$_2$ (3-79].

The structural results reported for CF$_3$-substituted phosphorus pentafluorides seem to be less unambiguous. The molecular configurations are indicated in Fig. 3-67. However, it is necessary to discuss the available data in terms of the experimental

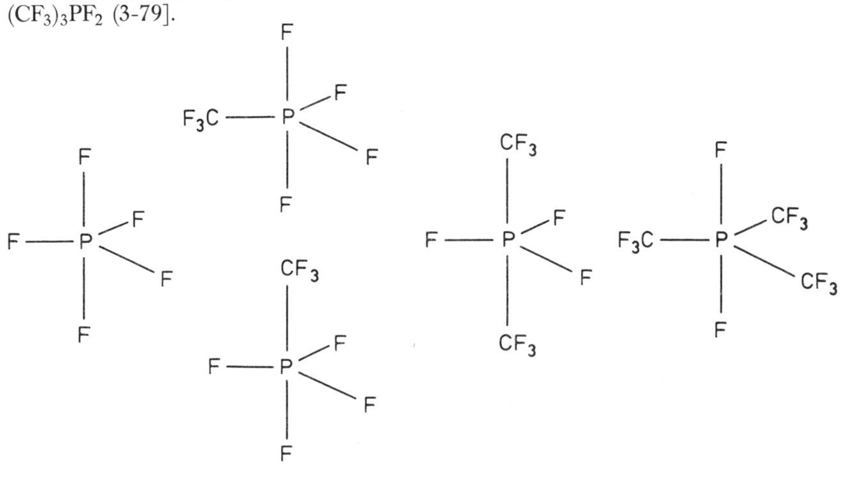

129

methods employed. According to electron diffraction [3-79], there is an equilibrium between two configurations of the mono-CF_3-substituted phosphorus pentafluoride, CF_3PF_4. The interpretations of the N.M.R. spectra were contradictory depending on the nucleus whose magnetic resonance was recorded. The F-19 data indicated an axial CF_3 group [3-80] whereas the C-13 data, an equatorial one [3-81]. Oberhammer et al. [3-79] suggested that both configurations are present and are rapidly interconverting. This may resolve the apparent contradiction in the N.M.R. information. A rapid interconversion between two configurations would be revealed as an apparent mixture by electron diffraction as the interaction time in this experiment is much shorter than the time of the interconversion, or than the N.M.R. interaction time, for that matter. The configuration of the $(CF_3)_3PF_2$ molecule is again in complete agreement with the VSEPR model. On the other hand, there is a puzzling discrepancy for the $(CF_3)_2PF_3$ configuration for which electron pair repulsion and ligand electronegativity considerations would suggest equatorial positions for the CF_3 groups. The experimentally determined structure has a higher symmetry than the one predicted by the VSEPR model would have. The $(CF_3)_2PF_3$ structure is analogous to that of $(CF_3)_2PCl_3$. For the latter, however, there is no discrepancy with the model as the CF_3 electronegativity is slightly greater than that of chlorine. As has been seen before, the methyl groups of $(CH_3)_2PF_3$ occur in equatorial positions, see Fig. 3-65, in agreement with the VSEPR model.

All six electron pairs are equivalent in the AX_6 molecule and so the symmetry is unambiguously O_h. An example is SF_6. The IF_5 molecule, however, corresponds to AX_5E and its square pyramidal configuration has C_{4v} symmetry. There is no question here as to the preferred position for the lone pair, as any of the six equivalent sites may be selected. When, however, a second lone pair is introduced, then the favored arrangement is that in which the two lone pairs find themselves at the maximum distance apart. Thus for XeF_4, i.e. AX_4E_2, the bond configuration is square planar, point group D_{4h}. The above mentioned three structures are depicted in Fig. 3-68.

Figure 3-68.
Molecules with octahedral and related configurations.

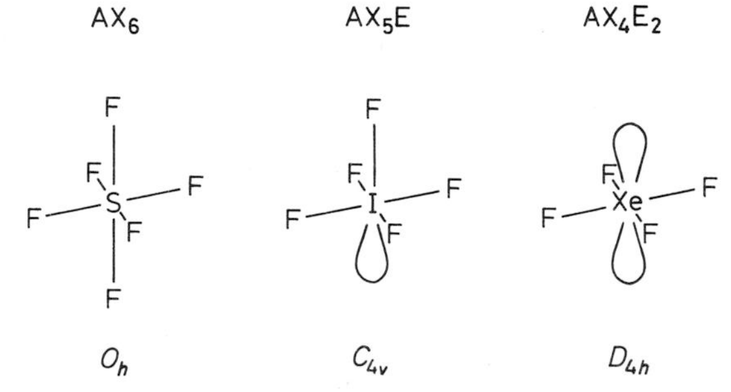

AX_6 $\quad\quad$ AX_5E $\quad\quad$ AX_4E_2

O_h $\quad\quad$ C_{4v} $\quad\quad$ D_{4h}

The difficulties encountered in the discussion of the five electron pair valence shells are intensified in the case of the seven electron pair case. Here again the ligand arrangements are less favorable than for the nearest coordination neighbors, i.e. six and eight. It is not possible to arrange seven equivalent points in a regular polyhedron, while the number of nonisomorphic polyhedra with seven vertices is large, viz. 34 [3-82]. No single one of them is distinguished, however, from the others on the basis of relative stability. Some of the possible arrangements are shown in Fig. 3-69. There may be quite rapid rearrangements among the various configurations.

Figure 3-69.
Configurations with 7 electron pairs in the valence shell of the central atom.

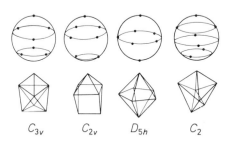

C_{3v} C_{2v} D_{5h} C_2

One of the early successes of the VSEPR model was that it correctly predicted a nonregular structure for XeF_6 by considering it as a seven-coordination case, AX_6E. Some possible distorted octahedral configurations for XeF_6 are shown in Fig. 3-70. They have been found to be in agreement with experimental evidence.

Figure 3-70.
Distorted octahedral structures of xenon hexafluoride. Bartell: "An electron pair isn't like a baseball bat that's round; it's like a British cricket bat that's flattened out." [3-83].

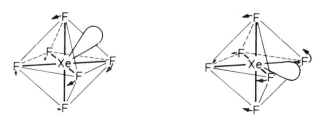

3.7.5.3 Generalized Applicability Test

From the very beginning of the applications of the VSEPR model, the predictions have usually been only for bond angle variations. While the influence of the lone pairs on the bond angles has been correctly assessed, the fact that the bond angles represent only a part of the geometrical characterization of the entire valence shell configuration has been largely ignored.

Some bond angle variations seemingly incompatible with the model have been noted, e.g., [3-84 – 3-87]. These discrepancies were the more puzzling in that they occurred among simple molecules, for which the VSEPR model was certainly expected to be applicable. Some examples will be discussed below. First, however, a more generalized approach for testing the applicability of the VSEPR model is formulated. This formulation stems directly from the basic idea of the VSEPR model.

As the shape and the geometry of a molecule is assumed to be determined by the repulsions among all electron pairs of the valence shell,

> the compatibility of a structure or structural variations with the VSEPR model has to be tested by examining the variations of *all angles of all electron pairs* rather than only those of the bond angles [3-88].

Indeed, the variations in the bond angles are usually the only ones considered as they are directly determined from the experiment. Sometimes, the angles made by the lone pairs may be deduced from the experimental data on the bond angles, by virtue of the molecular symmetry. For example, the E-P-F angle of the PF_3 molecule can be calculated from the F-P-F bond angle by virtue of the C_{3v} symmetry of the trifluorophosphine molecule. On the other hand, the angles E-S-E and E-S-F of the SF_2 molecule with C_{2v} symmetry cannot be calculated from the F-S-F bond angle. However, in many cases where the angles made by the lone pairs can easily be calculated from the experimentally determined bond angles, or when they may be obtained from the results of quantum chemical calculations, they are often ignored. The proper application of the VSEPR model, however, should direct at least as much attention to the angles of the lone pairs and their variations as to those of the bond angles themselves.

Let us consider first the experimental bond angle variations in the examples of the AX_4, EBX_3, E_2CX_2 series of molecules shown in Fig. 3-71. Originally it was stated that "... in the series CH_4, NH_3, and H_2O the bond angle decreases ... as the number of non-bonding pairs increases." [3-62]. While it was invariably observed that in going from AX_4 to EBX_3 the bond angles decreased, the replacement of another bond by a second lone pair did not lead to a further decrease in the bond angle in E_2CX_2, except for the case of the hydride molecules.

As not all the angles made by the lone pairs in these tetrahedral systems were attainable from the experimental data, *ab initio* molecular orbital calculations have been carried out for a series of molecules [3-66, 3-89]. The position of the lone pair was characterized by the center of its charge distribution. In the series, SiF_4, PF_3, SF_2, ClF, and Ar, for example, all the angles are listed in Table 3-6. This series of molecules may be expressed by the following general formulae, AX_4E_0, AX_3E_1, AX_2E_2, AX_1E_3, AX_0E_4, respectively. It is noted that the angular differences within each structure are in accordance with the VSEPR model. The calculated bond angles in the series well parallel the changes observed in the experimental values which are shown in Fig. 3-71.

Figure 3-71.
The bond angles (experimental results) in a series of AX_4, BX_3, CX_2 molecules; 0, 1, and 2 indicate the number of lone pairs in the valence shell of the central atom, for sources see [3-86].

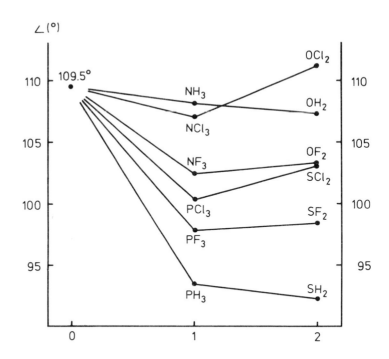

Table 3-6. Angles Made by Bonds and Lone Pairs in a Series of Isoelectronic Molecules

General formulae	AX_4E_0	AX_3E_1	AX_2E_2	AX_1E_3	AX_0E_4
Molecules	SiF_4	PF_3	SF_2	ClF	Ar
Symmetry	T_d	C_{3v}	C_{2v}	$C_{\infty v}$	–
Bond angles	F-Si-F 109.5°	F-P-F 96.9°	F-S-F 98.1°	–	
Other angles		E-P-F 120.2°	E-S-F 104.3°	E-Cl-F 101.6°	
			E-S-E 135.8°	E-Cl-E 116.1°	E-Ar-E 109.5°
Reference	–	[3-66]	[3-66]	[3-89]	–

There is a decrease in going from the F-Si-F to the F-P-F angle and the latter is smaller than the F-S-F angle. On the other hand, the E-P-F angle is much larger than the E-S-F angle. The origin of this difference is decisive as the relative strength of the repulsive interactions decreases in the order

$$E/E > E/b > b/b.$$

Also, there are four E/b interactions and only one b/b interaction in the sulfur valence shell of SF_2. The situation here, however, is even more complicated as there is also a strong E/E interaction.

133

Figure 3-72.
All angles in the series of SF$_2$, HSF, SH$_2$ molecules from *ab initio* molecular orbital calculations [3-66].

Another example is provided by the experimental bond angles of SF$_2$, 98.0° [3-69] and SH$_2$, 92.2° [3-67]. The bond angle difference here is in the opposite direction from what would be expected from the electronegativity subrule. However, the structural changes in the rest of the valence shell configuration should not be ignored. The calculated angles – all angles, regardless whether they involve bonds or lone pairs – are shown in Fig. 3-72. Firstly, it can be noted that the E-S-E, E-S-b, and b-S-b angles are related to each other in each molecule exactly as would be predicted by the VSEPR model, considering the

Table 3-7. Triple Average Angles for Bonds and Lone Pairs from *ab initio* Molecular Orbital Calculations [3-66]

Orbital	Molecule	Triple Average Angle (°)
S-F	HSF	102.2
	SF$_2$	102.4
	SOF$_2$	102.9[a]
S-H	HSF	103.8
	SH$_2$	103.1
	SOH$_2$	103.1
	SO$_2$H$_2$	104.7
S=O	SOH$_2$	113.6
	SO$_2$H$_2$	113.3
	SOF$_2$	113.7[a]
S-E	SH$_2$	114.2
	HSF	114.2
	SOH$_2$	114.8
	SF$_2$	114.7
	SOF$_2$	114.9[a]
O-H	OH$_2$	107.1
O-F	OF$_2$	103.5
O-E	OH$_2$	111.6
	OF$_2$	114.1
N-F	NF$_3$	106.9[a]
N-H	NH$_3$	108.0
N-E	NF$_3$	115.9[a]
	NH$_3$	113.3
P-F	PF$_3$	104.7[a]
P-H	PH$_3$	103.5
P-E	PF$_3$	120.2[a]
	PH$_3$	122.4

[a] An spd basis was used in the calculations except where this superscript indicates sp basis.

134

different space requirements of lone pairs and bonds. Furthermore, in agreement with the electronegativity subrule, the E-S-H angle is larger than the E-S-F angle. In both molecules there are four stronger E/b interactions and only one weaker b/b interaction. The first effect obviously prevails. The molecule HSF, not yet studied experimentally, is also shown among the computed structures. It is interesting to note that the E-S-F angle is somewhat smaller than the E-S-H angle in this molecule as well.

The general space requirements of various bonds and lone pairs can be conveniently characterized by the so-called triple average angles [3-66]. The triple average angle is the mean of the three angles made by a bond or a lone pair in a tetrahedral configuration. For a series of bonds and lone pairs the values of the triple average angles are collected in Table 3-7. The triple average angles for a bond or for a lone pair in various molecules appear to be rather constant. The space requirements of the fluorine bonds are somewhat smaller than those of the respective hydride bonds. The space requirement of the $S = O$ double bond is considerably larger than those of the single bonds and only slightly smaller than those of the lone pairs. The remarkable constancy of the general space requirements further facilitates an understanding of the bond angle changes displayed, e.g., by the SF_2 and SH_2 molecules, or by the series shown in Fig. 3-71. The sulfur bond angle variations in analogous sulfones/sulfoxides/sulfides as shown in Fig. 3-73 are similar to the angular variations given in Fig. 3-71. The difference is that here first one $S = O$ double bond, then two $S = O$ double bonds are replaced by lone electron pairs. A detailed discussion is given in [3-66].

Figure 3-73.
The experimental bond angles X-S-X in XSO$_2$X, XSOX, and XSX molecules.

X	sulfones	sulfoxides	sulfides
(CH$_3$)$_2$N	3-90	3-91	3-92
CH$_3$	3-93	3-94	3-95
CF$_3$	3-96	3-98	3-97
Cl	3-98	3-99	3-70
F	3-100	3-101	3-69

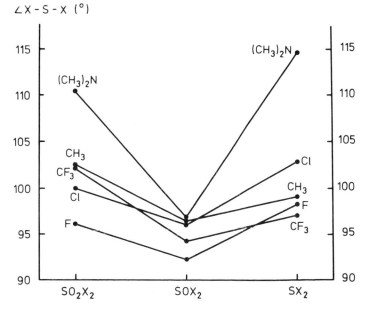

∠X - S - X (°)

135

Figure 3-74.

All angles in SF_4 and $S(CF_3)_2F_2$ molecules. Experimental bond angles from [3-102] and [3-87], respectively.

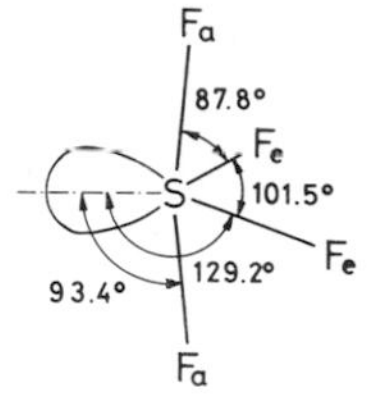

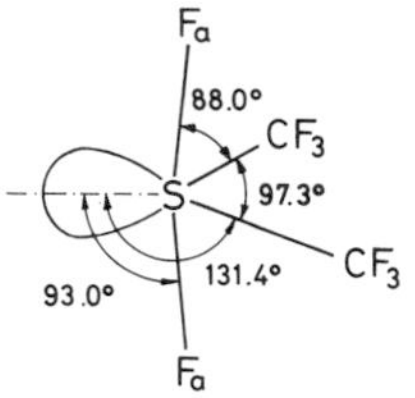

A comparison of the SF_4 [3-102] and $S(CF_3)_2F_2$ molecular geometries, Fig. 3-74, in terms of their bond angles only, would again suggest an incompatibility with the VSEPR model [3-87]. The general configuration of these molecules is unambiguously predicted by the VSEPR model to be trigonal bipyramidal. For the bis(trifluoromethyl) derivative it is also predicted correctly that the less electronegative CF_3 ligands should be found in the equatorial positions. According to the electronegativity subrule, the C-S-C bond angle of $S(CF_3)_2F_2$ would be expected to be larger than the F_e-S-F_e bond angle of SF_4. This would be the expected result if the other interactions were ignored. Incidentally, if steric effects rather than electron pair repulsions were the determining factor, then again the bulky CF_3 groups would be expected to cause an increase in the C-S-C bond angle as compared to the F_e-S-F_e bond angle. However, as can be seen from Fig. 3-74, the C-S-C bond angle is actually smaller than the F_e-S-F_e bond angle.

Luckily, the angles involving the lone pairs can be easily calculated from the bond angles in these structures by virtue of the C_{2v} symmetry of the sulfur bond configuration. There are two kinds of interactions in the equatorial plane, viz. E/b_Y and b_Y/b_Y in one, and E/b_X and b_X/b_X in the other molecule. The stronger E/b interaction occurs twice and the weaker b/b interaction only once in both structures. Both the E-S-b and b-S-b angles are in the equatorial plane. As the stronger and twice occurring E/b interaction obviously prevails over the b/b interaction, the real question is whether the difference in the E-S-b angles of the two molecules is consistent with the VSEPR model or not? The E-S-C angle is larger than the E-S-F_e angle, exactly as expected from the VSEPR model if all interactions are properly considered. The observed change in the bond angles then is consistent with the changes in the prevailing interactions between the two molecules. In the present comparison the angles involving the bonds to the axial fluorine ligands are ignored as they are equal in the two structures within experimental error.

It is also instructive to consider the so-called quadruple average angles in the trigonal bipyramidal molecules for characterizing the general space requirements [3-103]. The quadruple average angle is the mean of the four angles made by Q-A in QAX_4 where Q may be a ligand or a lone pair and the X ligands may all be the same or they may be different. The quadruple average angles of the lone pairs in the two molecules considered above are

SF_4	111.4°
$S(CF_3)_2F_2$	112.2°.

Although the difference is small, its direction is in complete agreement with the prediction of the VSEPR model postulating the E/b repulsions to be stronger when involving bonds to less electronegative ligands. It is again assumed that the $E/S\text{-}F_a$ interactions are equal in the two molecules.

It is of interest to compare the triple average angles of lone pairs and double bonds in the tetrahedral systems and the corresponding quadruple average angles in the trigonal bipyramidal systems. It has been noted that the triple average angles of the $S=O$ double bond are only slightly smaller than those of the lone pairs. According to a recent electron diffraction reinvestigation of the structure of thionyl tetrafluoride [3-104], its geometry corresponds to a quadruple average angle of 110.65° for the $S=O$ bond. In fact, of the four models suggested earlier [3-105] as being compatible with the electron diffraction data but differing considerably in the bond angles, consideration of the quadruple average angles led to the choice of the model [3-103] that consequently proved to be the preferred structure. A remarkable constancy of the quadruple average angles in QSF$_4$ derivatives has also been noted and contrasted with the rather large variation in the equatorial bond angles [3-103]. The data are shown in Fig. 3-75. Even in such $XN=SF_4$ derivatives

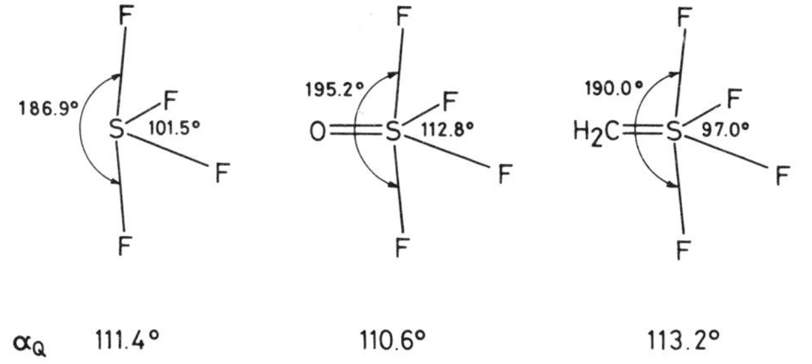

Figure 3-75.
Bond angles and quadruple average angles (α_Q) in some QSF$_4$ derivatives; Refs: SF$_4$ [3-102], OSF$_4$ [3-104], H$_2$CSF$_4$ [3-106].

α_Q 111.4°   110.6°   113.2°

where the molecular symmetry is strongly distorted by the axial orientation of the X ligand, the $N=S$ double bond retains a similar quadruple average angle and it remains remarkably constant. This is illustrated in Fig. 3-76.

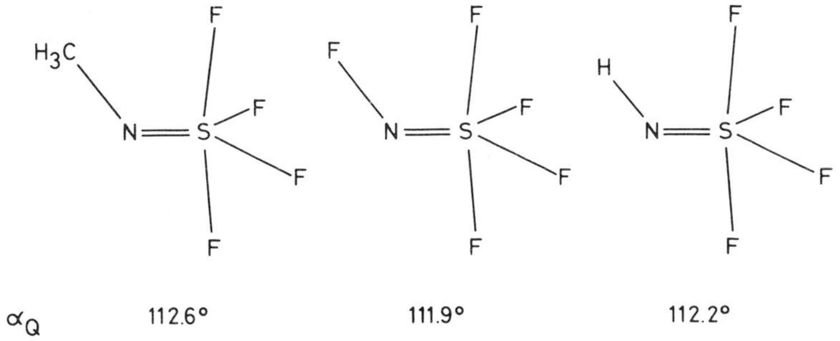

Figure 3-76.
Molecular configurations of the nitrogen-sulfur double bond in $XN=SF_4$ derivatives: CH$_3$NSF$_4$ [3-107], FNSF$_4$ [3-108], HNSF$_4$ [3-109].

α_Q 112.6°   111.9°   112.2°

137

3.7.5.4 Directional Repulsion Effects

Christe and Oberhammer [3-109] called attention to the directional repulsion effects of lone pairs and double bonds in trigonal bipyramidal structures. Fig. 3-77 shows the ideal angles in the AX_5 molecule and the deviations from these ideal angles in SF_4, OSF_4, and H_2CSF_4. If the lone pair of SF_4 can be assumed to have cylindrical symmetry, then the different angular deviations from the ideal values in the axial and equatorial directions point to directional differences in the repulsion strengths. The difference in the deviations observed in the sulfur tetrafluoride structure suggests that the equatorial direction is "softer" and the axial direction is "harder". The $C=S$ double bond seems to have similar, although somewhat greater effects than the lone pair. On the other hand, the $O=S$ double bond has not only a smaller overall repulsion, but shows directional preference in the axial direction.

Figure 3-77.
The "ideal" bond angles in the trigonal bipyramidal bond configurations and deviations from the "ideal" angles in SF_4, OSF_4, and H_2CSF_4 molecules.

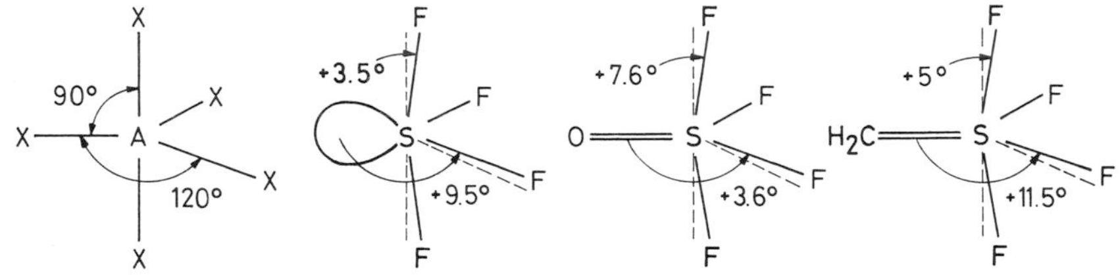

A striking example that demonstrates the importance of these directional effects is the molecular geometry of XeO_2F_2 as determined by neutron diffraction [3-110]. The molecule is of EAX_4 type and the lone pair and the two double bonds are found in equatorial positions as predicted by the VSEPR model and as shown in Fig. 3-78. The $O=Xe=O$ bond angle is considerably smaller than $120°$, even though this angle is between two double bonds. Obviously the two $E/Xe=O$ interactions are prevailing over the $Xe=O/Xe=O$ interaction in the equatorial plane. It is then somewhat surprising that the axial Xe-F bonds are bent *towards* the xenon lone pair of electrons rather than away from it. This fact indicates, however, that the repulsions in the axial directions are dominated by the directional effect from the $Xe=O$ double bonds.

Figure 3-78.
The structure of XeO_2F_2; molecular configuration, bond angles and deviations from the "ideal" angles, neutron crystallographic results of Peterson et al. [3-110].

138

It is worth mentioning that XeO_2F_2 is the only compound in the present discussion of the VSEPR model for which a crystal-phase molecular structure is considered. For all the others the structures of the free molecules were available. XeO_2F_2 has a layer structure in the crystal resulting from the Xe ... O inter-molecular bridging as shown in Fig. 3-79. These contacts might be thought to decrease somewhat the repulsive strength of the xenon lone pair as well as that of the Xe=O double bonds. In the absence of vapor-phase data for comparison, we have no way to judge the extent of such an effect. What is important for our present discussion is the *opposite sign* in the deviations from the "ideal" angles in the equatorial and axial directions. This result certainly points to the difference in the directional repulsion strengths of the xenon lone pair and the Xe=O double bond.

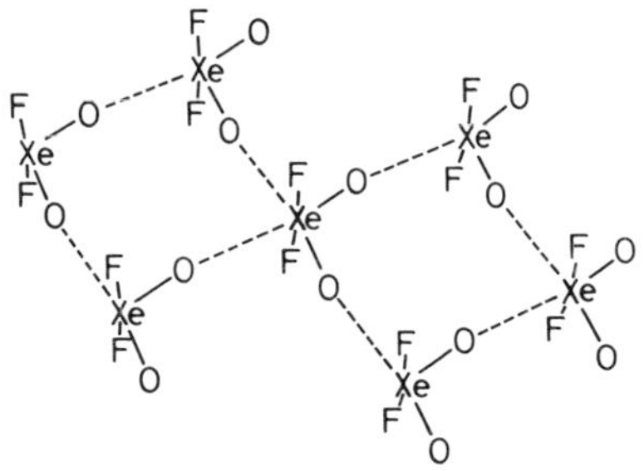

Figure 3-79.
From the layer structure of the XeO_2F_2 crystal [3-110].

3.7.5.5 Historical Remarks

The structural peculiarities described above which can be explained by the directional effects go beyond the original VSEPR model. Such additional considerations are useful or necessary to explain particular structural variations, but their introduction makes the original simple model more complex. It is a matter of opinion whether one remains within the scope of the original model and applies it strictly observing its limitations. Alternatively, the scope of the model may be expanded at the expense of its original simplicity.

However, this simplicity of the VSEPR model is one of its primary strengths. In addition, the model provides a continuity in the development of the qualitative ideas about the nature of the chemical bond and its correlation with molecular structure. Abegg's octet rule [3-111] and Lewis' theory of the shared electron pair [3-112] may be considered as the direct fore-runners of the model.

Lewis' cubical atom [3-112] deserves special mention. It was instrumental in shaping the concept of the shared electron pair. It also permitted a resolution of the apparent contradiction between the two distinctly different bonding types, viz. the

139

Figure 3-80.
Lewis' cubical atoms and
some molecules built from
such atoms (cf. [3-112]).

shared electron pair and the ionic electron-transfer bond. In terms of Lewis' theory the two bonding types could be looked at as mere limiting cases. Lewis' cubical atoms are illustrated in Fig. 3-80. They are also noteworthy as an example of a certainly useful though not necessarily correct application of a polyhedral model.

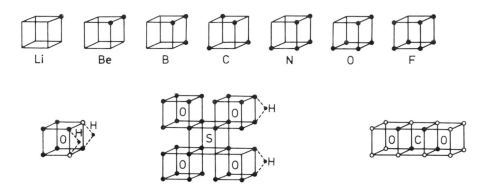

Sidgwick and Powell [3-113] were first to correlate the number of electron pairs in the valence shell of a central atom and its bond configuration. Then Gillespie and Nyholm [3-114] introduced allowances for the difference between the effects of bonding pairs and lone pairs and applied the model to large classes of inorganic compounds. The name VSEPR was coined by Gillespie (cf. [3-115]), who has also very effectively popularized the model over the years (e.g. [3-116, 3-117, 3-62]). The model has found its way into most introductory chemistry texts in addition to being a research tool. Although the primary applications of the model are in the area of molecular geometries it is currently also contributing to molecular force field studies [3-118, 3-119].

There have been attempts to provide quantum-mechanical foundations for the VSEPR model (cf. e.g. [3-118]). Roughly speaking, these attempts have developed along two lines. One was concerned with assigning a rigorous theoretical basis to the model, primarily involving the Pauli exclusion principle, to the extent that it was even suggested that the application of the model be named "Pauli mechanics" [3-63]. However, it should be remembered that the VSEPR model is a qualitative one. It overemphasizes some interactions and ignores many others. It cannot therefore be expected that a rigorous quantum-mechanical treatment will parallel it in its entirety. On the other hand, numerous quantum chemical calculations (e.g. [3-66, 3-120]) have already produced a large amount of structural data which are in complete agreement with the VSEPR predictions, demonstrating that the model captures some very important effects which appear to be dominant in some structural classes.

For example, the model emphasizes electron pair repulsions while ignoring ligand-ligand interactions. With large central atoms and small ligands, these emphases work well. However, with increasing ligand size relative to the size of the central atom, the non-bonded interactions become gradually more important. Obviously, the two effects may be commensurable in some structures and eventually the ligand-ligand interactions will become dominant. Another basic assumption of the VSEPR model is that the shape of the valence shell of the central atom is spherical. In cases where this assumption is invalid, the applicability of the predictions from the model will be less useful. Further work is expected to make the model more reliable, by investigating and establishing its limitations.

3.7.6 Consequences of Intramolecular Motion

Henri Matisse's Dance is reproduced in Fig. 3-81 [3-121]. The picture radiates dynamism. We can readily imagine that we are really observing such a dance from a distance. According to the choreography, one of the dancers jumps and is thus out of the plane of the other four. As soon as this dancer returns into the plane of the others, it is now the role of the next to jump, and so on. The exchange of roles from one dancer to another

Figure 3-81.
Henri Matisse: "Dance". – The Hermitage, Leningrad. Reproduced by permission.

throughout the five-member group is so quick that if we take a normal photograph, we shall have a blurred picture of the five dancers. However, if we have a very sensitive film, we may be able to use such a short exposure that a well-defined configuration of the dancers at a particular moment can be identified.

141

The above description well simulates the *pseudorotation* of the cyclopentane molecule, although on a different time scale. The cyclopentane, $(CH_2)_5$, molecule has a special degree of freedom when the out-of-plane carbon atom exchanges roles with one of its two neighboring carbon atoms (and their hydrogen ligands). The effect is also equivalent to a rotation by $2\pi/5$ about the axis perpendicular to the plane of the four in-plane carbons [3-122].

In discussing molecular structure, an extreme approach is to disregard intramolecular motion and to consider the molecule to be motionless. A completely rigid molecule is a hypothetical state corresponding to the minimum position of the potential energy function for the molecule. Such a motionless structure has an important and well-defined physical meaning; it even has been given a special name "equilibrium structure". It is this equilibrium structure that emerges from quantum chemical calculations. On the other hand, real molecules are never motionless. Furthermore, the various physical measurement techniques determine the structures of *real* molecules. As our discussion of Matisse's Dance has illustrated, the relationship between the lifetime of the configuration under investigation and the time-scale of the investigating technique is of crucial importance [3-123]. The structural properties of mono-CF_3-substituted phosphorus pentafluoride, CF_3PF_4, provide a good example as previously discussed. Techniques with different interaction times came up with different positions for the CF_3 substituent. Clearly, the relationship of these interaction times and the permutation time of the ligands in the molecule resulted in these different conclusions.

Large-amplitude, low-frequency intramolecular vibrations may lower the molecular symmetry of the average structure from the higher symmetry of the equilibrium structure. Some examples from metal halide molecules are shown in Fig. 3-82. In fact it is not yet definitely known whether all the equilibrium structures indeed correspond to those shown in Fig. 3-82. No

Figure 3-82.
Equilibrium versus average structures of metal halide molecules with low-frequency, large-amplitude deformation vibrations.

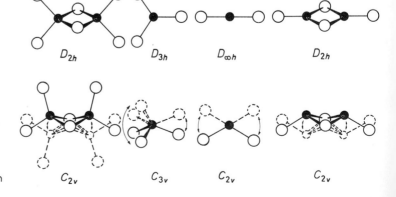

● metal
○ halogen

142

permutations of the nuclei are considered to result from these intramolecular vibrations on the time-scale involved. The structures determined so far include $MnCl_2$ [3-124], $MnBr_2$, Mn_2Br_4 [3-125], $FeCl_2$ [3-126], $FeBr_2$, Fe_2Br_4 [3-127], $CoCl_2$ [3-128], $CoBr_2$, Co_2Br_4 [3-129], $NiCl_2$ [3-130], $NiBr_2$ [3-131], $AlCl_3$ [3-132], $FeCl_3$ [3-133], Al_2Cl_6 [3-134], and Fe_2Cl_6 [3-135]. They invariably indicate the consequences of large-amplitude deformation vibrations in the lowered symmetry of the average structures as compared with the equilibrium geometries. Of course, not all metal dihalides have linear equilibrium and bent average structures. There are genuinely bent equilibrium geometries even among the dihalides of first-row transition metals [3-136].

A rapid interconversion of the nuclei takes place in the bullvalene molecule under very mild conditions in fluid media. This process involves making and breaking bonds, but this is accompanied by very small shifts in the nuclear positions. The molecular formula is $(CH)_{10}$ and the carbon skeleton is shown in Fig. 3-83. There are only four different kinds of carbon positions (and hydrogen positions, accordingly) and all four positions are being interconverted simultaneously [3-137]. Hypostrophene is another $(CH)_{10}$ hydrocarbon whose trivial name was chosen to reflect its behavior [3-138]. The Greek hypostrophe means turning about, a recurrence. The molecule is ceaselessly undergoing the intramolecular rearrangements indicated in Fig. 3-84. The atoms have a complete time-averaged equivalence yet hypostrophene could not be converted into pentaprismane (cf. [3-29]).

Permutational isomerism among inorganic substances was discovered by Berry [3-139] for trigonal bipyramidal structures. Although the trigonal bipyramid and the square pyramid have very different symmetries, D_{3h} versus C_{4v}, they are easily inter-

Figure 3-83.
The interconversion of bull-valene [3-137].

Figure 3-84.
The interconversion of hypo-strophene [3-138].

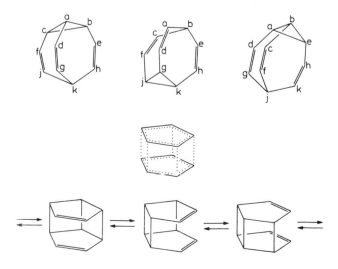

143

converted by means of bending vibrations as is illustrated in Fig. 3-85. The possible change in the potential energy during this structural reorganization is also shown. The permutational isomerism of an AX_5 molecule, e.g. PF_5, is easy to visualize as the two axial ligands replace two of the three equatorial ones, while the third equatorial ligand becomes the axial ligand in the transitional square pyramidal structure. The rearrangements quickly follow one another, without any position being constant for any significant time period. The C_{4v} form originates from a D_{3h} structure and yields then again to another D_{3h} form. A somewhat similar pathway was established [3-140] for the $(CH_3)_2NPF_4$ molecule in which the dimethylamine group is permanently locked to an equatorial position whereas the fluorines exchange in pairs all the time.

Figure 3-85.
Berry-pseudorotation of PF_5-type molecules [3-139].

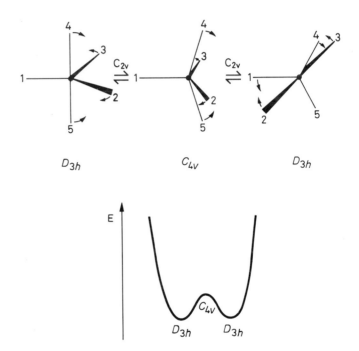

Iodine heptafluoride, IF_7, has an interesting structure in that several pieces of experimental evidence [3-141] suggested a slightly deformed D_{5h} symmetry for its geometry. Representations of three proposed structures are shown in Fig. 3-86. The possible structures with C_2 and C_s symmetry result from a static deformation from the D_{5h} symmetry structure. Describing iodine heptafluoride in terms of these structures is analogous to describing cyclopentane by the so-called "half-chair" and "envelope" conformations shown in Fig. 3-87. Conversely, the dynamic structure of iodine heptafluoride may be described by pseudorotation.

Figure 3-86.
Representations of D_{5h}, C_2, and C_s structures for iodine heptafluoride after Adams et al. [3-142].

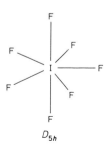

D_{5h}

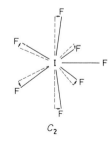

C_2

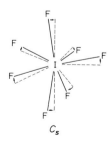

C_s

Figure 3-87.
The "half-chair" (C_2) and "envelope" (C_s) conformations of cyclopentane.

Of particular interest for the IF$_7$ structure is that the VSEPR model and corresponding points-on-the-sphere calculations predicted the equatorial sites to be less advantageous than the axial ones from point of view of spatial possibilities. While in the trigonal bipyramidal configuration the equatorial positions are more spacious, i.e. suffering relatively less repulsion from neighboring electron pairs than the axial ones, the opposite seems to be the case in the pentagonal bipyramidal and related configurations. A direct consequence is that the axial I-F bonds are shorter than the equatorial ones [3-142]. The more crowded equatorial arrangement results in stiffer equatorial in-plane bends than the other bending modes. Recognition of these facts led to a reassignment of the vibrational spectrum of iodine heptafluoride [3-141].

The rearrangement that characterizes the PF$_5$ molecule also describes well the permutation of the atomic nuclei in five-atom polyhedral boron skeletons in borane molecules [3-143].

Lipscomb [3-144] has elaborated a general concept for the rearrangements of polyhedral boranes. According to this concept, two common triangulated faces are stretched to a square face in the borane polyhedra. There is an intermediate polyhedral structure with square faces. In the final step of the rearrangement, the intermediate configuration may revert to the

145

original polyhedron with no net change, but it may as well turn into a different arrangement. The arrangement has rectangular faces with an orthogonal linkage with respect to the bonding situation in the original polyhedron [3-143]. This is illustrated in Fig. 3-88. There are many practical examples, among which is the rearrangements of dicarba-*closo*-dodecaboranes, illustrated in Fig. 3-89. There are three isomers of this beautiful carborane molecule:

> 1,2-dicarba-*closo*-dodecaborane, or o-$C_2B_{10}H_{12}$,
> 1,7-dicarba-*closo*-dodecaborane, or m-$C_2B_{10}H_{12}$, and
> 1,12-dicarba-*closo*-dodecaborane, or p-$C_2B_{10}H_{12}$.

Whereas the ortho isomer easily transforms into the meta isomer in agreement with the above mentioned model, the para isomer is obtained only under more drastic conditions and only in a small amount [3-144]. More recently a very similar model has been proposed [3-146] for the so-called carbonyl scrambling mechanism in molecules like $Co_4(CO)_{12}$, $Rh_4(CO)_{12}$, and $Ir_4(CO)_{12}$.

Figure 3-88.
(a) The Lipscomb model of the rearrangement in polyhedral boranes [3-144].
(b) An example of icosahedron/cuboctahedron/icosahedron rearrangement [3-144].

Figure 3-89.
Ortho-, meta-, and para-dicarba-*closo*-dodecaborane structures [3-145]. Whereas the ortho isomer easily transforms into the meta, the para isomer is obtained only under drastic conditions and only in small amounts (cf. [3-144]).

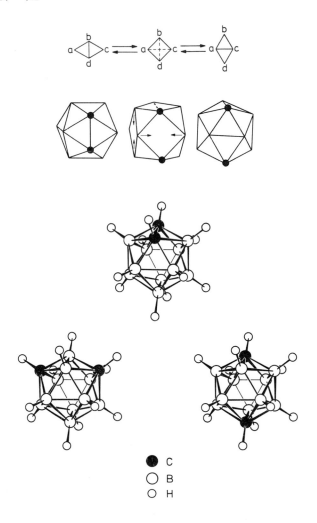

● C
○ B
○ H

Figure 3-90.
The structure of $[Co_6(CO)_{14}]^{4-}$ in two representations after Benfield et al. [3-148].
(a) The octahedron of the cobalt cluster possesses six terminal and eight triply bridging carbonyl groups.

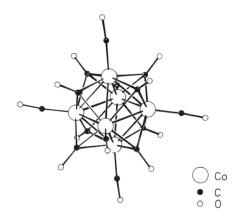

Co
C
O

(b) An omnicapped cube of the carbonyl oxygen envelopes the cobalt octahedron.

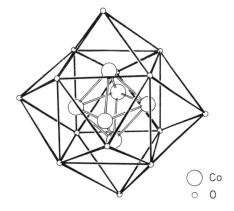

Co
O

Incidentally, the carbonyl ligands can have several modes of coordination, viz. terminal and a variety of bridging possibilities. Rapid interconversion between the different coordination modes is possible, even in the solid state [3-147]. The above mentioned metal-carbonyl molecules belong to a large class of compounds whose general formula is $M_m(CO)_n$, where M is a transition metal. The usually small m-atomic metal cluster polyhedron is enveloped by another polyhedron whose vertices are occupied by the carbonyl oxygens [3-148]. An attractive example is the structure of $[Co_6(CO)_{14}]^{4-}$ in which the octahedral metal cluster has six terminal and eight triply bridging carbonyl groups, as shown in Fig. 3-90a. This structure may also be represented by an omnicapped cube enveloping an octahedron as shown in Fig. 3-90b after [3-148]. These models are reminiscent of another model in which, also, polyhedra were enveloping other polyhedra. That model was Kepler's planetary system [3-149] cited in Fig. 2-73.

References

[3-1]   I. Hargittai, J. Chem. Educ. **60**, 94 (1983).

[3-2]   Full color reproductions of the original are available in editions of Degas' work.
(a) The original itself is in the Louvre, Musée de l'Impressionisme, Paris.
(b) The original itself is in The Hermitage, Leningrad.

[3-3]   W. D. Hounshell, D. A. Dougherty, and K. Mislow, J. Am. Chem. Soc. **100**, 3149 (1978).

[3-4]   N. F. M. Henry and K. Lonsdale, eds., *International Tables for X-ray Crystallography,* Vol. I., *Symmetry Groups,* Kynoch Press, Birmingham, 1969.

[3-5]   F. A. Cotton, *Chemical Applications of Group Theory,* Second Edition, Wiley-Interscience, New York, 1971.

[3-6]   M. Orchin and H. H. Jaffe, J. Chem. Educ. **47**, 372 (1970).

[3-7]   (a) B. Baer Capitman, *American Trademark Design,* Dover Publications, New York, 1976.
(b) K. L. Wolf and R. Wolff, *Symmetrie,* Böhlau-Verlag, Münster/Köln, 1956.

[3-8]   B. Rozsondai, B. Zelei, and I. Hargittai, J. Mol. Struct. **95**, 187 (1982).

[3-9]   The origins of the structural data for the XC_6H_4X molecules, $X=F$, A. Domenicano, G. Schultz, and I. Hargittai, J. Mol. Struct. **78**, 97 (1982); $X=NO_2$, N. P. Penionzkevich, N. I. Sadova, M. I. Popik, L. V. Vilkov, and Yu. A. Pankrushev, Zh. Strukt. Khim. **20**, 603 (1979); $X=CN$, M. Colapietro, A. Domenicano, G. Portalone, G. Schultz, and I. Hargittai, J. Mol. Struct. **112**, 141 (1984); $X=Cl$, G. Schultz, I. Hargittai, and A. Domenicano, J. Mol. Struct. **68**, 281 (1980); $X=CH_3$, A. Domenicano, G. Schultz, M. Kolonits, and I. Hargittai, J. Mol. Struct. **53**, 197 (1979); $X=Si(CH_3)_3$, [3-8].

[3-10]   M. Hargittai and I. Hargittai, *The Molecular Geometries of Coordination Compounds in the Vapour Phase,* Akadémiai Kiadó, Budapest and Elsevier, Amsterdam and New York, 1977.

[3-11]   M. Hargittai, I. Hargittai, and V. P. Spiridonov, J. Chem. Soc. Chem. Commun. 750 (1973).

[3-12]   H. S. M. Coxeter, *Regular Polytopes,* Third Edition, Dover Publications, New York, 1973.

[3-13]   R. J. Ternansky, D. W. Balogh, and L. A. Paquette, J. Am. Chem. Soc. **104**, 4503 (1982).

[3-14]   H. P. Schultz, J. Org. Chem. **30**, 1361 (1965).

[3-15]   E. L. Muetterties, in *Boron Hydride Chemistry,* E. L. Muetterties, ed., Academic Press, New York, San Francisco, London, 1975. The passage on p. 98 is quoted with permission from Academic Press.

[3-16]   C. H. MacGillavry, *Symmetry Aspects of M. C. Escher's Periodic Drawings,* Bohn, Scheltema and Holkema, Utrecht, 1976.

[3-17]   V. P. Spiridonov and G. I. Mamaeva, Zh. Strukt. Khim. **10**, 113 (1969).

[3-18]   V. Plato and K. Hedberg, Inorg. Chem. **10**, 590 (1970).

[3-19]   W. N. Lipscomb, in *Boron Hydride Chemistry,* E. L. Muetterties, ed., Academic Press, New York, San Francisco, London, 1975.

[3-20]   R. W. Rudolph, Acc. Chem. Res. **9**, 446 (1976).

[3-21]   R. E. Williams, Inorg. Chem. **10**, 210 (1971).

[3-22]   R. W. Rudolph and W. R. Pretzer, Inorg. Chem. **11**, 1974 (1972).

[3-23]   A. Greenberg and J. F. Liebman, *Strained Organic Molecules,* Academic Press, New York, 1978.

[3-24]   J. F. Liebman and A. Greenberg, Chem. Rev. **76**, 311 (1976).

[3-25]   L. N. Ferguson, J. Chem. Educ. **46**, 404 (1969).

[3-26]   G. Maier, S. Pfriem, U. Schafer, and R. Matush, Angew. Chem. Int. Ed. Engl. **17**, 520 (1978).

[3-27]   P. E. Eaton and T. J. Cole, J. Am. Chem. Soc. **86**, 3157 (1964).

[3-28]   T. J. Katz and N. Acton, J. Am. Chem. Soc. **95**, 2738 (1973); V. Ramamurthy and T. J. Katz, Nouv. J. Chim. **1**, 363 (1977).

[3-29]   P. E. Eaton, Y. S. Or, and S. J. Branca, J. Am. Chem. Soc. **103**, 2134 (1981).

[3-30]   W. T. Wipke's computations are cited in [3-21].

[3-31]   D. Farcasin, E. Wiskott, E. Osawa, W. Thielecke, E. M. Engler, J. Slutsky and P. v. R. Schleyer, J. Am. Chem. Soc. **96**, 4669 (1974).

[3-32]   C. A. Cupas and L. Hodakowski, J. Am. Chem. Soc. **96**, 4668 (1974).

[3-33]   L. F. Fieser, J. Chem. Educ. **42**, 408 (1965).

[3-34]   C. Cupas, P. v. R. Schleyer, and D. J. Trecker, J. Am. Chem. Soc. **87**, 917 (1965); T. M. Gund, E. Osawa, V. Z. Williams, and P. v. R. Schleyer, J. Org. Chem. **39**, 2979 (1974).

[3-35]   I. Hargittai and K. Hedberg, in *Molecular Structures and Vibrations*, S. J. Cyvin, ed., Elsevier, Amsterdam, London, New York, 1972.

[3-36]   E. Boelema, J. Strating, and H. Wynberg, Tetrahedron Lett. 1175 (1972); W. D. Graham and P. v. R. Schleyer, Tetrahedron Lett. 1179 (1972).

[3-37]   W. D. Graham, P. v. R. Schleyer, E. W. Hagaman, and E. Wenkert, J. Am. Chem. Soc. **95**, 5785 (1973).

[3-38]   W. Z. Williams, P. v. R. Schleyer, G. J. Gleicher, and L. B. Rodewald, J. Am. Chem. Soc. **88**, 3862 (1966).

[3-39]   W. Burns, T. R. B. Mitchell, M. A. McKervey, J. J. Rooney, G. Ferguson, and P. Roberts, J. Chem. Soc. Chem. Commun. 893 (1976).

[3-40]   M. Hargittai, Kém. Közlem. **50**, 371 (1978); **50**, 489 (1978).

[3-41]   E. Vajda, I. Hargittai, and J. Tremmel, Inorg. Chim. Acta **25**, L143 (1977).

[3-42]   G. Rauscher, T. Clark, D. Poppinger, and P. v. R. Schleyer, Angew. Chem. Int. Ed. Engl. **17**, 276 (1978).

[3-43]   E. Weiss and G. Hencken, J. Organometal. Chem. **21**, 265 (1970).

[3-44]   A. Haaland and J. E. Nilsson, Acta Chem. Scand. **22**, 2653 (1968).

[3-45]   F. A. Cotton and R. A. Walton, *Multiple Bonds Between Metal Atoms,* Wiley-Interscience, New York, Chichester etc., 1982.

[3-46]   F. A. Cotton and C. B. Harris, Inorg. Chem. **4**, 330 (1965); V. G. Kuznetsov and P. A. Koz'min, Zh. Strukt. Khim. **4**, 55 (1963).

[3-47]   M. H. Kelly and M. Fink, J. Chem. Phys. **76**, 1407 (1982).

[3-48]   E. H. Hahn, H. Bohm, and D. Ginsburg, Tetrahedron Lett. 507 (1973).

[3-49]   K. B. Wiberg and F. H. Walker, J. Am. Chem. Soc. **104**, 5239 (1982).

[3-50]   J. E. Jackson and L. C. Allen, J. Am. Chem. Soc. **106**, 591 (1984).

[3-51]   See, e.g., R. Hoffmann, Science **211**, 995 (1981).

[3-52]   V. Plato, W. D. Hartford, and K. Hedberg, J. Chem. Phys. **53**, 3488 (1970).

[3-53]   L. S. Bartell, J. Chem. Phys. **32**, 827 (1960).

[3-54]   L. S. Bartell, J. Chem. Educ. **45**, 754 (1968).

[3-55]   L. Pauling, *The Nature of the Chemical Bond,* Third Edition, Cornell University Press, Ithaca, N. Y., 1960.

[3-56]   C. Glidewell, Inorg. Chim. Acta **20**, 113 (1976).

[3-57]   I. Hargittai, J. Mol. Struct. **54**, 287 (1979).

[3-58]   I. Hargittai, Z. Naturforsch. **34a**, 755 (1979).

[3-59]   I. Hargittai, *The Structure of Volatile Sulphur Compounds,* Akadémiai Kiadó, Budapest and Reidel Publ. Co., Dordrecht, Boston, Lancaster, 1985.

[3-60]   R. L. Kuczkowski, R. D. Suenram, and F. J. Lovas, J. Am. Chem. Soc. **103**, 2561 (1981).

[3-61]   K. P. Petrov, V. V. Ugarov, and N. G. Rambidi, Zh. Strukt. Khim. **21**, 159 (1980).

[3-62]   R. J. Gillespie, *Molecular Geometry,* Van Nostrand Reinhold Co., London, 1972.

[3-63]   L. S. Bartell, Kém. Közlem. **43**, 497 (1975).

[3-64]   H. R. Jones and R. B. Bentley, Proc. Chem. Soc. 438, 1961.

[3-65]   G. Niac and C. Florea, J. Chem. Educ. **57**, 429 (1980).

[3-66]   A. Schmiedekamp, D. W. J. Cruickshank, S. Skaarup, P. Pulay, I. Hargittai, and J. E. Boggs, J. Am. Chem. Soc. **101**, 2002 (1979).

[3-67]   Landolt-Börnstein, *Numerical Data and Functional Relation ships in Science and Technology (New Series).* Vol. 7: *Structure Data of Free Polyatomic Molecules,* K.-H. Hellwege and A. M. Hellwege, eds., Springer, Berlin-Heidelberg-New York, 1976.

[3-68]   J. Brunvoll, O. Exner, and I. Hargittai, J. Mol. Struct. **73**, 99 (1981).

[3-69]   Y. Endo, S. Saito, E. Hirota, and T. Chikaraishi, J. Mol. Spectrosc. **77**, 222 (1979).

[3-70]   R. W. Davis and M. C. L. Gerry, J. Mol. Spectrosc. **65**, 455 (1977).

[3-71]   R. J. French, K. Hedberg, and J. M. Shreeve, *Tenth Austin Symposium on Molecular Structure,* Abstracts, p. 51, Austin, Texas, 1984.

[3-72]   L. S. Bartell and K. W. Hansen, Inorg. Chem. **4**, 1775 (1965).

[3-73]   H. Yow and L. S. Bartell, J. Mol. Struct. **15**, 209 (1973).

[3-74]   R. J. Gillespie, Inorg. Chem. **5**, 1634 (1966).

[3-75]   W. J. Adams and L. S. Bartell, J. Mol. Struct. **8**, 23 (1971).

[3-76]   J. E. Griffiths, J. Chem. Phys. **41**, 3510 (1964).

[3-77]   J. E. Griffiths and A. L. Beach, J. Chem. Phys. **44**, 2686 (1966).

[3-78]   H. Oberhammer and J. Grobe, Z. Naturforsch. **30b**, 506 (1975).

[3-79]   H. Oberhammer, J. Grobe, and D. Le Van, Inorg. Chem. **21**, 275 (1982).

[3-80]   E. L. Muetterties, W. Mahler, and R. Schmutzler, Inorg. Chem. **2**, 613 (1963).

[3-81]   R. G. Cavell, J. A. Gibson, and K. I. The, J. Am. Chem. Soc. **99**, 7841 (1977).

[3-82]   R. Hoffmann, B. F. Beier, E. L. Muetterties, and A. Rossi, Inorg. Chem. **16**, 511 (1977).

[3-83]   L. S. Bartell, Trans. Am. Crystallogr. Assoc. **2**, 134 (1966).

[3-84]   I. Hargittai, Természet Világa **104**, 78 (1973).

[3-85]   I. Hargittai and A. Baranyi, Acta Chim. Acad. Sci. Hung. **93**, 279 (1977).

[3-86]   I. Hargittai, *Sulphone Molecular Structures*. Lecture Notes in Chemistry, Vol. 6. Springer, Berlin-Heidelberg-New York, 1978.

[3-87]   H. Oberhammer, R. C. Kumar, G. D. Knerr, and J. M. Shreeve, Inorg. Chem. **20**, 3871 (1981).

[3-88]   I. Hargittai, Inorg. Chem. **21**, 4334 (1982).

[3-89]   P. Scharfenberg, L. Harsányi, and I. Hargittai, unpublished calculations, 1984.

[3-90]   I. Hargittai, E. Vajda, and A. Szőke, J. Mol. Struct. **18**, 381 (1973).

[3-91]   I. Hargittai and L. V. Vilkov, Acta Chim. (Budapest) **63**, 143 (1970).

[3-92]   I. Hargittai and M. Hargittai, Acta Chim. (Budapest) **75**, 129 (1973).

[3-93]   M. Hargittai and I. Hargittai, J. Mol. Struct. **20**, 283 (1974).

[3-94]   V. Typke, Z. Naturforsch. **33a**, 842 (1978).

[3-95]   T. Iijima, S. Tsuchiya, and M. Kimura, Bull. Chem. Soc. Japan **50**, 2564 (1977); T. Radnai, I. Hargittai, M. Kolonits, and D. C. Gregory, unpublished results, 1975.

[3-96]   H. Oberhammer, G. D. Knerr, and J. M. Shreeve, J. Mol. Struct. **82**, 143 (1982).

[3-97]   H. Oberhammer, H. Willner, and W. Gombler, J. Mol. Struct. **70**, 273 (1981).

[3-98]   M. Hargittai and I. Hargittai, J. Mol. Struct. **73**, 253 (1981).

[3-99]   I. Hargittai, Acta Chim. Acad. Sci. Hung. **60**, 231 (1969).

[3-100]   K. Hagen, V. R. Cross, and K. Hedberg, J. Mol. Struct. **44**, 187 (1978).

[3-101]   I. Hargittai and F. C. Mijlhoff, J. Mol. Struct. **16**, 69 (1973).

[3-102]   M. W. Tolles and W. D. Gwinn, J. Chem. Phys. **36**, 1119 (1962).

[3-103]   I. Hargittai, J. Mol. Struct. **56**, 301 (1979).

[3-104]   L. Hedberg and K. Hedberg, J. Phys. Chem. **86**, 598 (1982).

[3-105]   G. Gundersen and K. Hedberg, J. Chem. Phys. **51**, 2500 (1969).

[3-106]   H. Oberhammer and J. E. Boggs, J. Mol. Struct. **56**, 107 (1979).

[3-107]   H. Gunther, H. Oberhammer, R. Mews, and I. Stahl, Inorg. Chem. **21**, 1872 (1982).

[3-108]   D. D. DesMarteau, H. H. Eysel, H. Oberhammer, and H. Gunther, Inorg. Chem. **21**, 1607 (1982).

[3-109]   K. O. Christe and H. Oberhammer, Inorg. Chem. **20**, 297 (1981).

[3-110]   S. W. Peterson, R. D. Willett, and J. L. Huston, J. Chem. Phys. **59**, 453 (1973).

[3-111]   See, e.g., W. B. Jensen, J. Chem. Educ. **61**, 191 (1984).

[3-112]   G. N. Lewis, J. Am. Chem. Soc. **38**, 762 (1916); G. N. Lewis, *Valence and the Structure of Atoms and Molecules,* Chemical Catalog Co., New York, 1923.

[3-113]   N. V. Sidgwick and H. M. Powell, Proc. R. Soc. London, Ser. A, **176**, 153 (1940).

[3-114]   R. J. Gillespie and R. S. Nyholm, Quart. Rev. Chem. Soc. **11**, 339 (1957).

[3-115]   R. J. Gillespie, Current Contents, **24**, 14 (1984).

[3-116]   R. J. Gillespie, J. Chem. Educ. **40**, 295 (1963).

[3-117]   R. J. Gillespie, Angew. Chem. Int. Ed. Engl. **6**, 819 (1967).

[3-118]   L. S. Bartell, Croatica Chem. Acta **57**, 927 (1984).

[3-119]   L. S. Bartell and Y. Z. Barshad, J. Am. Chem. Soc. **106**, 7700 (1984).

[3-120]   R. F. Bader, P. J. MacDougall, and C. D. H. Lau, J. Am. Chem. Soc. **106**, 1594 (1984).

[3-121]   Henri Matisse: Dance; The Hermitage, Leningrad.

[3-122]   See, e.g., R. S. Berry, in *Quantum Dynamics of Molecules. The New Experimental Challenge to Theorists,* R. G. Wooley, ed., Plenum Press, New York and London, 1980.

[3-123]   E. L. Muetterties, Inorg. Chem. **4**, 769 (1965).

[3-124]   I. Hargittai, J. Tremmel, and G. Schultz, J. Mol. Struct. **26**, 116 (1975).

[3-125]   M. Hargittai, I. Hargittai, and J. Tremmel, Chem. Phys. Lett. **83**, 207 (1981).

[3-126]   M. Hargittai and I. Hargittai, J. Mol. Spectrosc. **108**, 155 (1984).

[3-127]   E. Vajda, J. Tremmel, and I. Hargittai, J. Mol. Struct. **44**, 101 (1978).

[3-128]   J. Tremmel, A. A. Ivanov, G. Schultz, I. Hargittai, S. J. Cyvin, and A. Eriksson, Chem. Phys. Lett. **23**, 533 (1973).

[3-129]   M. Hargittai, O. V. Dorofeeva, and J. Tremmel, Inorg. Chem. **24**, 245 (1985).

[3-130]   L. Eddy, Dissertation, Oregon State University, Corvallis, Oregon, 1973.

[3-131]   Z. Molnár, G. Schultz, J. Tremmel, and I. Hargittai, Acta Chim. (Budapest) **86**, 223 (1975).

[3-132]   V. P. Spiridonov, A. G. Gershikov, E. Z. Zasorin, N. I. Popenko, A. A. Ivanov, and L. I. Ermolaeva, High Temp. Sci. **14**, 285 (1981).

[3-133]   See, e.g., I. Hargittai and M. Hargittai, Magy. Kém. L. **39**, 80 (1984).

[3-134]   Q. Shen, Dissertation, Oregon State University, Corvallis, Oregon, 1973.

[3-135]   M. Hargittai, J. Tremmel, and I. Hargittai, J. Chem. Soc. Dalton Trans. 87 (1980).

[3-136]   M. Hargittai, O. V. Dorofeeva, and J. Tremmel, Inorg. Chem. **24**, 3963 (1985).

[3-137]   W. v. E. Doering and W. R. Roth, Tetrehadron **19,** 720 (1963); G. Schroeder, Angew. Chem. Int. Ed. Engl. **2,** 481 (1963); M. Saunders, Tetrahedron Lett. 1699 (1963).

[3-138]   J. S. McKennis, L. Brener, J. S. Ward, and R. Pettit, J. Am. Chem. Soc. **93,** 4957 (1971).

[3-139]   R. S. Berry, J. Chem. Phys. **32,** 933 (1960).

[3-140]   G. M. Whitesides and H. L. Mitchell, J. Am. Chem. Soc. **91,** 5384 (1969).

[3-141]   L. S. Bartell, M. J. Rothman, and A. Gavezzotti, J. Chem. Phys. **76,** 4136 (1982) and references therein.

[3-142]   W. J. Adams, H. B. Thompson, and L. S. Bartell, J. Chem. Phys. **53,** 4040 (1970).

[3-143]   E. L. Muetterties and W. H. Knoth, *Polyhedral Boranes,* Marcel Dekker, New York, 1968.

[3-144]   W. N. Lipscomb, Science **153,** 373 (1966).

[3-145]   R. K. Bohn and M. D. Bohn, Inorg. Chem. **10,** 350 (1971).

[3-146]   B. F. G. Johnson and R. E. Benfield, J. Chem. Soc. Dalton Trans. 1554 (1978).

[3-147]   B. E. Hanson, M. J. Sullivan, and R. J. Davis, J. Am. Chem. Soc. **106,** 251 (1984).

[3-148]   R. E. Benfield and B. F. G. Johnson, J. Chem. Soc. Dalton Trans. 1743 (1980).

[3-149]   J. Kepler, *Mysterium cosmographicum,* 1595.

4 Helpful Mathematical Tools

4.1 Groups

So far our discussion has been non-mathematical. Ignoring mathematics, however, does not make things necessarily easier. Group theory is the mathematical apparatus for describing symmetry operations. It facilitates the understanding and the use of symmetries. It may not even be possible to successfully attack some complex problems without the use of group theory. Besides, groups are fascinating.

This introductory chapter gives the reader the tools necessary to understand the next three chapters in which molecular vibrations, electronic structure, and chemical reactions are discussed. Further reading is recommended for broader knowledge of the subject [4-1 – 4-5].

A mathematical group is a very general idea. It is a special case when the elements of the group are symmetry operations. When the symmetries of molecules are characterized by Schoenflies symbols, for example, C_{2v}, C_{3v} or C_{2h}, these symbols represent well-defined groups of symmetry operations. Let us consider first the C_{2v} point group. It consists of a two-fold rotation, C_2, and two reflections through mutually perpendicular symmetry planes, σ_v, and σ_v', whose intersection coincides with the rotation axis. All the corresponding elements are shown in Fig. 4-1. One more operation can be added to these, called the identity operation, E. Its application leaves the molecule unchanged. The set of the operations C_2, σ_v, σ_v' and E together make a mathematical group.

A mathematical group is a set of elements related by certain rules. They will be illustrated on the symmetry operations.

1. The product of any two elements of a group is also an element of the group. The product here means consecutive application of the elements rather than common multiplication. Thus, for example, the product $\sigma_v \cdot C_2$ means that first a two-fold rotation is applied to an operand[1] and then reflection is applied to the new operand. Let us perform

Figure 4-1.

Symmetry operations in the C_{2v} point group.

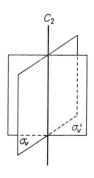

---

1 Shortly, we shall use a wide range of operands related to molecular structure.

these operations on the atomic positions of a sulfuryl chloride molecule as is shown in Fig. 4-2a. The same final result is obtained by simply applying the symmetry plane σ_v' as is also shown in Fig. 4-2b. Thus

$$\sigma_v \cdot C_2 = \sigma_v'.$$

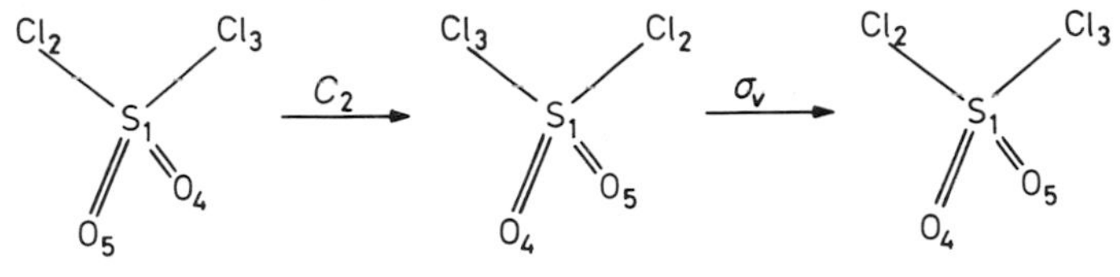

Figure 4-2.
(a) Consecutive application of two symmetry operations, C_2 and σ_v to the nuclear positions of the SO_2Cl_2 molecule.

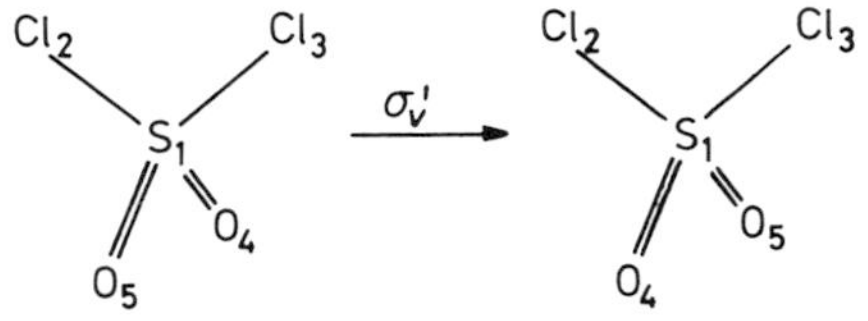

(b) Application of σ_v' to SO_2Cl_2.

2. The products of the elements in a group are generally not commutative. That means that the result of the consecutive application of the symmetry operations depends on the order in which they are applied. Fig. 4-3 gives an example for the ammonia molecule, which belongs to the C_{3v} point group. Whether the C_3 operation is applied first and then the σ_v'', or vice versa, the effect is different. The product of the identity operation with any other element in the group is commutative by definition. Thus, for example,

$$C_3 \cdot E = E \cdot C_3 \text{ and } \sigma_v \cdot E = E \cdot \sigma_v.$$

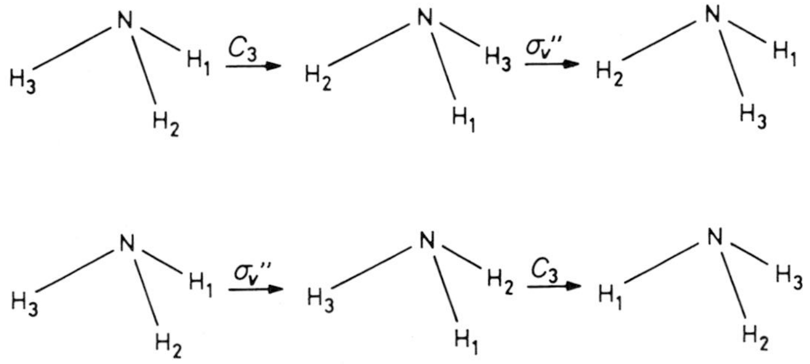

Figure 4-3.
Illustration for the non-commutative character of the symmetry operations.

The C_{2v} point group is special in that all the possible products of its elements are commutative. Thus in Fig. 4-2a we could get the same result first applying the σ_v reflection and then the two-fold rotation.

3. The products of the elements in a group are always associative. That means that if there is a consecutive application of several symmetry operations, their application may be grouped in any way without changing the final result as long as the order of the application remains the same. Thus, for example,

$$C_2 \cdot \sigma_v \cdot \sigma_v{}' = C_2 \cdot (\sigma_v \cdot \sigma_v{}') = (C_2 \cdot \sigma_v) \cdot \sigma_v{}'.$$

4. For each element X in a group, there is an inverse or reciprocal operation which satisfies the following condition:

$$X \cdot X^{-1} = X^{-1} \cdot X = E.$$

For example,

$$C_2 \cdot C_2{}^{-1} = C_2{}^{-1} \cdot C_2 = E$$

or

$$\sigma_v \cdot \sigma_v{}^{-1} = \sigma_v{}^{-1} \cdot \sigma_v = E.$$

The symmetry operation corresponding to an inverse operation can be found in group multiplication tables. These tables contain the products of the elements of a group. An example is shown in Table 4-1, for the C_{2v} point group. Here each element of the group, i.e. each symmetry operation is listed only once in the initial row at the top and in the initial column at the far left. In forming the product of any two elements, one belonging to the row and the other to the column, the order of the application of the elements is strictly defined. First the element in the top row is applied, followed by the application of the element in the far left column. The result is found at the intersection of the corresponding column and row. Any one of the results is also a symmetry operation belonging to the C_{2v} point group. In fact, each row and each column in the field of the results is a rearranged list of the initial operations but no two rows or two columns may be identical. From the C_{2v} multiplication table it is seen that the inverse operation of C_2 is C_2, since their intersection is E; similarly, the inverse operation of σ_v is σ_v in this group.

Table 4-1. Group Multiplication Table for the C_{2v} Point Group

C_{2v}	E	C_2	σ_v	$\sigma_v{}'$
E	E	C_2	σ_v	$\sigma_v{}'$
C_2	C_2	E	$\sigma_v{}'$	σ_v
σ_v	σ_v	$\sigma_v{}'$	E	C_2
$\sigma_v{}'$	$\sigma_v{}'$	σ_v	C_2	E

Table 4-2. Group Multiplication Table for the C_{3v} Point Group

C_{3v}	E	C_3	$C_3{}^2$	σ_v	$\sigma_v{}'$	$\sigma_v{}''$
E	E	C_3	$C_3{}^2$	σ_v	$\sigma_v{}'$	$\sigma_v{}''$
C_3	C_3	$C_3{}^2$	E	$\sigma_v{}''$	σ_v	$\sigma_v{}'$
$C_3{}^2$	$C_3{}^2$	E	C_3	$\sigma_v{}'$	$\sigma_v{}''$	σ_v
σ_v	σ_v	$\sigma_v{}'$	$\sigma_v{}''$	E	C_3	$C_3{}^2$
$\sigma_v{}'$	$\sigma_v{}'$	$\sigma_v{}''$	σ_v	$C_3{}^2$	E	C_3
$\sigma_v{}''$	$\sigma_v{}''$	σ_v	$\sigma_v{}'$	C_3	$C_3{}^2$	E

Similarly to Table 4-1, the multiplication table of the C_{3v} point group is compiled in Table 4-2. Here

$$C_3 \cdot C_3 = C_3{}^2$$

means two successive applications of the three-fold rotation. Applying it once yields a 120° rotation, while $C_3{}^2$ corresponds to a 240° rotation altogether. Accordingly, for example, the meaning of $C_5{}^2$ is a rotation by $2 \cdot (360°/5) = 144°$.

The number of elements in a group is called *the order of the group*. Its conventional symbol is h. The group multiplication tables show that $h=4$ for the C_{2v} point group and $h=6$ for C_{3v}.

A group may be divided into two kinds of subunits: subgroups and classes. A *subgroup* is a smaller group within a group which still possesses the four fundamental properties of a group. The identity operation, E, is always a subgroup by itself and it is also a member of all other possible subgroups.

A *class* is a complete set of the elements, in our case symmetry operations, of the group which are conjugate to one another. Conjugate means that if A and B are elements of the same class then there is some element, Z, in the group for which

$$B=Z^{-1} \cdot A \cdot Z$$

This kind of operation is called a *similarity transformation*. B is the similarity transform of A by Z, or in other words A and B are conjugates. The inverse operation can be applied with the aid of the multiplication table and rule 4. given above,

$$Z^{-1} \cdot Z = Z \cdot Z^{-1} = E.$$

To find out what operations belong to the same class within a group, all possible similarity transformations in the group have to be performed. Let us work this out for the C_{3v} point group and begin with the identity operation:

$$E^{-1} \cdot (E \cdot E) = E^{-1} \cdot E = E$$
$$C_3^{-1} \cdot (E \cdot C_3) = C_3^{-1} \cdot C_3 = C_3^2 \cdot C_3 = E$$
$$(C_3^2)^{-1} \cdot (E \cdot C_3^2) = (C_3^2)^{-1} \cdot C_3^2 = C_3 \cdot C_3^2 = E$$
$$\sigma_v^{-1} \cdot (E \cdot \sigma_v) = \sigma_v^{-1} \cdot \sigma_v = \sigma_v \cdot \sigma_v = E$$
$$\sigma_v'^{-1} \cdot (E \cdot \sigma_v') = \sigma_v'^{-1} \cdot \sigma_v' = \sigma_v' \cdot \sigma_v' = E$$
$$\sigma_v''^{-1} \cdot (E \cdot \sigma_v'') = \sigma_v''^{-1} \cdot \sigma_v'' = \sigma_v'' \cdot \sigma_v'' = E.$$

The conclusion is that E is always a class by itself; it commutes with all other elements in the group and leaves them unchanged. Consequently, it is not conjugate with any other element. This is true for all other point groups as well.

Consider now σ_v:

$$E^{-1} \cdot (\sigma_v \cdot E) = E^{-1} \cdot \sigma_v = E \cdot \sigma_v = \sigma_v$$
$$C_3^{-1} \cdot (\sigma_v \cdot C_3) = C_3^{-1} \cdot \sigma_v' = C_3^2 \cdot \sigma_v' = \sigma_v''$$
$$(C_3^2)^{-1} \cdot (\sigma_v \cdot C_3^2) = (C_3^2)^{-1} \cdot \sigma_v'' = C_3 \cdot \sigma_v'' = \sigma_v'$$
$$\sigma_v^{-1} \cdot (\sigma_v \cdot \sigma_v) = \sigma_v^{-1} \cdot E = \sigma_v \cdot E = \sigma_v$$
$$\sigma_v'^{-1} \cdot (\sigma_v \cdot \sigma_v') = \sigma_v'^{-1} \cdot C_3 = \sigma_v' \cdot C_3 = \sigma_v''$$
$$\sigma_v''^{-1} \cdot (\sigma_v \cdot \sigma_v'') = \sigma_v''^{-1} \cdot C_3^2 = \sigma_v'' \cdot C_3^2 = \sigma_v'.$$

We have performed all possible similarity transformations for the operation σ_v. As a result it is seen that the three operations expressing vertical mirror symmetry belong to the same class. We could reach the same conclusion by similarity transformations on either of the other two σ_v operations.

Next let us examine C_3:

$$E^{-1} \cdot (C_3 \cdot E) = E^{-1} \cdot C_3 = E \cdot C_3 = C_3$$
$$C_3^{-1} \cdot (C_3 \cdot C_3) = C_3^{-1} \cdot C_3^2 = C_3^2 \cdot C_3^2 = C_3$$
$$(C_3^2)^{-1} \cdot (C_3 \cdot C_3^2) = (C_3^2)^{-1} \cdot E = C_3 \cdot E = C_3$$
$$\sigma_v^{-1} \cdot (C_3 \cdot \sigma_v) = \sigma_v^{-1} \cdot \sigma_v'' = \sigma_v \cdot \sigma_v'' = C_3^2$$
$$\sigma_v'^{-1} \cdot (C_3 \cdot \sigma_v') = \sigma_v'^{-1} \cdot \sigma_v = \sigma_v' \cdot \sigma_v = C_3^2$$
$$\sigma_v''^{-1} \cdot (C_3 \cdot \sigma_v'') = \sigma_v''^{-1} \cdot \sigma_v' = \sigma_v'' \cdot \sigma_v' = C_3^2.$$

According to these transformations, C_3 and C_3^2 are conjugates and thus belong to the same class.

The *order of a class* is defined as the number of elements in the class. For example, the order of the class of the reflection operations in C_{3v} is 3 and the order of the class of the rotation operations is 2. Generally the order of a class or a subgroup is an integral divisor of the order of the group.

The mathematical handling of the symmetry operations is done by means of matrices.

4.2 Matrices

A matrix is a rectangular array of numbers, or symbols for numbers. These elements are put between square brackets. A numerical example of a matrix is shown below:

$$\begin{bmatrix} 3 & 1 & 0 & 2 \\ 5 & 7 & 0 & -3 \\ 0 & 0 & -2 & 1 \end{bmatrix}$$

Generally a matrix has m rows and n columns:

$$\begin{bmatrix} a_{11} & a_{12} & \cdots & a_{1n} \\ a_{21} & a_{22} & \cdots & a_{2n} \\ \cdot & & & \cdot \\ \cdot & & & \cdot \\ \cdot & & & \cdot \\ a_{m1} & a_{m2} & \cdots & a_{mn} \end{bmatrix}$$

The above matrix may be represented by a capital letter A. Another representation is: $[a_{ij}]$, where subscript i denotes the number of rows, $0 \leq i \leq m$, and subscript j the number of columns, $0 \leq j \leq n$.

There are some special matrices which are important for our discussion. A *square matrix* is a matrix which has equal numbers of rows and columns. According to the general notation a matrix $[a_{ij}]$ is a square matrix if $m = n$. The following matrix:

$$\begin{bmatrix} 0 & 1 & 2 \\ 3 & 4 & 5 \\ 6 & 7 & 8 \end{bmatrix}$$

is a square matrix.

A special square matrix is the *unit matrix* in which all elements along the top-left-to-bottom-right diagonal are 1 and all the other elements are zero. The short notation for a unit matrix is E. Some unit matrices are presented here:

$$\begin{bmatrix} 1 & 0 \\ 0 & 1 \end{bmatrix} \qquad \begin{bmatrix} 1 & 0 & 0 \\ 0 & 1 & 0 \\ 0 & 0 & 1 \end{bmatrix} \qquad \begin{bmatrix} 1 & 0 & 0 & 0 & 0 \\ 0 & 1 & 0 & 0 & 0 \\ 0 & 0 & 1 & 0 & 0 \\ 0 & 0 & 0 & 1 & 0 \\ 0 & 0 & 0 & 0 & 1 \end{bmatrix}$$

A *column matrix* consists of only one column, e.g.,

$$\begin{bmatrix} 1 \\ 2 \\ 3 \end{bmatrix}$$

Figure 4-4.
Representation of a vector in three-dimensional space.

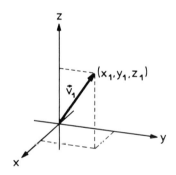

Column matrices are used to represent vectors. A *vector* is characterized by its length *and* direction. A vector in three-dimensional space is shown in Fig. 4-4. If one end of the vector is at the origin of the Cartesian coordinate system, then the three coordinates of its other end fully describe the vector. These three Cartesian coordinates can be written as a column matrix:

$$\begin{bmatrix} x_1 \\ y_1 \\ z_1 \end{bmatrix}$$

Thus this column matrix represents a vector.

While column matrices are used to represent vectors, square matrices are used to *represent symmetry operations.* Performing a symmetry operation on a vector is actually a geometrical transformation. How can these geometrical transformations be translated into matrix "language"? Consider a specific example and see how the symmetry operations of the C_s symmetry group can be applied to the vector of Fig. 4-4. For a matrix representation we first write (or usually just imagine) the coordinates of the original vector in the top row and the coordinates of the vector resulting from the symmetry operation in the left hand column:

$$x_1 \quad y_1 \quad z_1 \quad \leftarrow \quad \text{original vector}$$

resultant $\quad x_1'$

vector $\quad y_1'$

$\qquad z_1'$

$$\begin{bmatrix} \\ \\ \\ \end{bmatrix}$$

Then we examine the effect of the symmetry operation in detail. If a coordinate is transformed into itself, 1 is placed into the intersection position, while if it is transformed into its negative self, –1 is put into the intersection. Both these positions will be along the diagonal of the matrix. If a coordinate is transformed into another one or into the negative of this other coordinate, 1 or –1 is placed into the intersection position, respectively. These intersection positions will be off the matrix diagonal.

There are two symmetry operations in the C_s point group, E and σ_h. The identity operation, E, does not change the position of the vector, so it can be represented by a unit matrix:

$$
\begin{array}{c}
\begin{array}{ccc} x_1 & y_1 & z_1 \end{array} \\
\begin{array}{c} x_1' \\ y_1' \\ z_1' \end{array}
\begin{bmatrix} 1 & 0 & 0 \\ 0 & 1 & 0 \\ 0 & 0 & 1 \end{bmatrix}
\cdot
\begin{bmatrix} x_1 \\ y_1 \\ z_1 \end{bmatrix}
=
\begin{bmatrix} x_1 \\ y_1 \\ z_1 \end{bmatrix}
\end{array}
$$

Accordingly,

$$ E \quad \cdot \quad \vec{v_1} \; = \; \vec{v_1}. $$

If the matrix elements are a_{ij} and the vector components are b_j, then the components of the product vector, c_i, are given by

$$ c_i = \sum_j a_{ij} \cdot b_j. $$

To get the first member of the resulting matrix, all the elements of the first row of the square matrix are multiplied by the consecutive members of the column matrix and then added together. To get the second member, the same procedure is followed with the second row of the square matrix, and so on, as shown below:

$$
\begin{bmatrix} 1 & 0 & 0 \\ 0 & 1 & 0 \\ 0 & 0 & 1 \end{bmatrix}
\cdot
\begin{bmatrix} x_1 \\ y_1 \\ z_1 \end{bmatrix}
=
\begin{bmatrix} 1 \cdot x_1 + 0 \cdot y_1 + 0 \cdot z_1 \\ 0 \cdot x_1 + 1 \cdot y_1 + 0 \cdot z_1 \\ 0 \cdot x_1 + 0 \cdot y_1 + 1 \cdot z_1 \end{bmatrix}
=
\begin{bmatrix} x_1 \\ y_1 \\ z_1 \end{bmatrix}
$$

Figure 4-5.
Reflection of a vector by a
horizontal mirror plane.

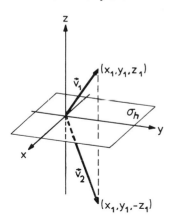

The other symmetry operation of the C_s point group is the
horizontal reflection (see Fig. 4-5). In matrix language this
operation can be written as follows:

$$x_1 \quad y_1 \quad z_1$$

$$
\begin{matrix} x_1' \\ y_1' \\ z_1' \end{matrix}
\begin{bmatrix} 1 & 0 & 0 \\ 0 & 1 & 0 \\ 0 & 0 & -1 \end{bmatrix}
\cdot
\begin{bmatrix} x_1 \\ y_1 \\ z_1 \end{bmatrix}
=
\begin{bmatrix} 1 \cdot x_1 + 0 \cdot y_1 + 0 \cdot z_1 \\ 0 \cdot x_1 + 1 \cdot y_1 + 0 \cdot z_1 \\ 0 \cdot x_1 + 0 \cdot y_1 + (-1)z_1 \end{bmatrix}
=
\begin{bmatrix} x_1 \\ y_1 \\ -z_1 \end{bmatrix}
$$

$$\sigma_h \qquad \cdot \qquad \vec{v_1} \qquad\qquad\qquad\qquad = \quad \vec{v_2}$$

It often happens that the coordinates are not transformed
simply into each other by a symmetry operation. Trigonometric
relations must be used to express, for instance, the consequences of three-fold rotation.

Fig. 4-6 illustrates a vector rotated by an angle α in the xy
plane. The coordinates of the rotated vector are related to the
coordinates of the original vector in the following way (β is an
auxiliary angle shown in Fig. 4-6):

Figure 4-6.
Rotation of a vector by an
angle α in the xy plane.

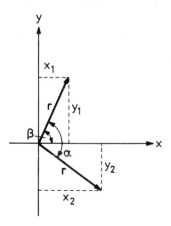

$$x_1 = r \cdot \cos \beta \quad \text{and} \quad y_1 = r \cdot \sin \beta \tag{4-1}$$
$$x_2 = r \cdot \cos(\alpha\text{-}\beta) \quad \text{and} \quad y_2 = -r \cdot \sin(\alpha\text{-}\beta) \tag{4-2}$$

Utilizing the trigonometric expressions:

$$\cos(\alpha\text{-}\beta) = \cos \alpha \cdot \cos \beta + \sin \alpha \cdot \sin \beta$$
$$\sin(\alpha\text{-}\beta) = \sin \alpha \cdot \cos \beta - \cos \alpha \cdot \sin \beta \tag{4-3}$$

and substituting Eqs. (4-3) and (4-1) into Eq. (4-2), we get:

$$x_2 = r \cdot \cos\alpha \cdot \cos\beta + r \cdot \sin\alpha \cdot \sin\beta = x_1 \cdot \cos\alpha + y_1 \cdot \sin\alpha$$
$$y_2 = -r \cdot \sin\alpha \cdot \cos\beta + r \cdot \cos\alpha \cdot \sin\beta = -x_1 \cdot \sin\alpha + y_1 \cdot \cos\alpha \tag{4-4}$$

The same equations in matrix formulation:

$$
\begin{bmatrix} \cos\alpha & \sin\alpha \\ -\sin\alpha & \cos\alpha \end{bmatrix}
\cdot
\begin{bmatrix} x_1 \\ y_1 \end{bmatrix}
=
\begin{bmatrix} x_2 \\ y_2 \end{bmatrix}
$$

The square matrix above is the matrix representation of a rotation through an angle α.

Since matrices can be used to represent symmetry operations, the set of matrices representing all symmetry operations
of a point group will be the representation of that group.
Moreover, if a set of matrices forms a representation of a symmetry group, it will obey all the rules of a mathematical group. It
will also obey the group multiplication table. Let the SO_2Cl_2
molecule serve as an example again. This molecule belongs to
the C_{2v} point group and some of its symmetry operations have
already been illustrated in Fig. 4-2. To construct the corre-

sponding matrices the same procedure can be applied as used before with a vector. The original nuclear positions of the molecule can be written at the top row and the nuclear positions resulting from the symmetry operation at the far left column.

There are four operations in the C_{2v} point group. E leaves the molecule unchanged, so the corresponding matrix will be a unit matrix:

$$E = \begin{array}{c} \\ S_1' \\ Cl_2' \\ Cl_3' \\ O_4' \\ O_5' \end{array} \begin{array}{ccccc} S_1 & Cl_2 & Cl_3 & O_4 & O_5 \\ \left[\begin{array}{ccccc} 1 & 0 & 0 & 0 & 0 \\ 0 & 1 & 0 & 0 & 0 \\ 0 & 0 & 1 & 0 & 0 \\ 0 & 0 & 0 & 1 & 0 \\ 0 & 0 & 0 & 0 & 1 \end{array}\right] \end{array}$$

The two-fold rotation changes the positions of the two chlorine atoms and also the positions of the two oxygen atoms. The sulfur atom remains in place.

$$C_2 = \begin{array}{c} \\ S_1' \\ Cl_2' \\ Cl_3' \\ O_4' \\ O_5' \end{array} \begin{array}{ccccc} S_1 & Cl_2 & Cl_3 & O_4 & O_5 \\ \left[\begin{array}{ccccc} 1 & 0 & 0 & 0 & 0 \\ 0 & 0 & 1 & 0 & 0 \\ 0 & 1 & 0 & 0 & 0 \\ 0 & 0 & 0 & 0 & 1 \\ 0 & 0 & 0 & 1 & 0 \end{array}\right] \end{array}$$

σ_v changes the positions of the two chlorines and leaves the other three atoms in place (the aiding top-row and left-hand-column will no longer be indicated):

$$\sigma_v = \left[\begin{array}{ccccc} 1 & 0 & 0 & 0 & 0 \\ 0 & 0 & 1 & 0 & 0 \\ 0 & 1 & 0 & 0 & 0 \\ 0 & 0 & 0 & 1 & 0 \\ 0 & 0 & 0 & 0 & 1 \end{array}\right]$$

Finally σ_v' changes the positions of the two oxygen atoms, and leaves the sulfur and the two chlorines in their original positions:

$$\sigma_v' = \begin{bmatrix} 1 & 0 & 0 & 0 & 0 \\ 0 & 1 & 0 & 0 & 0 \\ 0 & 0 & 1 & 0 & 0 \\ 0 & 0 & 0 & 0 & 1 \\ 0 & 0 & 0 & 1 & 0 \end{bmatrix}$$

Since each of these four 5 x 5 matrices represent one of the symmetry operations of the C_{2v} point group, the set of these four 5 x 5 matrices will be a representation of this group. They will also obey the C_{2v} multiplication table. As was shown in Fig. 4-2

$$\sigma_v \cdot C_2 = \sigma_v'.$$

The corresponding matrix representations are the following:

$$\begin{bmatrix} 1 & 0 & 0 & 0 & 0 \\ 0 & 0 & 1 & 0 & 0 \\ 0 & 1 & 0 & 0 & 0 \\ 0 & 0 & 0 & 1 & 0 \\ 0 & 0 & 0 & 0 & 1 \end{bmatrix} \cdot \begin{bmatrix} 1 & 0 & 0 & 0 & 0 \\ 0 & 0 & 1 & 0 & 0 \\ 0 & 1 & 0 & 0 & 0 \\ 0 & 0 & 0 & 0 & 1 \\ 0 & 0 & 0 & 1 & 0 \end{bmatrix} =$$

$$\sigma_v \qquad\qquad\qquad C_2$$

$$= \begin{bmatrix} 1\cdot1+0\cdot0+0\cdot0+0\cdot0+0\cdot0 & 1\cdot0+0\cdot0+0\cdot1+0\cdot0+0\cdot0 \\ 0\cdot1+0\cdot0+1\cdot0+0\cdot0+0\cdot0 & 0\cdot0+0\cdot0+1\cdot1+0\cdot0+0\cdot0 \\ 0\cdot1+1\cdot0+0\cdot0+0\cdot0+0\cdot0 & 0\cdot0+1\cdot0+0\cdot1+0\cdot0+0\cdot0\ldots \\ 0\cdot1+0\cdot0+0\cdot0+1\cdot0+0\cdot0 & 0\cdot0+0\cdot0+0\cdot1+1\cdot0+0\cdot0 \\ 0\cdot1+0\cdot0+0\cdot0+0\cdot0+1\cdot0 & 0\cdot0+0\cdot0+0\cdot1+0\cdot0+1\cdot0 \end{bmatrix} =$$

$$= \begin{bmatrix} 1 & 0 & 0 & 0 & 0 \\ 0 & 1 & 0 & 0 & 0 \\ 0 & 0 & 1 & 0 & 0 \\ 0 & 0 & 0 & 0 & 1 \\ 0 & 0 & 0 & 1 & 0 \end{bmatrix}$$

$$\sigma_v'$$

The multiplication is shown here in detail only for the first two columns of the resulting matrix. The elements of the product matrix are given by:

$$c_{ik} = \sum_j a_{ij} \cdot b_{jk}.$$

To get the first member of the first row, all elements of the first row of the first matrix are multiplied by the corresponding elements of the first column of the second matrix and then the results added. To get the second member of the first row, all elements of the first row of the first matrix are multiplied by the corresponding members of the second column of the second matrix and then the results are added, and so on. To get the second row members, the same procedure is repeated with the second row members of the first matrix, and so on. It is also possible to visualize the second matrix as a series of column matrices and then consider the multiplication of each of these column matrices, one by one, by the first matrix.

4.3 Representation of Groups

Any collection of quantities which obey the multiplication table of a group is a *representation* of the group [4-2]. In our examples these quantities show how certain characteristics of a molecule behave under the symmetry operations of the group. The symmetry operations may be applied to various characteristics or descriptions of the molecule. The particular description to which the symmetry operations are applied, forms the *basis* for a representation of the group. Generally speaking, any set of algebraic functions or vectors may be the basis for a representation of a group [4-1]. Our choice of a suitable basis depends on the particular problem we are studying. After choosing the basis set, the task is to construct the matrices which transform the basis or its components according to each symmetry operation. The most common basis sets in chemical applications are summarized in Sect. 4.11. Some of them will be used in the following discussion.

Let us now work out the representation of a point group for a very simple basis. Choose just the *changes*, Δr_1 and Δr_2, of the two N-H bond lengths of the diimide molecule, N_2H_2:

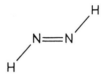

These two vectors may be used in the description of the stretching vibrations of the molecule. The molecular symmetry is C_{2h}. Fig. 4-7 helps to visualize the effects of the symmetry operations of this group on the selected basis. There are four symmetry operations in the C_{2h} point group, E, C_2, i, and σ_h. E leaves the basis unchanged, so the corresponding matrix representation is a unit matrix:

$$E \cdot \begin{bmatrix} \Delta r_1 \\ \Delta r_2 \end{bmatrix} = \begin{bmatrix} 1 & 0 \\ 0 & 1 \end{bmatrix} \cdot \begin{bmatrix} \Delta r_1 \\ \Delta r_2 \end{bmatrix}$$

Both C_2 and i interchanges the two vectors, Δr_1 "goes into" Δr_2 and *vice versa*

$$C_2 \cdot \begin{bmatrix} \Delta r_1 \\ \Delta r_2 \end{bmatrix} = \begin{bmatrix} 0 & 1 \\ 1 & 0 \end{bmatrix} \cdot \begin{bmatrix} \Delta r_1 \\ \Delta r_2 \end{bmatrix}$$

$$i \cdot \begin{bmatrix} \Delta r_1 \\ \Delta r_2 \end{bmatrix} = \begin{bmatrix} 0 & 1 \\ 1 & 0 \end{bmatrix} \cdot \begin{bmatrix} \Delta r_1 \\ \Delta r_2 \end{bmatrix}$$

finally, σ_h leaves the molecule unchanged.

$$\sigma_h \cdot \begin{bmatrix} \Delta r_1 \\ \Delta r_2 \end{bmatrix} = \begin{bmatrix} 1 & 0 \\ 0 & 1 \end{bmatrix} \cdot \begin{bmatrix} \Delta r_1 \\ \Delta r_2 \end{bmatrix}$$

With this basis the representation consists of four 2 x 2 matrices.

Let us take now a more complicated basis, and consider all the nuclear coordinates of HNNH shown in Fig. 4-8a. These are the so-called Cartesian displacement vectors and will be dis-

Figure 4-7.
The four symmetry operations of the C_{2h} point group applied to the two N-H bond length changes of the HNNH molecule.

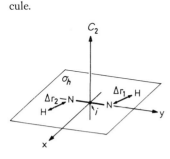

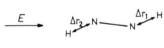

Figure 4-8.
(a) Cartesian coordinates as basis for a representation.

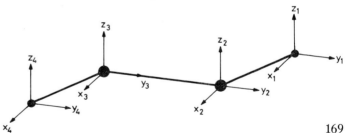

cussed in Chapter 5 on molecular vibrations. Let us find the matrix representation of the σ_h operation (see Fig. 4-8b). The horizontal mirror plane leaves all x and y coordinates unchanged while all z coordinates will "go" into their negative selves. In matrix notation this is expressed in the following way:

$$
\sigma_h \cdot
\begin{bmatrix}
x_1 \\ y_1 \\ z_1 \\ x_2 \\ y_2 \\ z_2 \\ x_3 \\ y_3 \\ z_3 \\ x_4 \\ y_4 \\ z_4
\end{bmatrix}
=
\begin{bmatrix}
1 & 0 & 0 & 0 & 0 & 0 & 0 & 0 & 0 & 0 & 0 & 0 \\
0 & 1 & 0 & 0 & 0 & 0 & 0 & 0 & 0 & 0 & 0 & 0 \\
0 & 0 & -1 & 0 & 0 & 0 & 0 & 0 & 0 & 0 & 0 & 0 \\
0 & 0 & 0 & 1 & 0 & 0 & 0 & 0 & 0 & 0 & 0 & 0 \\
0 & 0 & 0 & 0 & 1 & 0 & 0 & 0 & 0 & 0 & 0 & 0 \\
0 & 0 & 0 & 0 & 0 & -1 & 0 & 0 & 0 & 0 & 0 & 0 \\
0 & 0 & 0 & 0 & 0 & 0 & 1 & 0 & 0 & 0 & 0 & 0 \\
0 & 0 & 0 & 0 & 0 & 0 & 0 & 1 & 0 & 0 & 0 & 0 \\
0 & 0 & 0 & 0 & 0 & 0 & 0 & 0 & -1 & 0 & 0 & 0 \\
0 & 0 & 0 & 0 & 0 & 0 & 0 & 0 & 0 & 1 & 0 & 0 \\
0 & 0 & 0 & 0 & 0 & 0 & 0 & 0 & 0 & 0 & 1 & 0 \\
0 & 0 & 0 & 0 & 0 & 0 & 0 & 0 & 0 & 0 & 0 & -1
\end{bmatrix}
\cdot
\begin{bmatrix}
x_1 \\ y_1 \\ z_1 \\ x_2 \\ y_2 \\ z_2 \\ x_3 \\ y_3 \\ z_3 \\ x_4 \\ y_4 \\ z_4
\end{bmatrix}
$$

(b) Effect of σ_h.

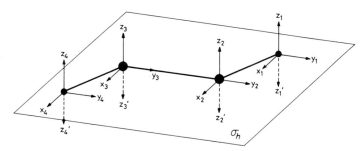

Take one more operation, the C_2 rotation (Fig. 4-8c). This operation introduces the following changes:

x_1, y_1, and z_1 to $-x_4$, $-y_4$, and z_4,
x_2, y_2, and z_2 to $-x_3$, $-y_3$, and z_3,
x_3, y_3, and z_3 to $-x_2$, $-y_2$, and z_2, and
x_4, y_4, and z_4 to $-x_1$, $-y_1$, and z_1.

(c) Effect of C_2.

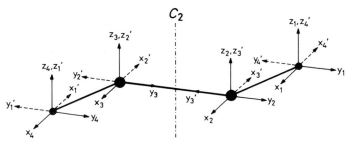

170

In matrix notation:

$$C_2 \cdot \begin{bmatrix} x_1 \\ y_1 \\ z_1 \\ x_2 \\ y_2 \\ z_2 \\ x_3 \\ y_3 \\ z_3 \\ x_4 \\ y_4 \\ z_4 \end{bmatrix} = \begin{bmatrix} 0 & 0 & 0 & 0 & 0 & 0 & 0 & 0 & 0 & -1 & 0 & 0 \\ 0 & 0 & 0 & 0 & 0 & 0 & 0 & 0 & 0 & 0 & -1 & 0 \\ 0 & 0 & 0 & 0 & 0 & 0 & 0 & 0 & 0 & 0 & 0 & 1 \\ 0 & 0 & 0 & 0 & 0 & 0 & -1 & 0 & 0 & 0 & 0 & 0 \\ 0 & 0 & 0 & 0 & 0 & 0 & 0 & -1 & 0 & 0 & 0 & 0 \\ 0 & 0 & 0 & 0 & 0 & 0 & 0 & 0 & 1 & 0 & 0 & 0 \\ 0 & 0 & 0 & -1 & 0 & 0 & 0 & 0 & 0 & 0 & 0 & 0 \\ 0 & 0 & 0 & 0 & -1 & 0 & 0 & 0 & 0 & 0 & 0 & 0 \\ 0 & 0 & 0 & 0 & 0 & 1 & 0 & 0 & 0 & 0 & 0 & 0 \\ -1 & 0 & 0 & 0 & 0 & 0 & 0 & 0 & 0 & 0 & 0 & 0 \\ 0 & -1 & 0 & 0 & 0 & 0 & 0 & 0 & 0 & 0 & 0 & 0 \\ 0 & 0 & 1 & 0 & 0 & 0 & 0 & 0 & 0 & 0 & 0 & 0 \end{bmatrix} \cdot \begin{bmatrix} x_1 \\ y_1 \\ z_1 \\ x_2 \\ y_2 \\ z_2 \\ x_3 \\ y_3 \\ z_3 \\ x_4 \\ y_4 \\ z_4 \end{bmatrix}$$

Considering all four symmetry operations of the C_{2h} point group, the complete representation of the displacement coordinates of HNNH as basis consists of four 12 x 12 matrices. Working with such big matrices is awkward and time-consuming. Fortunately, they can be simplified. We shall not go into the details of how this is done since only the easiest and quickest methods utilizing matrix representations will be used in the next chapters. We shall merely outline the procedure leading from the big unpleasant representations of symmetry operations to simpler tools [4-1]. With the help of suitable similarity transformations, matrices can be turned into so-called *block-diagonal matrices*. A block-diagonal matrix has nonzero values only in square blocks along the diagonal from the top left to the bottom right. For example, the following is a typical block-diagonal matrix:

$$\begin{bmatrix} 1 & 2 & 3 & 0 & 0 & 0 \\ 4 & -1 & 2 & 0 & 0 & 0 \\ 3 & 1 & -2 & 0 & 0 & 0 \\ 0 & 0 & 0 & 5 & 0 & 0 \\ 0 & 0 & 0 & 0 & 1 & -1 \\ 0 & 0 & 0 & 0 & 1 & 2 \end{bmatrix}$$

The merits of block-diagonal matrices are best illustrated in their multiplication. Suppose, for example, that two 5 x 5 matrices are to be multiplied, as follows:

$$
\begin{bmatrix} 2 & 3 & 0 & 0 & 0 \\ 1 & 2 & 0 & 0 & 0 \\ 0 & 0 & 1 & 1 & 0 \\ 0 & 0 & 1 & 1 & 0 \\ 0 & 0 & 0 & 0 & 2 \end{bmatrix} \cdot \begin{bmatrix} 1 & 2 & 0 & 0 & 0 \\ 2 & 1 & 0 & 0 & 0 \\ 0 & 0 & 2 & 2 & 0 \\ 0 & 0 & 1 & 2 & 0 \\ 0 & 0 & 0 & 0 & 1 \end{bmatrix} = \begin{bmatrix} 8 & 7 & 0 & 0 & 0 \\ 5 & 4 & 0 & 0 & 0 \\ 0 & 0 & 3 & 4 & 0 \\ 0 & 0 & 3 & 4 & 0 \\ 0 & 0 & 0 & 0 & 2 \end{bmatrix}
$$

The determination of the first row is already quite complicated:

$$2 \cdot 1 + 3 \cdot 2 + 0 \cdot 0 + 0 \cdot 0 + 0 \cdot 0 = 8$$
$$2 \cdot 2 + 3 \cdot 1 + 0 \cdot 0 + 0 \cdot 0 + 0 \cdot 0 = 7$$
$$2 \cdot 0 + 3 \cdot 0 + 0 \cdot 2 + 0 \cdot 1 + 0 \cdot 0 = 0$$
$$2 \cdot 0 + 3 \cdot 0 + 0 \cdot 2 + 0 \cdot 2 + 0 \cdot 0 = 0$$
$$2 \cdot 0 + 3 \cdot 0 + 0 \cdot 0 + 0 \cdot 0 + 0 \cdot 1 = 0$$

Notice that the product of two equally block-diagonalized matrices – as those two above – is another similarly block-diagonalized matrix. What is especially important, this resulting matrix can be obtained simply by multiplying the corresponding individual blocks of the original matrices. Check this on the above example:

$$
\begin{bmatrix} 2 & 3 \\ 1 & 2 \end{bmatrix} \cdot \begin{bmatrix} 1 & 2 \\ 2 & 1 \end{bmatrix} = \begin{bmatrix} 2 \cdot 1 + 3 \cdot 2 & 2 \cdot 2 + 3 \cdot 1 \\ 1 \cdot 1 + 2 \cdot 2 & 1 \cdot 2 + 2 \cdot 1 \end{bmatrix} = \begin{bmatrix} 8 & 7 \\ 5 & 4 \end{bmatrix}
$$

$$
\begin{bmatrix} 1 & 1 \\ 1 & 1 \end{bmatrix} \cdot \begin{bmatrix} 2 & 2 \\ 1 & 2 \end{bmatrix} = \begin{bmatrix} 1 \cdot 2 + 1 \cdot 1 & 1 \cdot 2 + 1 \cdot 2 \\ 1 \cdot 2 + 1 \cdot 1 & 1 \cdot 2 + 1 \cdot 2 \end{bmatrix} = \begin{bmatrix} 3 & 4 \\ 3 & 4 \end{bmatrix}
$$

$$
\begin{bmatrix} 2 \end{bmatrix} \cdot \begin{bmatrix} 1 \end{bmatrix} = \begin{bmatrix} 2 \end{bmatrix}
$$

Generally, if two matrices A and B can be transformed by similarity transformation into identically shaped block-diagonalized matrices, their product matrix C will also have the same block-diagonal form:

$$
\begin{bmatrix} A_1 & & \\ & A_2 & \\ & & A_3 \end{bmatrix} \cdot \begin{bmatrix} B_1 & & \\ & B_2 & \\ & & B_3 \end{bmatrix} = \begin{bmatrix} C_1 & & \\ & C_2 & \\ & & C_3 \end{bmatrix}
$$

The multiplication will also be valid for the individual blocks:

$$
A_1 \cdot B_1 = C_1
$$
$$
A_2 \cdot B_2 = C_2
$$
$$
A_3 \cdot B_3 = C_3
$$

Since the blocks themselves will obey the same multiplication table that the big matrices do, each block will be a new representation for an operation of the group. Thus if the above A and B matrices are representations for the respective symmetry operations σ_v and $\sigma_v{}'$ in the C_{2v} point group, so will be matrices A_1, A_2, A_3 and B_1, B_2, and B_3, respectively. The C_{2v} multiplication table (Table 4-1) shows that

$$
\sigma_v \cdot \sigma_v{}' = C_2,
$$

and accordingly, not only the big C matrix but also the small matrices C_1, C_2, C_3, will be representations of the C_2 operation. This way the big matrices *reduce* into smaller ones which are more convenient to handle. Let us suppose that the above big matrices A, B, and C together with the E matrix constitute a representation for the C_{2v} point group. This is called then a *reducible representation* of the group indicating that it is possible to find a similarity transformation to reduce it. Then we take each individual block and try to find again a similarity transformation which reduces it into a simpler form yet. This procedure is repeated until a set of the simplest possible matrices is reached along the diagonal of each big matrix. This is called the *irreducible representation*. Suppose now that in the example above the small matrices along the diagonal cannot be reduced further by a similarity transformation. In this case each set of the small matrices along the diagonal of the big ones will be an irreducible representation of the C_{2v} point group. The set of A_1, B_1, C_1, and E_1 will be an irreducible representation, so will be the set of A_2, B_2, C_2, and E_2, and yet another the set of A_3, B_3, C_3, and E_3. Thus the reducible representation was reduced to three irreducible representations. Since the symmetry operations can be applied to all kinds of bases for a molecule, there may be countless numbers of reducible representations. The important thing is, that all these representations reduce into a *small and finite number* of irreducible representations for practically all point groups. These irreducible representations, often called *symmetry species,* are then used in many areas of chemistry to describe symmetry properties.

4.4 The Character of a Representation

Considering the sizes of the initial matrices, the irreducible representations are a great improvement. Fortunately, even further simplification is possible. Instead of working with irreducible representations we can use simply their *characters.* The utility of this approach will be amply demonstrated later. The *character* (or trace) *of a matrix* is the sum of its diagonal elements. For the following matrix

$$\begin{bmatrix} 1 & 2 & 0 & 3 \\ 0 & 7 & 1 & 1 \\ 1 & 2 & 0 & 0 \\ 1 & -2 & 3 & -4 \end{bmatrix}$$

the character is:

$$1 + 7 + 0 + (-4) = 4$$

Since a representation – reducible or irreducible – is a set of matrices corresponding to all symmetry operations of a point group, the *character of the representation* is the set of characters of all these matrices. For the simple basis of Δr_1 and Δr_2 used before for the HNNH molecule in the C_{2h} point group, the representation consisted of four 2 x 2 matrices:

characters

$$E \;=\; \begin{bmatrix} 1 & 0 \\ 0 & 1 \end{bmatrix} \qquad 1+1=2$$

$$C_2 \;=\; \begin{bmatrix} 0 & 1 \\ 1 & 0 \end{bmatrix} \qquad 0+0=0$$

$$i \;=\; \begin{bmatrix} 0 & 1 \\ 1 & 0 \end{bmatrix} \qquad 0+0=0$$

$$\sigma_h \;=\; \begin{bmatrix} 1 & 0 \\ 0 & 1 \end{bmatrix} \qquad 1+1=2$$

Thus the character of this representation is the following set:

$$2 \quad 0 \quad 0 \quad 2$$

We don't know yet, however, whether this representation is reducible or irreducible? To answer this question, first we have to know the characters of the irreducible representations of the C_{2h} point group.

4.5 Character Tables

The characters of irreducible representations are collected in the so-called *character tables*. We shall not discuss here how to find the characters of a given irreducible representation. The character tables are always available in textbooks and handbooks and some of them are also given in the subsequent chapters of this book. Table 4-3 shows the character table for the C_{2h} point group. Here the top row contains the complete set of symmetry operations of this group. The left column shows, for the time being, some temporary names. Γ is the generally used label for the representations. The main body of the character table contains the characters themselves. Thus each row constitutes the characters of an irreducible representation, and the number of rows tells us how many irreducible representations does that particular point group have.

Table 4-3. A Preliminary Character Table for the C_{2h} Point Group

C_{2h}	E	C_2	i	σ_h
Γ_1	1	1	1	1
Γ_2	1	−1	1	−1
Γ_3	1	1	−1	−1
Γ_4	1	−1	−1	1

Some relationships and terminology must be known in order to use the character tables. First of all there are *classes* of irreducible representations, to which three rules are applicable:
(1) Usually symmetry operations of the same kind belong to the same class (e.g., C_3 and $C_3{}^2$, or, the three vertical mirror planes of the C_{3v} point group); (2) The number of irreducible representations of a group is equal to the number of classes in that group; and (3) The character of any irreducible representation is the same for all operations in the same class.
Let us check these rules on the C_{2h} character table given above. All the four symmetry elements stand here by themselves, one element in each class. The number of irreducible representations of the C_{2h} point group is 4 just as is the number of classes.

Table 4-4. A Preliminary Character Table for the C_{3v} Point Group

C_{3v}	E	C_3	$C_3{}^2$	σ_v	$\sigma_v{}'$	$\sigma_v{}''$
Γ_1	1	1	1	1	1	1
Γ_2	1	1	1	−1	−1	−1
Γ_3	2	−1	−1	0	0	0

Table 4-4 shows a kind of preliminary character table for the C_{3v} point group. The complete set of operations is listed in the upper row. Clearly, some of them must belong to the same class since the number of irreducible representations is 3 and the number of operations is 6. A closer look at this table reveals that the characters of all irreducible representations are equal in C_3 and C_3^2 and also in σ_v, σ_v', and σ_v'', respectively. Indeed, both three-fold rotation operations have the same effect on the ammonia molecule as seen in Fig. 4-9. The same is true for the

Figure 4-9.
Rotation and reflection of the ammonia molecule.

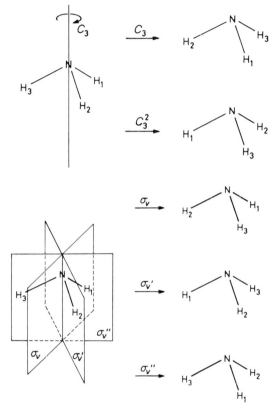

three vertical reflection planes; all of them leave one of the hydrogens in place and exchange the other two. Accordingly, C_3 and C_3^2 form one class and σ_v, σ_v' and σ_v'' together form another class. This way the number of classes in C_{3v} is three as well as the number of irreducible representations. The compact C_{3v} character table is shown in Table 4-5.

Table 4-5. Compact Character Table for the C_{3v} Point Group

C_{3v}	E	$2C_3$	$3\sigma_v$
Γ_1	1	1	1
Γ_2	1	1	−1
Γ_3	2	−1	0

177

The number of operations in a given class is always indicated in the upper row of the character table. The identity operation, E, is always a class by itself, and the same is true for the inversion operation, i.

Another important feature of an irreducible representation is its *dimension*. This is simply the dimension of any of its matrices which in turn is the number of rows or columns of the matrix. Since the identity operation always leaves the molecules unchanged, its representation is a unit matrix. The character of a unit matrix is equal to the number of rows or columns of that matrix, as is demonstrated below:

$$E = \begin{bmatrix} 1 & 0 & 0 \\ 0 & 1 & 0 \\ 0 & 0 & 1 \end{bmatrix} \qquad \text{character} = 1+1+1=3$$

$$E = \begin{bmatrix} 1 & 0 \\ 0 & 1 \end{bmatrix} \qquad \text{character} = 1+1=2$$

$$E = \begin{bmatrix} 1 \end{bmatrix} \qquad \text{character} = 1$$

From this follows that the *character under E is always the dimension of the given irreducible representation*. The one-dimensional representations are non-degenerate and the two- or more dimensional representations are degenerate. The meaning of degeneracy will be discussed in Chapter 6.

Now it is time to see a complete character table. Table 4-6 shows one for the C_{3v} point group. Consider now the symbols used for the names of the irreducible representations. These are the so-called Mulliken symbols and their meaning is described below together with other Mulliken symbols collected in Table 4-7.

Letters A and B are used for one-dimensional irreducible representations, depending on whether they are symmetric or antisymmetric with respect to rotation around the principal

Table 4-6. Complete Character Table for the C_{3v} Point Group

C_{3v}	E	$2C_3$	$3\sigma_v$		
A_1	1	1	1	z	$x^2+y^2,\ z^2$
A_2	1	1	-1	R_z	
E	2	-1	0	$(x,\ y)(R_x,\ R_y)$	$(x^2-y^2,\ xy)(xz,\ yz)$

Table 4-7. Symbols for Irreducible Representations of Finite Groups

Dimension of Representation	Character under E C_n i σ_h C_2[a] or σ_v	Symbols
1	1 1 1 −1	A B
2	2	E
3	3	T

	1
	 −1
	    1
	    −1
	       1
	       −1
		A_g B_g E_g T_g
		 A_u B_u E_u T_u
		 A' B'
		 A'' B''
		 A_1 B_1
		 A_2 B_2

[a] C_2 axis perpendicular to the principal axis

axis of the point group. An antisymmetric behavior here means changing sign or direction[2]. The character for a symmetric representation is +1 and this is designated by letter A. An antisymmetric behavior is represented by letter B and has −1 character. E is the symbol[3] for two-dimensional, and T (sometimes F) for three-dimensional representations. The subscripts g and u indicate whether the representation is symmetric or antisymmetric with respect to inversion. The German *gerade* means even and *ungerade* means odd. The superscripts ' and " are used for irreducible representations which are symmetric and antisymmetric to a horizontal mirror plane, respectively. The subscripts 1 and 2 with A and B refer to symmetric (1) and

---

2 Antisymmetry will be discussed in the next section.
3 Not to be confused with the symbol of the identity operation which is also E.

179

antisymmetric (2) behavior with respect to either a C_2 axis perpendicular to the principal axis, or, in its absence, to a vertical mirror plane. The meaning of subscripts 1 and 2 with E and T is more complicated, and will not be discussed here. The character tables of the infinite groups, $C_{\infty v}$ and $D_{\infty h}$, use Greek rather than Latin letters: Σ stands for one-dimensional representations and Π, Δ, Φ, etc., for two-dimensional representations.

It is always possible to find such a behavior which remains unchanged under any of the symmetry operations of the given point group. Thus there is always an irreducible representation which has only +1 characters. This is the *totally symmetric irreducible representation,* and it is always the first one in any character table.

The character tables usually consist of four main areas (sometimes three if the last two are merged), as is seen in Table 4-6 for the C_{3v} and in Table 4-8 for the C_{2h} group. The first area contains the symbol of the group (in the upper left hand corner) and the Mulliken symbols referring to the dimensionality of the representations and their relationship to various symmetry operations. The second area contains the classes of symmetry operations (in the upper row) and the characters of the irreducible representations of the group.

Table 4-8. C_{2h} Character Table

C_{2h}	E	C_2	i	σ_h		
A_g	1	1	1	1	R_z	x^2, y^2, z^2, xy
B_g	1	−1	1	−1	R_x, R_y	xz, yz
A_u	1	1	−1	−1	z	
B_u	1	−1	−1	1	x, y	

Figure 4-10.
(a) Applying the identity and the C_3 operation to a rotating spinning top.

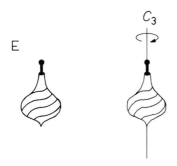

The third and fourth areas of the character table contain some chemically important basis functions for the group. The third area contains six symbols: x, y, z, R_x, R_y, and R_z. The first three are the Cartesian coordinates that we have already used before as bases for a representation of the C_{2h} point group. The symbols R_x, R_y, and R_z stand for rotations around the x, y, and z axes, respectively. A popular toy, the spinning top, is helpful in visualizing the consequences of symmetry operations on rotation. Let us work out the characters for rotation around the z axis in the C_{3v} point group (Fig. 4-10a). Obviously, the identity operation leaves the rotating spinning top unchanged (character 1). So does the rotation around the same axis since the rotational symmetry axis is indistinguishable from the axis of

(b) Illustration of the effect of mirror planes on the rotating spinning top.

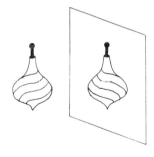

rotation of the toy. The corresponding character is again 1. Now place a mirror next to the rotating toy (Fig. 4-10b). No matter where the mirror is, the rotation of the mirror image will always be in the opposite direction with respect to the real rotation. Accordingly, the character will be −1.

Thus the characters of the rotation around the z axis in the C_{3v} point group will be:

$$1 \quad 1 \quad -1$$

Indeed, R_z belongs to the irreducible representation A_2 in the C_{3v} character table. In other words, R_z *transforms* as A_2, or, it *forms a basis* for A_2.

The fourth area of the character table contains all the squares and binary products of the coordinates according to their behavior under the symmetry operations. All the coordinates and their products listed in the third and fourth areas of the character table are important basis functions. They have the same symmetry properties as the atomic orbitals under the same names; z corresponds to p_z, x^2-y^2 to $d_{x^2-y^2}$ and so on. We shall meet them again in the discussion of the properties of atomic orbitals.

Antisymmetry has occurred several times above and it is a whole new idea in our discussion. It is again a point where chemistry and other fields meet in a uniquely important symmetry concept.

4.6 Antisymmetry

Figure 4-11.
Identity operation (top) and antiidentity operation (bottom).

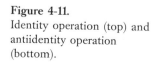

"Operations of antisymmetry transform objects possessing two possible values of a given property from one value to the other" [4-6]. The simplest demonstration of an antisymmetry operation is by color change. Fig. 4-11 shows an identity operation and an antiidentity operation. Nothing changes, of course, in the former, whereas merely the black-and-white coloring reverses in the latter. Antimirror symmetry along with mirror symmetry can be found in Fig. 4-12, and further antimirror symmetries are presented in Fig. 4-13.

Not only a symmetry plane but other symmetry elements may also serve as antisymmetry elements. Thus, for example, two-fold, four-fold, and six-fold antirotation axes appear in Fig. 4-14 after Shubnikov [4-8]. The four-fold antirotation axis includes a two-fold rotation axis and the six-fold antirotation axis includes a three-fold rotation axis. The antisymmetry elements have the same notation as the ordinary ones except that they are underlined. Antimirror rotation axes characterize

Figure 4-12.
Mirror symmetries and anti-mirror symmetries.
1-2 and 3-4 mirror symmetries
1-4 and 2-3 antimirror symmetries.

the rosettes in the second row of Fig. 4-14. The antirotation axes appear in combination with one or more symmetry planes perpendicular to the plane of the drawing in the third and fourth rows of Fig. 4-14. Finally, the ordinary rotation axes are combined with one or more antisymmetry planes in the bottom row of this figure. In fact, symmetry $1 \cdot \underline{m}$ here is the symmetry illustrated also in Figs. 4-12 and 4-13. The examples above are all point groups and only from among the simplest ones.

Figure 4-13.
Illustration of antimirror symmetry.
(a) Eastern Orthodox church in Zagorsk, photograph and courtesy of Dr. A. A. Ivanov, Moscow.

(b) Batik design after [4-7].

(c) Logo of the TUNGSRAM Works, Budapest.

(d) Reproduction of an original work by Victor Vasarely. Used by permission.

182

Figure 4-14.
Antisymmetry operations: antirotation axes 2̱, 4̱, 6̱; antimirror rotation axes 2̄, 4̄, 6̄; antirotation axes combined with ordinary mirror planes 2̱ · m, 4̱ · m̈, 6̱ · m; ordinary rotation axes combined with antimirror planes 1 · m̱, 3 · m̱; after Shubnikov [4-8]. Reproduced with permission from Nauka Publ. Co., Moscow.

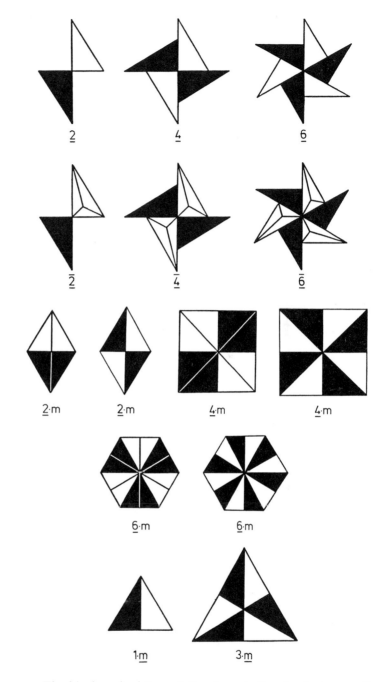

The black-and-white variation is again the simplest case of what is color symmetry. This area is vast and becomes very complicated, and its importance is increasingly being recognized [4-8 – 4-11]. Our single example indicative of the complexity of color symmetry involves the Rubik's cube. In its monocolor version the cube has many symmetry elements, among them four-fold axes going through the midpoints of opposite faces. In the unscrambled starting position of the

Rubik's cube each side has a different color. Thus the original four-fold axis of the cube is no longer a symmetry element for the Rubik's cube. However, it is possible to specify this axis in such a way that it corresponds to the color changes of the Rubik's cube. To do so, the coloring of the cube must be known. Consider the coloring indicated in Fig. 4-15. Now it is possible to imagine a four-fold rotation axis, for example going through the midpoints of the yellow and white faces. Then the four-fold color-rotation axis will change the colors every quarter of rotation in the following order: red/blue/orange/green/red, or, red/green/orange/blue/red, depending on the sense of the rotation.

Figure 4-15.
The application of a four-fold color-rotation axis to the Rubik's cube. It involves the following color changes during a complete rotation, red → blue → orange → green → red or *vice versa*. The color arrangement of the cube is shown.

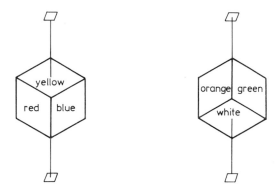

The above examples applied to point groups. Such distinctions and further coloring, of course, may be introduced in space-group symmetries as well [4-8].

The black-and-white variation is perhaps the simplest version of antisymmetry. The general definition of antisymmetry at the beginning of this section, however, calls for a much broader interpretation and application. The relationship between matter and antimatter is a conspicuous example of antisymmetry. There is no limit to down-to-earth examples, especially if, again, symmetry is considered rather loosely.

Another literary example is taken from Karinthy's writing, this time to illustrate antisymmetry. It is from a short story entitled "Two Diagnoses" [4-12]. The same person, Dr. Same, goes to see a physician at two different places. At the recruiting station he would obviously like to avoid being drafted, while at the insurance society he would like to acquire the best possible terms for his policy. His answers to the identical questions of the two physicians are related by antisymmetry.

184

Two Diagnoses

Physician

Dr. Same	Dr. Same
At the recruiting station	At the insurance society

Broken-looking, sad, ruined human wreckage, feeble masculinity, haggard eyes, wavering movement. He steps forward, with humped back and coughs in front of the doctor. His speech is soft, whispering.	*Young athlete with straightened back, flashing eyes. He crosses the room with banging steps. He has a sonorous baritone, swinging hips.*

How old are you?

Old ... very old, indeed.	*Coyly.* Oh, my gosh, I'm almost ashamed of it ... I'm so silly ...

Your I. D. says you're thirty two.

With pain. To be old is not to be far from the cradle – but near the coffin.	To be young is not to be near the cradle, but far from the coffin.

Your parents alive?

Alive, just alive, if life it is, rolling about in qualms of conscience day and night for giving life to the pitiable, sick bastard I am, to my own suffering and mankind's.	My father is chairman by seniority of the Methuselah club of very old men.

Examines the heart. Take a breath.

Takes a breath like a small humming bird.	*Takes a breath enough for staying at the bottom of the sea for two weeks.*

Show me your tongue.

Waves his hand. Don't even look at it, Doctor. You look to me, Sir, like a sensitive soul and my tongue is a sad sight, that's what it is, to tell you the truth.

Hanging out his tongue half a meter. Look at this tongue, Doctor – don't be afraid to feel it – you won't see the like of it again soon, Sir, in this city – it's so fresh and appetizing that sometimes I have the urge to bite and eat it. *He's patting his tongue.*

Have you had any organic diseases?

Who, me? Not to my knowledge, unless this one-sided brain-cancer counts as an organic disease; my physician has always suspected I had it.

Sighs. Alas ... I have to admit that I have had. When I was a child, my nose was severely itching at one point. However, I scratched it in good time and it was gone. Since then I have been fine.

Are you ever dizzy?

Don't mention dizziness, please, Doctor, or else I'll collapse at once. I always have to walk in the middle of the street, because if I look down from the curb, I become dizzy at once.

Quite often, sorry to say. Every time I'm aboard an airplane and it's up-side-down, and breaking to pieces. Otherwise, not.

No complaints, otherwise?

None, otherwise. If I'm holding my head on one side, I feel this dull pain in my back. If I'm turning around, I feel the same dull pressure in my head. Sometimes I have this feeling in my chest as if my aching leg were throbbing, and sometimes I can't breathe through my ears. I feel a stabbing pain around my kidney when I'm blowing my nose, and sometimes I have the urge to cry, and I don't even know why. In the evenings when I'm tired to death, I'm in such a state that I can barely find my bed. Otherwise, nothing, except for that rabid dog that bit me about an hour ago.

Oh, yes ... I don't dare to go near a hungry horse lest it mistake me for oats. I anguish at all times about what will happen to me in two or three hundred years, when I get tired of living, and shall not be able to die.

Figure 4-16.
(a) Reflection: two fast food chain restaurants side by side, downtown Washington, D.C. Photograph by the authors.
(b) Antireflection: an exclusive, one-of-a-kind restaurant and a fast food chain restaurant side by side, downtown Washington, D.C. Photograph by the authors.

The next example will be appreciated at once by the American reader but needs some clarification for the rest of us. Fig. 4-16 shows four restaurants in two pairs from Washington, D.C. Both Burger King and McDonald's are fast food chains whereas Sans Souci is (or at least used to be) a very exclusive, elegant, certainly one-of-a-kind restaurant. Thus Fig. 4-16a may be considered to be an ordinary reflection while Fig. 4-16b an antireflection. The symmetry and antisymmetry planes are imaginary, but in a way the vertical walls between the restaurants are their physical appearances.

The antireflection in Fig. 4-17a relates the identical Flemish and French texts in the holiday advertisement. Fig. 4-17b shows a pair of door handles. They are each other's mirror images except that the one on the inner side of the door looks brand new while the other exposed to harsh weather conditions is considerably corroded. The final example shows two military jets and a sea gull in Fig. 4-17c.

The above examples of antisymmetry may have implied at least as much abstraction as any chemical applications. The symmetric and antisymmetric behavior of orbitals describing electronic structure and vectors describing molecular vibrations may be perceived with greater ease after the preceding diversion. Before that, however, some more of group theory will have to be covered.

Figure 4-17.
Antireflections. Photographs by the authors.
(a) Holiday advertisement in Belgium in Flemish and in French.

(b) A pair of door handles, one of them inside, protected, the other outside, exposed to harsh weather conditions, corroded. On the island Harøy in the North Sea, off Molde, Norway.

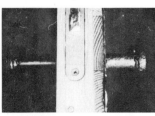

(c) Military jets and a sea gull, off Bodø, Norway.

188

4.7 Shortcut to Determine a Representation

It was quite easy to find the irreducible representation of R_z before as the representation we worked out appeared to be an irreducible representation itself. In most cases, however, a reducible representation is found when the symmetry operations are applied to a certain basis. Now a simpler way will be shown (1) to describe the representation on a given basis without generating the matrices themselves and (2) to reduce them, if reducible, to irreducible representations.

The diimide molecule is our example again, and the basis is the two N-H bond length changes (see Fig. 4-7). It is easy to generate the matrices corresponding to each operation using such a simple basis; however, even this may not be necessary. It has already been mentioned that instead of the representations themselves, we can work with their characters. For this particular case the characters of the representation have already been determined:

$$\Gamma_1 \quad 2 \quad 0 \quad 0 \quad 2$$

But how can we know the character of a matrix without writing down the whole matrix?

Looking back at the effect of the different symmetry operations on HNNH (Fig. 4-7) it is recalled, for example, that C_2 interchanges Δr_1 and Δr_2, so the diagonal elements of the matrix will all be 0. Consequently, these vectors do not contribute to the character.

This observation can be generalized so that those basis elements that are associated with an atom changing its position during the symmetry operation, will have zero contribution to the character. The basis element that is unchanged by a given operation contributes +1 to the character. Finally, the basis element that is transformed into its negative, contributes −1. The only complication arises with the rotational operations when the atom does not move during the symmetry operation but the basis element associated with it is rotated by a certain angle. Here the matrix of the rotation has to be constructed as shown in Sect. 4.2.

Returning to the diimide N-H bond length changes, let us see how the above simple rules work. The identity operation, E, leaves the molecule unchanged, so the two vectors, Δr_1 and Δr_2, will each contribute +1 to the character:

$$1 + 1 = 2$$

The effect of C_2 has already been looked at. Its character is 0. The effect of the inversion operation is the same as that of C_2, so the character will be

$$0 + 0 = 0$$

Finally, operation σ_h leaves the two bonds unchanged, so both of them contribute +1 to the character:

$$1 + 1 = 2$$

The result is the same as before:

$$\Gamma_1 \quad 2 \quad 0 \quad 0 \quad 2$$

Now check the rules with a larger basis set, the Cartesian displacement coordinates of the atoms of HNNH (see Fig. 4-8). Operation E leaves all the 12 vectors unchanged, so its character will be 12. C_2 brings each atom into a different position so their vectors will also be shifted. This means that all the vectors will have zero contribution to the character. The same applies to the inversion operation. Finally, as already worked out before, the horizontal reflection leaves all the x and y vectors unchanged and brings the four z vectors into their negative selves. The result is

$$8 + (-4) = 4$$

The whole representation of the displacement vectors is:

$$\Gamma_2 \quad 12 \quad 0 \quad 0 \quad 4$$

Both representations we constructed here are reducible since there are no 2 and 12 dimensional representations in the C_{2h} character table (Table 4-8). The next question is how to reduce these representations?

4.8 Reducing a Representation

It was discussed before that the irreducible representations can be produced from the reducible representations by suitable similarity transformations. Another important point is that the character of a matrix is not changed by any similarity transformation. From this follows that the sum of the characters of the irreducible representations is equal to the character of the original reducible representation from which they are obtained. We have seen that for each symmetry operation the matrices of the irreducible representations stand along the diagonal of the matrix of the reducible representation, and the character is just the sum of the diagonal elements. When reducing a representation, the simplest way is to look for the combination of the irreducible representations of that group – that is the sum of their characters in each class of the character table – which will produce the characters of the reducible representation.

First reduce the representation of the two N-H bond length changes of HNNH:

$$\Gamma_1 \quad 2 \quad 0 \quad 0 \quad 2$$

The C_{2h} character table shows that Γ_1 can be reduced to $A_g + B_u$:

C_{2h}	E	C_2	i	σ_h
A_g	1	1	1	1
B_g	1	−1	1	−1
A_u	1	1	−1	−1
B_u	1	−1	−1	1
$A_g + B_u$	2	0	0	2

It may be asked, of course, whether this is the only way of decomposing the Γ_1 representation? The answer is reassuring: *The decomposition of any reducible representation is unique.* If we find a solution just by inspection of the character table, it will be the only one. Often this is the fastest and simplest way to decompose a reducible representation.

A more general and more complicated way is to use a *reduction formula:*

$$a_i = (1/h) \sum_Q N \cdot \chi(R) \cdot \chi_i(R)$$

where

a_i is the number of times the ith irreducible representation appears in the reducible representation

h is the order of the group

Q is a class of the group

N is the number of operations in the class Q

R is an operation of the group

$\chi(R)$ is the character of R in the reducible representation[4]

$\chi_i(R)$ is the character of R in the ith irreducible representation

The summation extends over all classes of the group.

The reduction formula can be applied only to finite point groups. For the infinite point groups, $D_{\infty h}$ and $C_{\infty v}$, the usual practice is to reduce the representations by inspection of the character table.

For illustration, let us find the irreducible representations of the two examples used before. First on the basis of the two N-H distance changes of diimide (i.e., Γ_1):

C_{2h}	E	C_2	i	σ_h
A_g	1	1	1	1
B_g	1	−1	1	−1
A_u	1	1	−1	−1
B_u	1	−1	−1	1
Γ_1	2	0	0	2

The order of the group is 4. The number of times the irreducible representation A_g appears in the reducible representation is:

$$a_{A_g} = (1/4) \: [1 \cdot 2 \cdot 1 + 1 \cdot 0 \cdot 1 + 1 \cdot 0 \cdot 1 + 1 \cdot 2 \cdot 1] =$$
$$= (1/4) \: (2 + 0 + 0 + 2) = 4/4 = 1.$$

In the same way we can deduce the number of times the other irreducible representations appear in Γ_1:

$$a_{B_g} = (1/4) \: [1 \cdot 2 \cdot 1 + 1 \cdot 0 \cdot (-1) + 1 \cdot 0 \cdot 1 + 1 \cdot 2 \cdot (-1)] =$$
$$= (1/4) \: (2 + 0 + 0 - 2) = 0$$

---

4 Here and hereafter the short expression "character of R" stands for the character of the matrix corresponding to operation R, in accordance with our previous discussion.

$$a_{A_u} = (1/4) \, [1 \cdot 2 \cdot 1 + 1 \cdot 0 \cdot 1 + 1 \cdot 0 \cdot (-1) + 1 \cdot 2 \cdot (-1)] =$$
$$= (1/4) \, (2 + 0 + 0 - 2) = 0$$
$$a_{B_u} = (1/4) \, [1 \cdot 2 \cdot 1 + 1 \cdot 0 \cdot (-1) + 1 \cdot 0 \cdot (-1) + 1 \cdot 2 \cdot 1] =$$
$$= (1/4) \, (2 + 0 + 0 + 2) = 4/4 = 1$$

That is $\Gamma_1 = A_g + B_u$, and the result is the same as before.

With the twelve-dimensional reducible representation of the Cartesian displacement vectors of HNNH the inspection method probably does not work. However, the reduction formula can be used.
The reducible representation is:

$$\Gamma_2 \quad 12 \quad 0 \quad 0 \quad 4$$

$$a_{A_g} = (1/4) \, [1 \cdot 12 \cdot 1 + 1 \cdot 0 \cdot 1 + 1 \cdot 0 \cdot 1 + 1 \cdot 4 \cdot 1] =$$
$$= (1/4) \, (12 + 4) = 4$$
$$a_{B_g} = (1/4) \, [1 \cdot 12 \cdot 1 + 1 \cdot 0 \cdot (-1) + 1 \cdot 0 \cdot 1 + 1 \cdot 4 \cdot (-1)] =$$
$$= (1/4) \, (12 - 4) = 2$$
$$a_{A_u} = (1/4) \, [1 \cdot 12 \cdot 1 + 1 \cdot 0 \cdot 1 + 1 \cdot 0 \cdot (-1) + 1 \cdot 4 \cdot (-1)] =$$
$$= (1/4) \, (12 - 4) = 2$$
$$a_{B_u} = (1/4) \, [1 \cdot 12 \cdot 1 + 1 \cdot 0 \cdot (-1) + 1 \cdot 0 \cdot (-1) + 1 \cdot 4 \cdot 1] =$$
$$= (1/4) \, (12 + 4) = 4$$

$$\Gamma_2 = 4 \, A_g + 2 \, B_g + 2 \, A_u + 4 \, B_u$$

4.9 Auxiliaries

A few additional things need to be mentioned before embarking on chemical applications of group theoretical methods. For detailed descriptions and proofs we refer to [4-1 – 4-3].

4.9.1 Direct Product

Wave functions form bases for representations of the point group of the molecule [4-1]. Suppose that f_i and f_j are such functions, then the new set of functions, $f_i f_j$, called the *direct product* of f_i and f_j is also basis for a representation of the group. The characters of the direct product can be determined by the following rule: *the characters of the representation of a direct*

Table 4-9. Character Table and Some Direct Products for the C_{2v} Point Group

C_{2v}	E	C_2	σ_v	$\sigma_v{}'$	
A_1	1	1	1	1	
A_2	1	1	-1	-1	
B_1	1	-1	1	-1	
B_2	1	-1	-1	1	
$A_1 \cdot A_2$	1	1	-1	-1	$= A_2$
$A_2 \cdot B_1$	1	-1	-1	1	$= B_2$
$B_1 \cdot B_2$	1	1	-1	-1	$= A_2$

Table 4-10. Character Table and Direct Products for the C_{3v} Point Group

C_{3v}	E	$2C_3$	$3\sigma_v$	
A_1	1	1	1	
A_2	1	1	-1	
E	2	-1	0	
$A_2 \cdot A_2$	1	1	1	$= A_1$
$A_2 \cdot E$	2	-1	0	$= E$
$E \cdot E$	4	1	0	$= A_1 + A_2 + E$

product are equal to the products of the characters of the representations of the original functions. The direct product of two irreducible representations will be a new representation which is either an irreducible representation itself or which can be reduced into irreducible representations. Tables 4-9 and 4-10 show some examples for direct products with the C_{2v} and C_{3v} point groups, respectively.

4.9.2 Integrals of Product Functions

Integrals of product functions often occur in the quantum mechanical description of molecular properties and it is helpful to know their symmetry behavior. Why? The reason is that an integral whose integrand is the product of two or more functions will vanish unless the integrand is invariant under all symmetry operations of the point group. This means that the integral will be nonzero only if the integrand belongs to the totally symmetric irreducible representation of the molecular point group.

The representation of a product function can be determined by forming the direct product of the original functions. The representation of a direct product will contain the totally symmetric representation only if the original functions, whose product is formed, belong to the *same* irreducible representation of the molecular point group.

These rules can be extended to integrals of products of more than two functions. For a triple product the integral will be nonzero only if the representation of the product of any two functions is the same as or contains the representation of the third function. If the integral is

$$\int f_i \cdot f_j \cdot f_k \, d\tau$$

then the above condition is expressed by

$$\Gamma_{f_i} \cdot \Gamma_{f_k} \subset \Gamma_{f_j}$$

where Γ stands for the representation and $\subset$ means "is or contains". Very often f_j is a quantum chemical operator and then the relations are:

$$\int f_i \, \hat{o}p. \, f_k \, d\tau$$

or with other notation

$$\langle f_i | \hat{o}p. | f_k \rangle$$

and

$$\Gamma_{f_i} \cdot \Gamma_{f_k} \subset \Gamma_{\hat{o}p.}$$

This kind of condition appears in energy integrals, spectral selection rules, and in the discussion of chemical reactions.

4.9.3 Projection Operator

In applying group theory to chemical problems one of the most useful concepts is the *projection operator* [4-1, 4-2]. It is an operator which takes the non-symmetry adapted basis of a representation and projects it along new directions in such a way that it belongs to a specific irreducible representation of the group. The projection operator is represented by $\hat{P}$ in the following form:

$$\hat{P}^i = (1/h) \sum_R \chi_i(R) \cdot \hat{R}$$

where h is the order of the group,

i is an irreducible representation of the group,

R is an operation of the group,

$\chi_i(R)$ is the character of R in the i th irreducible representation, and

$\hat{R}$ means the application of the symmetry operation R to our basis component.

The summation extends over all operations of the group.

Consider now the construction of the A_1 symmetry group orbital of the hydrogen s atomic orbitals in ammonia as an example for the application of the projection operator. (The various kinds of orbitals will be discussed in detail in Chapter 6.)

The projection operator for the A_1 irreducible representation in the C_{3v} point group is

$$\hat{P}^{A_1} = (1/6) \sum_R \chi_{A_1}(R) \cdot \hat{R}$$

Applying this operator to the s orbital of one of the hydrogens (H1) of ammonia, we obtain

$$\hat{P}^{A_1} s_1 \approx 1 \cdot E \cdot s_1 + 1 \cdot C_3 \cdot s_1 + 1 \cdot C_3{}^2 \cdot s_1 +$$
$$+ 1 \cdot \sigma \cdot s_1 + 1 \cdot \sigma' \cdot s_1 + 1 \cdot \sigma'' \cdot s_1 =$$
$$= s_1 + s_2 + s_3 + s_1 + s_2 + s_3 \approx s_1 + s_2 + s_3.$$

The expression is an approximation here since the 1/6 numerical factor was omitted. The coefficient in the symmetry adapted linear combinations can be determined at a later stage by normalization. In an actual calculation this is necessary whereas here we are interested in the symmetry aspects only which are well represented by the relative values. In fact, the coefficients will be ignored throughout our discussions!

Application of the projection operator will also be demonstrated pictorially in forthcoming chapters. These representations will emphasize the results of summation of symmetry-sensitive properties while the absolute magnitudes will not be treated rigorously. Thus, for example, the directions of vectors will be summed in describing vibrations, and the signs of the angular components of the electronic wave functions will be summed in describing the electronic structure.

4.10 Dynamic Properties

Molecular properties can be of either static or dynamic nature. A static property remains unchanged by every symmetry operation carried out on the molecule. The geometry of the nuclear arrangement in the molecule is such a property: a symmetry operation transforms the nuclear arrangement into another which will be indistinguishable from the initial.[5] The mass and the energy of a molecule are also static properties.

The dynamic properties, on the other hand, may change under symmetry operations. Molecular motion itself is a most common dynamic property. In our previous discussions of molecular structure, the molecules were mostly assumed to be motionless, and only the symmetry of their nuclear arrangement was considered. However, real molecules are not motionless. On the contrary, their chemical behavior is determined by their motion to a great extent.

In order to appreciate the effects of symmetry operations on motion, an example from our macroscopic world is invoked here, following the idea of Orchin and Jaffe [4-13]. Suppose there exists a long wall of mirror, and one walks alongside this mirror (Fig. 4-18a). Our mirror image will be walking with us

Figure 4-18.
Symmetric (a) and antisymmetric (b) consequences of the "mirror operation" for two movements. Drawing and courtesy of György Doczi, Seattle, WA.

---

5 Unless, of course, identical atoms are distinguished by labels as, e.g., in Figs. 4-2 and 4-3.

When the road chosen is paralell to the mirror

When we chose a road perpendicular to the mirror

with the same speed and in the same direction (its velocity will be the same as ours). Walk now from a distance towards the mirror perpendicularly to it. In this case our mirror image will have a different velocity from ours: the speed will be the same again but the direction will be just the opposite. Both we and our mirror image will be walking towards the plane of the mirror and if we do not stop in time, we shall collide in that plane (Fig. 4-18b).

The consequences of the mirror operation were different for the two movements. One was symmetric and the other was antisymmetric.

There are analogous phenomena for all kinds of molecular motion which may be symmetric and antisymmetric with respect to the various symmetry operations of the molecular point group. The two main kinds of motion in a molecule is nuclear and electronic. The nuclear motion may be translational, rotational and vibrational (Chapter 5). The electronic motion is basically the changes in the electron density distribution (Chapter 6).

4.11 Where Is It Applied?

It is primarily the dynamic properties whose description is facilitated by group theoretical methods. This is in fact an understatement. The dynamic properties cannot be fully discussed without group theory. On the other hand, this theory need not be used to determine the point group symmetry of the nuclear arrangement of a molecule, as has been shown before (cf. Table 3-1).

The first step in the symmetry determination of the dynamic properties is the selection of the appropriate basis. Appropriate here means the correct representation of the changes in the properties examined. In the investigation of the molecular vibrations (Chapter 5), either Cartesian displacement vectors or internal coordinate vectors are used. In the description of the molecular electronic structure (Chapter 6), the angular components of the atomic orbitals are frequently used bases. Since the angular wave function changes its "sign" under certain symmetry operations, its behavior will be characteristic of the spatial symmetry of a particular orbital. Molecular orbitals can also be used as basis of representation. The simple scheme below shows some important areas in chemistry where group theory is indispensable and the most convenient basis functions are also indicated:

Areas	Basis functions
construction of molecular orbitals	atomic orbitals
construction of hybrid orbitals	position vectors pointing towards the ligands
predicting the decrease of degeneracies of d orbitals under a ligand field	d atomic orbitals
predicting the allowedness of chemical reactions	molecular orbitals
determining the number and symmetries of molecular vibrations	Cartesian displacement vectors
normal coordinate analysis (symmetry coordinates)	internal coordinate displacements

Group theory is also used prior to calculations to determine whether a quantum mechanical integral of the type $\int \psi_i \, \hat{op}. \, \psi_j \, d\tau$ is different from zero or not. This is important in such areas as the selection rules for electronic transitions, chemical reactions, infrared and Raman spectroscopy, and other spectroscopies.

References

[4-1]   F. A. Cotton, *Chemical Applications of Group Theory*, Second Edition, Wiley-Interscience, New York, 1971.

[4-2]   A. Nussbaum, *Applied Group Theory for Chemists, Physicists and Engineers*, Prentice-Hall, Inc., Englewood Cliffs, NJ, 1971.

[4-3]   L. H. Hall, *Group Theory and Symmetry in Chemistry*, McGraw-Hill Book Company, New York, St. Louis, San Francisco, etc., 1969.

[4-4]   G. Burns, *Introduction to Group Theory with Applications*, Material Science Series, A. M. Alper and A. S. Nowich, eds., Academic Press, New York, 1977.

[4-5]   A. Vincent, *Molecular Symmetry and Group Theory. A Programmed Introduction to Chemical Applications*, Wiley-Interscience, New York, 1977.

[4-6]   A. L. Mackay, Acta Cryst. **10;** 543 (1957).

[4-7]   Gy. Lengyel, *Handiwork. New techniques – new solutions* (in Hungarian), Kossuth, Budapest, 1975.

[4-8]   A. V. Shubnikov, *Simmetriya i antisimmetriya konechnykh figur*, Izd. Akad. Nauk S.S.S.R., Moskva, 1951.

[4-9]   A. Loeb, *Color and Symmetry*, Wiley-Interscience, New York, 1971.

[4-10]   A. Loeb, in *Patterns of Symmetry*, M. Senechal and G. Fleck, eds., University of Massachusetts Press, Amherst, MA, 1977.

[4-11]   M. Senechal, Acta Cryst. **A39,** 505 (1983).

[4-12]   F. Karinthy, *Selected Works* (in Hungarian), Szépirodalmi, Budapest, 1962. Our English translation was kindly checked by Dr. R. B. Wilkenfeld, Professor of English, The University of Connecticut, 1984.

[4-13]   M. Orchin and H. H. Jaffe, *Symmetry, Orbitals, and Spectra (S.O.S.)*, Wiley-Interscience, New York, 1971.

5 Molecular Vibrations

Vibration is a special kind of motion: the atoms of every molecule are permanently changing their relative positions at every temperature (even at absolute zero) without changing the position of the molecular center of mass. In terms of the molecular geometry these vibrations amount to continuously changing bond lengths and bond angles. Symmetry considerations will be applied to the molecular vibrations in this chapter following primarily Refs. [5-1 to 5-3]. Our brief discussion is only an indication of yet another important application of symmetry considerations. The mentioned references and two other fundamental monographs [5-4, 5-5] on vibrational spectroscopy are suggested for further reading. Our primary concern will be to examine in simple terms the following question. What kind of information can be deduced about the internal motion of the molecule from the mere knowledge of its point-group symmetry?

5.1 Normal Modes

The seemingly random motion of molecular vibrations can always be decomposed into the sum of relatively simple components, called *normal modes of vibration*. Each of the normal modes is associated with a certain frequency. Thus for a normal mode every atom of the molecule moves with the same frequency and in phase. Three characteristics will be examined for normal vibrations: their number, their symmetry, and their type.

5.1.1 Their Number

Since vibration is only one of the possible forms of motion, it has to be separated from the others, translation and rotation. Consider first a single atom. Its motion can be characterized by

Figure 5-1.
Three motional degrees of
freedom of an atom.

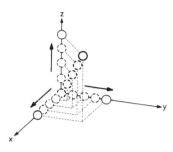

the three Cartesian coordinates of its instantaneous position as
shown in Fig. 5-1. In other words, the atom has three *degrees of
motional freedom*. Consider next a diatomic molecule. It will have
2 x 3 = 6 degrees of freedom. We might think again that the
three Cartesian coordinates of each atom describe the motion
of the molecule in space. However, this is not quite so. Since the
two atoms are not independent from each other, they must
move together in space. This means that three degrees of freedom will account, altogether, for the *translation* of a diatomic
molecule (see Fig. 5-2) – or of any polyatomic molecule, for that
matter. Two other degrees of freedom describe the *rotation* of
the diatomic molecule around the center of mass (see Fig. 5-3a).
The rotation around the z axis (Fig. 5-3b) need not be considered as it is the axis of the molecule and the rotation around it
does not change the position of the molecule.

Figure 5-2.
The three translational
degrees of freedom of a diatomic molecule.

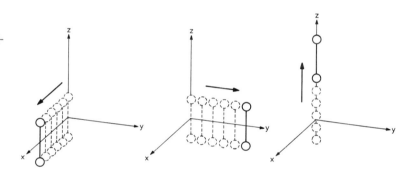

Figure 5-3.
Rotation of a diatomic molecule.
(a) Two rotational degrees of
freedom.

(b) Rotation around the molecular axis does not change the
position of the molecule.

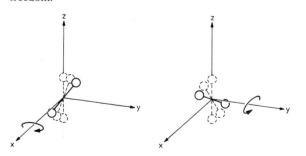
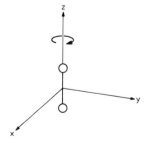

Thus of the six degrees of freedom five have been accounted
for. The sixth will describe the movement of the two atoms relative to each other without changing the center of the mass. This
is the *vibration* of the molecule.

The complete nuclear motion of an N atomic molecule can be described with $3N$ parameters; that is an N atomic molecule has $3N$ degrees of freedom. The translation of a molecule can always be described by 3 parameters. The rotation of a diatomic or any linear molecule will be described by 2 parameters and the rotation of a non-linear molecule by 3 parameters. This means that there are always 3 translational and 3 (for linear molecules 2) rotational degrees of freedom. The remaining $3N-6$ (for the linear case $3N-5$) degrees of freedom account for the vibrational motion of the molecule. They give the number of normal vibrations.

The translational and rotational degrees of freedom which do not change the relative positions of the atoms in the molecule, are often called *nongenuine modes*. The remaining $3N-6$ (or $3N-5$) degrees of freedom are called genuine vibrations or *genuine modes*.

5.1.2 Their Symmetry

The close relationship between symmetry and vibration is expressed by the following rule: *Each normal mode of vibration forms a basis for an irreducible representation of the point group of the molecule.*

Let us use the water molecule to illustrate the above statement. The normal modes of this molecule are shown in Fig. 5-4. The point group is C_{2v} and the character table is given in Table 5-1. It is seen that all operations bring v_1 and v_2 into themselves so their characters will be:

$$\begin{array}{ccccc} \Gamma_{v_1} & 1 & 1 & 1 & 1 \\ \Gamma_{v_2} & 1 & 1 & 1 & 1. \end{array}$$

Figure 5-4.
Normal modes of vibration for the water molecule. The lengths of the arrows indicate the relative displacements of the atoms.

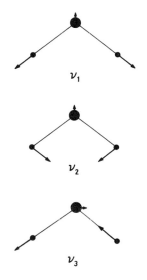

v_1

v_2

v_3

Table 5-1. The C_{2v} Character Table

C_{2v}	E	C_2	$\sigma_v(xz)$	$\sigma_v{'}(yz)$		
A_1	1	1	1	1	z	x^2, y^2, z^2
A_2	1	1	−1	−1	R_z	xy
B_1	1	−1	1	−1	x, R_y	xz
B_2	1	−1	−1	1	y, R_x	yz

The behavior of the third normal mode, v_3, is different. While E and σ' leave it unchanged, both C_2 and σ bring it into its negative self: each atom moves in the opposite direction after the operation. This means that v_3 is antisymmetric to these operations. The characters are:

$$\Gamma_{v_3} \quad 1 \quad -1 \quad -1 \quad 1.$$

Looking at the C_{2v} character table we can say that v_1 and v_2 belong to the totally symmetric irreducible representation A_1, and v_3 belongs to B_2.

It was easy to determine the symmetry of the normal modes of the water molecule because we already knew their forms. Can the symmetry of the normal modes of a molecule be determined without any previous knowledge of the actual forms of the normal modes? The answer is fortunately yes. From the symmetry group of the molecule the symmetry species of the normal modes can be determined without any additional information.

First an appropriate basis set has to be found. Considering that a molecule has $3N$ degrees of motional freedom, a system of $3N$ so-called *Cartesian displacement vectors* is a convenient choice. A set of such vectors is shown in Fig. 5-5 for the water molecule. A separate Cartesian coordinate system is attached to each atom of the molecule, with the atoms at the origin. The orientation of the axes is the same in each system. Any displacement of the atoms can be expressed by a vector and in turn this vector can be expressed as the vector sum of the Cartesian displacement vectors.

Next, the set of Cartesian displacement vectors is used as a basis for the representation of the point group. As discussed in Chapter 4, the vectors connected with atoms that change their position during an operation will not contribute to the character and thus they can be ignored.

Continuing with the water molecule as an example, the basis of the Cartesian displacement vectors will consist of 9 vectors (see Fig. 5-5). Operation E brings all of them into themselves, and the character is 9. Operation C_2 changes the position of the two hydrogen atoms, so only the three coordinates of the oxygen atom have to be considered. The corresponding block of the matrix representation is:

Figure 5-5.
Cartesian displacement vectors as basis for a representation of the water molecule.

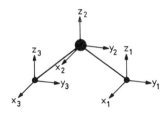

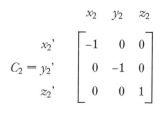

$$C_2 = \begin{array}{c} \\ x_2' \\ y_2' \\ z_2' \end{array} \begin{array}{ccc} x_2 & y_2 & z_2 \\ \begin{bmatrix} -1 & 0 & 0 \\ 0 & -1 & 0 \\ 0 & 0 & 1 \end{bmatrix} \end{array}$$

The character is: $(-1) + (-1) + 1 = -1$.

The next operation is σ. Again only the oxygen coordinates have to be considered. Reflection through the xz plane leaves x_2 and z_2 unchanged and brings y_2 into $-y_2$. The character is $1 + 1 + (-1) = 1$.

Finally, operation σ' leaves all three atoms in their place so all the 9 coordinates have to be taken into account. Reflection through the yz plane leaves all y and z coordinates unchanged and takes all x coordinates into their negative selves. The character will be:

$(-1) + 1 + 1 + (-1) + 1 + 1 + (-1) + 1 + 1 = 3$.

The representation is:

$$\Gamma_{\text{tot}} \quad 9 \quad -1 \quad 1 \quad 3.$$

This is, of course, a reducible representation. Reduce it now with the reduction formula (see Chapter 4):

$$a_{A_1} = (1/4) \ (1 \cdot 9 \cdot 1 + 1 \cdot (-1) \cdot 1 + 1 \cdot 1 \cdot 1 + 1 \cdot 3 \cdot 1) =$$
$$= (1/4) \ (9 - 1 + 1 + 3) = 3$$
$$a_{A_2} = (1/4) \ (1 \cdot 9 \cdot 1 + 1 \cdot (-1) \cdot 1 + 1 \cdot 1 \cdot (-1) + 1 \cdot 3 \cdot (-1)) =$$
$$= (1/4) \ (9 - 1 - 1 - 3) = 1$$
$$a_{B_1} = (1/4) \ (1 \cdot 9 \cdot 1 + 1 \cdot (-1) \cdot (-1) + 1 \cdot 1 \cdot 1 + 1 \cdot 3 \cdot (-1)) =$$
$$= (1/4) \ (9 + 1 + 1 - 3) = 2$$
$$a_{B_2} = (1/4) \ (1 \cdot 9 \cdot 1 + 1 \cdot (-1) \cdot (-1) + 1 \cdot 1 \cdot (-1) + 1 \cdot 3 \cdot 1) =$$
$$= (1/4) \ (9 + 1 - 1 + 3) = 3$$

The representation reduces to:

$$\Gamma_{\text{tot}} = 3A_1 + A_2 + 2B_1 + 3B_2.$$

These nine irreducible representations correspond to the nine motional degrees of freedom of the three-atomic water molecule. To obtain the symmetry of the genuine vibrations, the irreducible representations of the translational and rotational motion have to be separated. This can be done using some considerations described in Chapter 4. The translational motion always belongs to those irreducible representations where the three coordinates, x, y and z, belong. Rotations belong to the irreducible representations of the point group indicated by R_x, R_y, and R_z in the third area of the character tables. In the C_{2v} point group

$$\Gamma_{\text{tran}} = A_1 + B_1 + B_2$$

and

$$\Gamma_{\text{rot}} = A_2 + B_1 + B_2.$$

Subtracting these from the representation of the total motion, we get

$$
\begin{array}{lllllllll}
\Gamma_{\text{tot}} & = & 3A_1 & + & A_2 & + & 2B_1 & + & 3B_2 \\
-(\Gamma_{\text{tran}} & = & A_1 & & & + & B_1 & + & B_2) \\
-(\Gamma_{\text{rot}} & = & & & A_2 & + & B_1 & + & B_2) \\
\hline
\Gamma_{\text{vib}} & = & 2A_1 & & & & & + & B_2
\end{array}
$$

Thus, of the three normal modes of water, two will have A_1 and one B_2 symmetry. Let us stress again: this information could be derived purely from the molecular point-group symmetry.

5.1.3 Their Types

The normal modes can usually – though not always – be associated with a certain kind of motion. Those connected mainly with changes in bond lengths are the *stretching modes*. The ones connected mainly with changes of bond angles are the *deformation modes*. These may be mainly either in-plane or out-of-plane deformation modes. The simplest deformation mode is the *bending mode*.

Examine now the symmetries of these different types of vibration. For this purpose a new type of basis set is used. Since we are interested in the changes of the geometrical parameters, these changes are an obvious choice for basis set. The geometrical parameters are also called *internal coordinates,* and the basis is the displacement of these internal coordinates.

Let us continue with the water molecule and determine the symmetry of its stretching modes. The molecule has two O-H bonds, so the basis will be the *changes* of these O-H bonds. The representation of this basis set is

$$\Gamma_{\text{str}} \quad 2 \quad 0 \quad 0 \quad 2$$

and with inspection of the C_{2v} character table we see that it reduces to $A_1 + B_2$. This means that the stretching of the O-H bonds contributes to the normal modes of A_1 and B_2 symmetry (we shall later see that these are the symmetric and anti-symmetric stretches, respectively).

The third internal coordinate which can be considered in the water molecule is the bond angle, H-O-H. Its change will be the bending mode. All symmetry operations leave this basis unchanged, so the representation is:

$$\Gamma_{\text{bend}} \quad 1 \quad 1 \quad 1 \quad 1$$

and it belongs to the totally symmetric representation, A_1. What can we conclude? B_2 appears only in the stretching mode, so the B_2 normal mode will be a pure stretching mode. The A_1 symmetry mode, however, appears in both the stretching and the bending mode. At this point we cannot say whether one of the A_1 normal modes will be purely stretch and the other purely bend or they will be a mixture. This depends on the energy of these vibrations. If they are energetically close they can mix extensively. If they are separated by a large energy difference, they will not mix. In the case of H_2O, for example, the two A_1 symmetry modes are quite well separated while in Cl_2O they are completely mixed.

Modes of different symmetry never mix, even if they are close in energy. (This is a general rule which will have its analogous version for the transitions among electronic states as will be seen later in Chapters 6 and 7.)

The above analysis of the types of normal modes brings us to the limit where simple symmetry considerations can take us. Nothing yet has been said about the pictorial manifestation of the various normal modes. Above we deduced, e.g., that the B_2 normal mode of the water molecule is a pure stretch. The question may also be asked, how does it look? This question can be answered with the help of *symmetry coordinates*.

5.2 Symmetry Coordinates

The symmetry coordinates are symmetry adapted linear combinations of the internal coordinates. They always transform as one or another irreducible representation of the molecular point group.

The symmetry coordinates can be generated from the internal coordinates by the use of the projection operator introduced in Chapter 4. Both the symmetry coordinates and the normal modes of vibration belong to an irreducible representation of the point group of the molecule. A symmetry coordinate is always associated with one or another type of internal coordinates – that is pure stretch, pure bend, etc., – whereas a normal mode can be a mixture of different internal coordinate changes of the same symmetry. In some cases, as in H_2O, the symmetry coordinates are good representations of the normal vibrations. In other cases they are not. An example for the latter is Au_2Cl_6 [5-1] where the pure symmetry coordinate vibrations would be close in energy, so the real normal vibrations are mixtures of the different vibrations of the same symmetry type. The relationship between the symmetry coordinates and the normal vibrations can be established only by calculations called normal coordinate analysis [5-5, 5-6]. These calculations necessitate further data in addition to the knowledge of molecular symmetry, and are not pursued here.

Return now to the symmetry coordinates of the water molecule. They can be generated using the projection operator. As has been mentioned before, here we are interested only in the symmetry aspects of the symmetry coordinates. Thus the numerical factors are omitted and normalization is not considered. First work out the symmetry coordinate involving the stretching vibrations:

$$\hat{P}^{A_1} \Delta r_1 \approx 1 \cdot E \cdot \Delta r_1 + 1 \cdot C_2 \cdot \Delta r_1 + 1 \cdot \sigma \cdot \Delta r_1 + 1 \cdot \sigma' \cdot \Delta r_1 =$$
$$= \Delta r_1 + \Delta r_2 + \Delta r_2 + \Delta r_1 \approx \Delta r_1 + \Delta r_2$$
$$\hat{P}^{B_2} \Delta r_1 \approx 1 \cdot E \cdot \Delta r_1 + (-1) \cdot C_2 \cdot \Delta r_1 + (-1) \cdot \sigma \cdot \Delta r_1 +$$
$$+ 1 \cdot \sigma' \cdot \Delta r_1 =$$
$$= \Delta r_1 - \Delta r_2 - \Delta r_2 + \Delta r_1 \approx \Delta r_1 - \Delta r_2.$$

The same procedure is presented pictorially in Fig. 5-6. The bending mode of the water molecule stands alone so it will be a symmetry coordinate by itself:

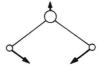

Since the symmetry coordinates of water are good approximations of the normal vibrations, the pictorial representations are applicable to them as well. Indeed, the three normal modes of Fig. 5-4 are the same as the symmetry coordinates we just derived. The A_1 symmetry stretching mode is called the symmetric stretch while the B_2 mode is the antisymmetric stretch.

Figure 5-6.
Generation of the symmetry coordinates representing bond stretching for H_2O.

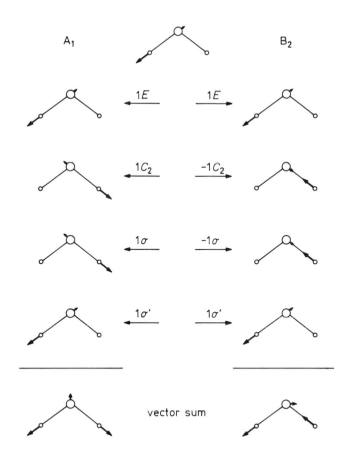

A_1

$1E$ $\quad$ $1E$

B_2

$1C_2$ $\quad$ $-1C_2$

1σ $\quad$ -1σ

$1\sigma'$ $\quad$ $1\sigma'$

vector sum

5.3 Selection Rules

The vibrational wave function, as any wave function, must form a basis for an irreducible representation of the molecular point group [5-2].

The total vibrational wave function, ψ_v, can be written as the product of the wave functions $\psi_i(n_i)$, where ψ_i is the wave function of the ith normal vibration ($i = 1$ through m) in the nth state.

$$\psi_v = \psi_1(n_1) \cdot \psi_2(n_2) \cdot \psi_3(n_3) \ldots\ldots \psi_m(n_m).$$

In general at any time each of the normal modes may be in any state. There is, however, a situation when all the normal modes are in their ground states and only one of them gets excited into the first excited state. Such a transition is called a *fundamental transition.* The intensity of the fundamental transitions is much higher than the intensity of the other kinds of transitions.[1] Therefore, these are of particular interest.

The vibrational wave function of the ground state belongs to the totally symmetric irreducible representation of the point group of the molecule [5-2]. The wave function of the first excited state will belong to the irreducible representation to which the normal mode undergoing the particular transition belongs.

A fundamental transition will occur only if one of the following integrals has nonzero value:

$$\langle \psi_v^{\,\circ} \,|\, x \,|\, \psi_v^{\,i} \rangle$$
$$\langle \psi_v^{\,\circ} \,|\, y \,|\, \psi_v^{\,i} \rangle$$
$$\langle \psi_v^{\,\circ} \,|\, z \,|\, \psi_v^{\,i} \rangle.$$

Here $\psi_v^{\,\circ}$ is the total vibrational wave function for the ground state, $\psi_v^{\,i}$ is the total vibrational wave function for the first excited state referring to the ith normal mode and x, y, and z are Cartesian coordinates.

The condition for an integral of product functions to have a nonzero value was given in Chapter 4. For the vibrational transitions this condition can be expressed in the following way:

$$\Gamma \psi_v^{\,\circ} \cdot \Gamma \psi_v^{\,i} \subset \Gamma_x \text{ or}$$
$$\Gamma \psi_v^{\,\circ} \cdot \Gamma \psi_v^{\,i} \subset \Gamma_y \text{ or}$$
$$\Gamma \psi_v^{\,\circ} \cdot \Gamma \psi_v^{\,i} \subset \Gamma_z.$$

The considerations on the symmetries of the ground and excited states and the above conditions lead to the selection rule for infrared spectroscopy: *A fundamental vibration will be infrared active if the corresponding normal mode belongs to the same irreducible representation as one or more of the Cartesian coordinates.*

The selection rule for Raman spectroscopy can also be derived by similar reasoning. It says: *A fundamental vibration will be Raman active if the normal mode undergoing the vibration belongs to the same irreducible representation as one or more of the components of the polarizability tensor of the molecule.* These components are the quadratic functions of the Cartesian coordinates given in the fourth area of the character tables. The Cartesian coordinates themselves are given in the third area. Thus the symmetry of the normal modes of a molecule is sufficient information to tell what transitions will be infrared and what transitions will be Raman active. The normal modes of the water molecule belong to the A_1 and B_2 irreducible representa-

---

1 Were the vibrations strictly harmonic, only fundamental transitions would be observable.

tion of the C_{2v} point group. Using merely the C_{2v} character table it is deduced that all three vibrational modes will be active in both the infrared and Raman spectra.

Since a particular normal mode may belong to different symmetry species in different point groups, its behavior depends strongly on the molecular symmetry. Just to mention one example, the v_1 symmetric stretching mode of an AX_3 molecule is not infrared active if the molecule is planar (D_{3h}). It is infrared active, however, if the molecule is pyramidal (C_{3v}). Vibrational spectroscopy is obviously one of the best experimental tools to determine the symmetry of molecules.

5.4 Examples

The utilization of the symmetry rules in the description of molecular vibrations will be further illustrated by a few examples.

Diimide, HNNH. This molecule belongs to the C_{2h} point group (see Fig. 4-7). The number of atoms is 4, so the number of normal vibrations is $3 \cdot 4 - 6 = 6$.

Our first task is to generate the representation of the Cartesian displacement vectors of the four atoms of the molecule (see Fig. 4-8a – 4-8c). As was shown in Chapter 4, the representation is:

$$\Gamma_{\text{tot}} \quad 12 \quad 0 \quad 0 \quad 4.$$

The reduction of this representation is also given in Chapter 4. The result is:

$$\Gamma_{\text{tot}} = 4A_g + 2B_g + 2A_u + 4B_u.$$

These 12 irreducible representations account for the 12 degrees of motional freedom of HNNH. Subtracting the irreducible representations corresponding to the translation and rotation of the molecule (see C_{2h} character table, Table 5-2) leaves us the symmetry species of the normal modes of vibration:

$$
\begin{aligned}
\Gamma_{\text{tot}} &= 4A_g + 2B_g + 2A_u + 4B_u \\
-(\Gamma_{\text{tran}} &= A_u + 2B_u) \\
-(\Gamma_{\text{rot}} &= A_g + 2B_g) \\
\hline
\Gamma_{\text{vib}} &= 3A_g + A_u + 2B_u
\end{aligned}
$$

Next we will see what kind of internal coordinate changes can account for each of these normal modes. There will have to be two N-H stretching modes and one N-N stretching mode. For deformation modes the two N-N-H angle bending modes are obvious choices, and they will be in-plane deformation modes. These constitute five normal vibrations so one is left to be accounted for. In deciding the nature of this normal mode, inspection of the character table may help. Of the above three different kinds of irreducible representations, A_g and B_u are symmetric with respect to σ_h so they must be vibrations within the molecular plane. The five vibrational modes suggested above then account for $3A_g + 2B_u$. The remaining A_u normal mode, however, is antisymmetric with respect to σ_h, so it must involve out-of-plane motion. Consequently, this normal mode will be an out-of-plane deformation mode.

Work out next the representations of the internal coordinates. The representation of the two N-H distance changes has been given in Chapter 4. This and the other representations are all shown in Table 5-2, together with the C_{2h} character table. The Γ_{NH} representation has been reduced to $A_g + B_u$ in Chapter 4. The reduction of the Γ_{NNH} representation is the same. Both the N-N stretching and the out-of-plane deformation are already irreducible representations by themselves. Since both A_g and B_u occur more than once, we cannot tell without calculation whether there will be three pure A_g modes, one N-H stretch, one N-N stretch and one N-N-H bend, or each of the three A_g modes will be a mixture of these three vibrations. The

Table 5-2. The C_{2h} Character Table and the Representations of the Internal Coordinates of Diimide.

C_{2h}	E	C_2	i	σ_h		
A_g	1	1	1	1	R_z	x^2, y^2, z^2, xy
B_g	1	–1	1	–1	R_x, R_y	xz, yz
A_u	1	1	–1	–1	z	
B_u	1	–1	–1	1	x, y	
Γ_{NH}	2	0	0	2	$= A_g + B_u$	
Γ_{NN}	1	1	1	1	$= A_g$	
Γ_{NNH}	2	0	0	2	$= A_g + B_u$	
Γ_{HNNH} [a]	1	1	–1	–1	$= A_u$	

[a] out-of-plane deformation mode

same is true for the B_u symmetry normal vibrations. They will be either pure N-H antisymmetric stretching and N-N-H bending modes or their mixtures. The only unambiguous assignment is that the A_u symmetry normal mode will be the out-of-plane deformation mode.

Generate now the symmetry coordinates of HNNH by means of the projection operator (α is the N-N-H angle):

$$\hat{P}^{A_g} \Delta r_1 \approx 1 \cdot E \cdot \Delta r_1 + 1 \cdot C_2 \cdot \Delta r_1 + 1 \cdot i \cdot \Delta r_1 + 1 \cdot \sigma_h \cdot \Delta r_1 =$$
$$= \Delta r_1 + \Delta r_2 + \Delta r_2 + \Delta r_1 \approx \Delta r_1 + \Delta r_2$$
$$\hat{P}^{B_u} \Delta r_1 \approx 1 \cdot E \cdot \Delta r_1 + (-1) \cdot C_2 \cdot \Delta r_1 + (-1) \cdot i \cdot \Delta r_1 + 1 \cdot \sigma_h \cdot \Delta r_1 =$$
$$= \Delta r_1 - \Delta r_2 - \Delta r_2 + \Delta r_1 \approx \Delta r_1 - \Delta r_2$$
$$\hat{P}^{A_g} \Delta \alpha_1 \approx 1 \cdot E \cdot \Delta \alpha_1 + 1 \cdot C_2 \cdot \Delta \alpha_1 + 1 \cdot i \cdot \Delta \alpha_1 + 1 \cdot \sigma_h \cdot \Delta \alpha_1 =$$
$$= \Delta \alpha_1 + \Delta \alpha_2 + \Delta \alpha_2 + \Delta \alpha_1 \approx \Delta \alpha_1 + \Delta \alpha_2$$
$$\hat{P}^{B_u} \Delta \alpha_1 \approx 1 \cdot E \cdot \Delta \alpha_1 + (-1) \cdot C_2 \cdot \Delta \alpha_1 + (-1) \cdot i \cdot \Delta \alpha_1 + 1 \cdot \sigma_h \cdot \Delta \alpha_1 =$$
$$= \Delta \alpha_1 - \Delta \alpha_2 - \Delta \alpha_2 + \Delta \alpha_1 \approx \Delta \alpha_1 - \Delta \alpha_2.$$

The same procedure is depicted in Fig. 5-7. The forms of the symmetry coordinates of HNNH are shown in Fig. 5-8. They might approximate well the normal modes of the molecule, and again, they might not.

Figure 5-7.
Generation of some symmetry coordinates of HNNH. (a) Symmetry coordinates corresponding to N-H bond stretches.

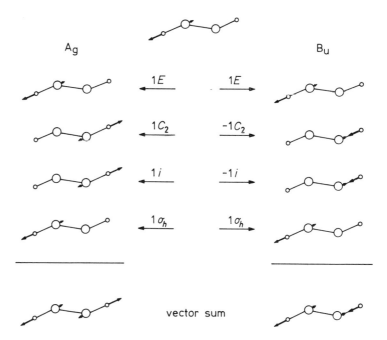

213

(b) Symmetry coordinates representing in-plane deformation.

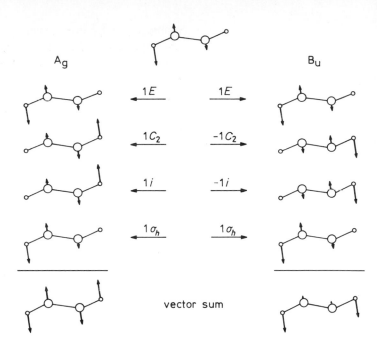

A_g B_u

1E   1E

1C_2 $-1C_2$

1i $-1i$

1σ_h 1σ_h

vector sum

Figure 5-8.
Symmetry coordinates for the HNNH molecule.

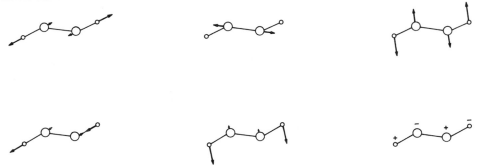

Finally, let us decide which normal mode will be infrared active and which one will be Raman active. The Cartesian coordinates belong to the A_u and B_u irreducible representation of the C_{2h} point group, while their binary products to A_g and B_g. Consequently, the selection rules are:

infrared active: A_u, B_u
Raman active: A_g.

This means that the A_g symmetry stretching modes and the A_g symmetry bending mode will be Raman active, while the B_u symmetry stretching and bending modes will be infrared active. Similarly, the A_u symmetry out-of-plane deformation mode will be infrared active.

Carbon dioxide, CO_2. The molecule is linear and belongs to the $D_{\infty h}$ point group. The number of atoms is 3, so the number of normal vibrations is: $3 \cdot 3 - 5 = 4$.

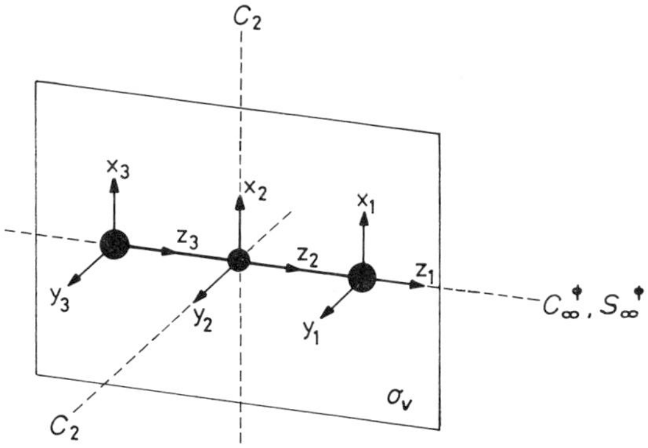

Figure 5-9.
Cartesian displacement vectors of CO_2.

The set of Cartesian displacement vectors as basis for a representation is shown in Fig. 5-9. The symmetry operations of the point group are also shown. The $D_{\infty h}$ character table is given in Table 5-3. Recall (Chapter 4) that the matrix of rotation by an angle Φ is

$$C^{\Phi} = \begin{bmatrix} \cos \Phi & \sin \Phi \\ -\sin \Phi & \cos \Phi \end{bmatrix}.$$

The rotation by an arbitrary angle Φ will leave the three z coordinates unchanged and the x and y coordinates mixed according to the above expression. The following matrix represents the C^{Φ} rotation:[2]

	x_1	y_1	z_1	x_2	y_2	z_2	x_3	y_3	z_3
x_1'	$c\Phi$	$s\Phi$	0	0	0	0	0	0	0
y_1'	$-s\Phi$	$c\Phi$	0	0	0	0	0	0	0
z_1'	0	0	1	0	0	0	0	0	0
x_2'	0	0	0	$c\Phi$	$s\Phi$	0	0	0	0
y_2'	0	0	0	$-s\Phi$	$c\Phi$	0	0	0	0
z_2'	0	0	0	0	0	1	0	0	0
x_3'	0	0	0	0	0	0	$c\Phi$	$s\Phi$	0
y_3'	0	0	0	0	0	0	$-s\Phi$	$c\Phi$	0
z_3'	0	0	0	0	0	0	0	0	1

---

2 cos is abbreviated by c and sin by s in the matrix.

Table 5-3. The $D_{\infty h}$ Character Table[a]

$D_{\infty h}$	E	$2C_{\infty}{}^{\Phi}$	...	$\infty \sigma_v$	i	$2S_{\infty}{}^{\Phi}$	...	∞C_2		
$\Sigma_g{}^{+}$	1	1	...	1	1	1	...	1		$x^2+y^2,\ z^2$
$\Sigma_g{}^{-}$	1	1	...	−1	1	1	...	−1	R_z	
Π_g	2	$2\,\mathrm{c}\Phi$	...	0	2	$-2\,\mathrm{c}\Phi$	...	0	$(R_x,\ R_y)$	$(xz,\ yz)$
Δ_g	2	$2\,\mathrm{c}2\Phi$	...	0	2	$2\,\mathrm{c}2\Phi$	...	0		$(x^2-y^2,\ xy)$
...	..	...	...	..	..	...	...	..		
$\Sigma_u{}^{+}$	1	1	...	1	−1	−1	...	−1	z	
$\Sigma_u{}^{-}$	1	1	...	−1	−1	−1	...	1		
Π_u	2	$2\,\mathrm{c}\Phi$	...	0	−2	$2\,\mathrm{c}\Phi$	...	0	$(x,\ y)$	
Δ_u	2	$2\,\mathrm{c}2\Phi$	...	0	−2	$-2\,\mathrm{c}2\Phi$	...	0		
...	..	...	...	..	..	...	...	..		

[a] c stands for cos

The character will be: $3 + 6 \cos \Phi$. The other relatively complicated operation is the mirror-rotation by an arbitrary angle, S^{Φ}. This operation means a rotation around the z axis by angle Φ, followed by reflection through the xy plane. This reflection interchanges the positions of the two oxygen atoms so they need not be considered. The block matrix of the S^{Φ} operation will be:

$$
\begin{array}{c}
\begin{array}{ccc} x_2 & y_2 & z_2 \end{array} \\
\begin{array}{c} x_2' \\ y_2' \\ z_2' \end{array}
\left[
\begin{array}{ccc}
\cos \Phi & \sin \Phi & 0 \\
-\sin \Phi & \cos \Phi & 0 \\
0 & 0 & -1
\end{array}
\right].
\end{array}
$$

The character is: $-1 + 2 \cos \Phi$.

Omitting the details of the determination of the remaining characters, the representation of the Cartesian displacement vectors is:

Γ_{tot} 9 $3 + 6 \cos \Phi$ 3   −3 $-1 + 2 \cos \Phi$ −1.

Subtract the characters of the translational and rotational representations. Remember that CO_2 is linear and the rotation around the molecular axis need not be taken into account.

Γ_{tot}	=	9	$3 + 6 \cos \Phi$	3	−3	$-1 + 2 \cos \Phi$	−1
$-(\Gamma_{\mathrm{tran}}$	=	3	$1 + 2 \cos \Phi$	1	−3	$-1 + 2 \cos \Phi$	−1)
$-(\Gamma_{\mathrm{rot}}$	=	2	$2 \cos \Phi$	0	2	$-2 \cos \Phi$	0)
Γ_{vib}	=	4	$2 + 2 \cos \Phi$	2	−2	$2 \cos \Phi$	0

The reduction formula cannot be applied to the infinite point groups (Chapter 4). Here inspection of the character table may help. Since $2 \cos \Phi$ at S^Φ appears with the Π_u irreducible representation, it is worth a try to subtract this one from Γ_{vib}:

$$
\begin{array}{rcccccc}
\Gamma_{vib} = & 4 & 2 + 2 \cos \Phi & 2 & -2 & 2 \cos \Phi & 0 \\
-(\Gamma_{\Pi_u} = & 2 & 2 \cos \Phi & 0 & -2 & 2 \cos \Phi & 0) \\
\hline
& 2 & 2 & 2 & 0 & 0 & 0
\end{array}
$$

This representation can be resolved as the sum of Σ_g and Σ_u.

$$
\begin{array}{rcccccc}
\Sigma_g = & 1 & 1 & 1 & 1 & 1 & 1 \\
\Sigma_u = & 1 & 1 & 1 & -1 & -1 & -1 \\
\hline
\Sigma_g + \Sigma_u = & 2 & 2 & 2 & 0 & 0 & 0
\end{array}
$$

Thus the normal modes of the CO_2 molecule will be:

$$\Gamma_{vib} = \Sigma_g + \Sigma_u + \Pi_u.$$

Since Π_u is a degenerate vibration, it counts as two and so we indeed have the four necessary normal vibrations.

The obvious choice for the three internal coordinate changes is the stretching of the two C=O bonds and the bending of the O=C=O angle. Using these as bases for representations we can build up the symmetry coordinates.

$$
\begin{array}{ccccccc}
\Gamma_{str} & 2 & 2 & 2 & 0 & 0 & 0
\end{array}
$$

We have already seen before that this representation reduces as $\Sigma_g + \Sigma_u$. The Π_u normal mode will correspond to the bending vibration.

Since each of the three symmetry species, Σ_g, Σ_u, and Π_u appears only once, the symmetry coordinates will be good representations of the normal modes. There is no possibility for mixing. Fig. 5-10 shows the forms of the normal vibrations of the CO_2 molecule. The two bending modes are degenerate; they are of equal energy.

Finally, apply the vibrational selection rules to CO_2:

infrared active: Σ_u, Π_u
Raman active: Σ_g.

Accordingly, the symmetric stretch C=O normal mode should appear in the Raman spectrum, while the antisymmetric stretch and the degenerate bend modes are expected to appear in the infrared.

Figure 5-10.
Normal modes of vibration of the CO_2 molecule.

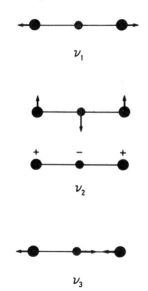

ν_1

ν_2

ν_3

References

[5-1]   D. C. Harris and M. D. Bertolucci, *Symmetry and Spectroscopy, An Introduction to Vibrational and Electronic Spectroscopy,* Oxford University Press, New York, 1978.

[5-2]   F. A. Cotton, *Chemical Applications of Group Theory,* Second Edition, Wiley-Interscience, New York, 1971.

[5-3]   M. Orchin and H. H. Jaffe, *Symmetry, Orbitals, and Spectra (S.O.S.),* Wiley-Interscience, New York, 1971.

[5-4]   G. Herzberg, *Infrared and Raman Spectra,* Van Nostrand Company, Princeton, NJ, 1959.

[5-5]   E. B. Wilson, Jr., J. C. Decius, and P. C. Cross, *Molecular Vibrations,* McGraw-Hill Book Co., New York, 1955.

[5-6]   K. Nakamoto, *Infrared Spectra of Inorganic and Coordination Compounds,* Second Edition, John Wiley and Sons, New York, 1970.

6 Electronic Structure of Atoms and Molecules

Everything that really counts in chemistry is related to the electronic structure of atoms and molecules. The formation of molecules from atoms, their behavior and reactivity all depend on their electronic structure. What is the role of symmetry? In various aspects of the electronic structure, symmetry can tell us a good deal; why certain bonds can form and others cannot, why certain electronic transitions are allowed and others are not, and why certain chemical reactions occur and others do not. Our discussion of these points is based primarily on some monographs listed in the references ([6-1] to [6-7]).

To describe the electronic structure, the electronic wave function $\psi(x,y,z,t)$ is used which depends in general on both space and time. Here, however, only its spatial dependence will be considered, $\psi(x,y,z)$. For detailed discussions of the nature of the electronic wave function we refer to texts on the principles of quantum mechanics ([6-1] to [6-3]). The physical meaning of the electronic wave function is expressed by the product of ψ with its complex conjugate ψ^*. This product $\psi \cdot \psi^*$ gives the probability of finding an electron in the volume $d\tau = dx\,dy\,dz$ about the point (x, y, z).

A many-electron system is described by a similar but multi-variable wave function

$$\psi(x_1,y_1,z_1,\ldots x_i,y_i,z_i,\ldots x_n,y_n,z_n).$$

The product $\psi \cdot \psi^*$ gives the probability of finding the first electron in $d\tau_1$, about the point (x_1,y_1,z_1), and the ith electron in $d\tau_i$ about the point (x_i, y_i, z_i), all at the same time.

The symmetry properties of the electronic wave function and the energy of the system are two determining factors in chemical behavior. The relationship between the wave function characterizing the behavior of the electrons and the energy of the system – atoms and molecules – is expressed by the Schrödinger equation. In its general and time-independent form, it is usually written as follows,

$$\hat{H}\psi = E\psi, \tag{6-1}$$

where $\hat{H}$ is the Hamiltonian operator and E is the energy of the system.

219

The Hamiltonian operator is an energy operator, which includes both kinetic and potential energy terms for all particles of the system. In our discussion only its symmetry behavior will be considered. With respect to the interchange of like particles (either nuclei or electrons) the *Hamiltonian must be unchanged* under a symmetry operation. A symmetry operation carries the system into an equivalent configuration, which is indistinguishable from the original. But, if nothing changes with the system, its energy must be the same before and after the symmetry operation. Thus the Hamiltonian of a molecule is *invariant* to any symmetry operation of the point group of the molecule. This means that it belongs to the totally symmetric representation of the molecular point group.

A fundamental property of the wave function is that it can be used as basis for irreducible representations of the point group of a molecule [6-4]. This property establishes the connection between the symmetry of a molecule and its wave function. The preceding statement follows from Wigner's theorem which says that all eigenfunctions of a molecular system belong to one of the symmetry species of the group [6-8].

In the expression of the energy of a system the following type of integral appears:

$$\int \psi_i \hat{H} \psi_j \, d\tau.$$

Depending on the system, ψ_i and ψ_j may be atomic orbitals used to construct molecular orbitals, or they may represent two different electronic states of the same atom or molecule, etc. The energy, then, expresses the extent of interaction between the two wave functions ψ_i and ψ_j. As was shown in Chapter 4, an integral will have a nonzero value only if the integrand is invariant to the symmetry operations of the point group, i.e., it belongs to the totally symmetric irreducible representation.

The above energy integral contains the $\hat{H}$ operator, which always belongs to the totally symmetric irreducible representation. Therefore, the symmetry of the whole integrand depends on the direct product of ψ_i and ψ_j. As also was shown in Chapter 4, the direct product of the representations of ψ_i and ψ_j belongs to the totally symmetric irreducible representation only if ψ_i and ψ_j belong to the same irreducible representation. Consequently, *the energy integral will be nonzero only if ψ_i and ψ_j belong to the same irreducible representation of the molecular point group*.

6.1 One-Electron Wave Function

Before discussing many-electron systems, the hydrogen atom (a one-electron system) will be described. This is essentially the only atomic system for which an exact solution of the wave function is available. The spherical symmetry of the hydrogen atom makes it convenient to express the wave func-

Figure 6-1.
The relationship of Cartesian coordinates and spherical polar coordinates.

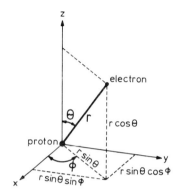

tion in a polar coordinate system. Such a system is shown in Fig. 6-1. Ignoring the translational motion of the hydrogen atom, the Schrödinger equation can be simplified as follows [6-5]:

$$\hat{H}_e \, \psi_e = E \, \psi_e \qquad (6\text{-}2)$$

where $\hat{H}_e$ depends only on the coordinates of the electron.

The electronic wave function can be represented as a product of a radial and an angular component:

$$\psi_e = R(r) \cdot A(\Theta, \Phi). \qquad (6\text{-}3)$$

The radial wave function $R(r)$ depends on two quantum numbers, n and l. The principal quantum number n determines the electron *shell*. The numbers $n = 1, 2, 3, 4,...$ correspond to the shells K, L, M, N, respectively. For the hydrogen atom n completely determines the energy of the shell, which is inversely proportional to n^2. Since this energy is negative, E is smallest for the first (K) shell, and increases with increasing n. The azimuthal quantum number l is associated with the total angular momentum of the electron and determines the shape of the *orbitals*. It may have integral values from 0 to n-1. The s, p, d, f,.... orbitals correspond to the azimuthal quantum numbers, $l = 0, 1, 2, 3,...$, respectively.

The angular wave function $A(\Theta, \Phi)$ depends also on two quantum numbers, l, and m_l. The magnetic quantum number, m_l, is associated with the component of angular momentum along

221

a specific axis in the atom. Since the hydrogen atom is spherically symmetrical, it is not possible to define a specific axis until the atom is placed in an external electric or magnetic field. This also means that the quantum number m_l has no effect on the energy and shape of the wave function of the hydrogen atom in the absence of such an external field. Generally, m_l may have values $-l$, $-l+1$,..., 0,..., $l-1$, l, altogether $2l+1$ of them, and the orbitals are subdivided accordingly.

For the one-electron wave functions, the most important orbitals are listed below:

n	l	m_l	shell	orbital	name
1	0	0	K	$1s$	$1s$
2	0	0	L	$2s$	$2s$
2	1	0		$2p$	$2p_z$
2	1	+1			$2p_x$
2	1	−1			$2p_y$
3	0	0	M	$3s$	$3s$
3	1	0		$3p$	$3p_z$
3	1	+1			$3p_x$
3	1	−1			$3p_y$
3	2	0		$3d$	$3d_{z^2}$
3	2	+1			$3d_{xz}$
3	2	−1			$3d_{yz}$
3	2	+2			$3d_{x^2-y^2}$
3	2	−2			$3d_{xy}$

Usually we refer to the energy of orbitals while what is really meant is the energy of an electron in that orbital. It was mentioned earlier that only the principal quantum number n influences the orbital energy in the hydrogen atom. This means that while $1s$ and $2s$ orbitals have different energies, the $2s$ and all three $2p$ orbitals have the same energy, i.e. these four $n = 2$ orbitals are degenerate.

In many-electron atoms the value of l also influences the energy of the orbitals; thus the $2s$ and $2p$ orbitals or the $3s$, $3p$, and $3d$ orbitals will no longer be degenerate. However, there are always three p orbitals and five d orbitals in each shell, and they differ only in the quantum number m_l, and these orbitals will be degenerate. As there are $2l+1$ values of m_l for an orbital with quantum number l, the p orbitals ($l=1$) will always be three-fold degenerate while the d orbitals ($l=2$) will always be five-fold degenerate.

Harris and Bertolucci [6-5] illustrated the relationship between symmetry and degeneracy of energy levels with a

Figure 6-2.
Illustration of the interrelation
of symmetry and degeneracy
after [6-5]. Used with permission.

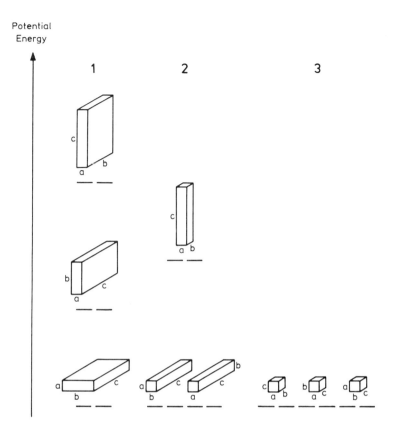

simple and attractive example. There are three parrallelepipeds in Fig. 6-2. Each of them has six stable resting positions. The potential energy of these positions depends on the height of the center of the mass above the supporting surface. This height, in turn, is determined by the choice of face on which the body rests. Three different positions are possible for the first parallelepiped (1) according to its three different kinds of faces. The potential energy of 1 will be largest when it stands on an *ab* face, since its center of mass is at the highest possible position. There are only two energetically different positions for 2 since its center of mass is at the same height when it rests on face *bc* or on face *ac*. Parallelepiped 3 is indeed a cube, and all possible positions will be energetically equivalent. Looking at the degeneracy of the most stable (lowest energy) position, it is two-fold degenerate for 1, four-fold degenerate for 2, and six-fold degenerate for the cube. Thus, with increasing symmetry, the degree of degeneracy increases. The connection between symmetry and degeneracy is strikingly obvious. *The greater the degree of symmetry, the smaller will be the number of different energy levels and the greater will be the degeneracy of these levels.*

This correlation between symmetry and degeneracy of energy levels is fundamental to understanding the electronic

223

structure of atoms and molecules. This relationship is valid not only when increasing symmetry renders the energy levels degenerate but also when energy levels are split as molecular symmetry decreases.

Let us return now to the wave function description of electronic structure. The separation of the wave function into two parts is convenient since these two parts relate to different properties. The radial part determines the energy of the system and is invariant to symmetry operations. The square of the radial function is related to probability. However, it has only significance for fixed values of the angular variables, Θ and Φ. These angular variables define a direction from the nucleus. The square of the radial function is proportional to the probability of finding the electron in a volume element along this direction. In order to determine the probability of finding the electron anywhere in a spherical shell surrounding the nucleus at a distance r from the nucleus, integration over both angular variables must be performed. The result is the radial distribution function.

Consider now the angular part of the one-electron wave function. It says nothing about the energy of the system but it can be altered by symmetry operations. Therefore we shall be dealing with this function in greater detail. The function $A(\Theta,\Phi)$ may have different signs ($+$ and $-$) in different spatial regions. A change in sign indicates a drastic change in the wave function. These signs might be thought of as signs of the amplitudes of the wave function; they certainly have nothing to do with electric charges. The places where the wave function changes sign are called *nodes*. The number of nodes is $n-1$, where n is the principal quantum number. Again, the squared function has physical significance, it is positive everywhere. The probability of finding an electron at a node is zero. However, as one proceeds in either direction from the nodes, the squared wave function has equal values relating to equal probabilities, to wit the probability of finding the electron on the "positive" or "negative" sides of the wave function is equal.

It generally helps to visualize and understand a problem in a pictorial way. However, since the wave function depends upon three variables, it can be represented only in four dimensions. To overcome this problem, symbolic representations are used to emphasize various properties of the wave function.

The angular wave function, $A(\Theta,\Phi)$, is shown for the H $1s$ and $2p_z$ orbitals in Fig. 6-3a. The H $1s$ orbital is positive everywhere, but the $2p_z$ orbital has one node, through which it changes sign. The $A^2(\Theta,\Phi)$ function is shown for the same orbitals in Fig. 6-3b. Their shape is similar to the shape of the $A(\Theta,\Phi)$ functions but they are positive everywhere. They represent the region in space where the electron can be found with a large probability (usually 90 % or more). The boundary

Figure 6-3.
Representations of the hydrogen $1s$ and $2p_z$ orbitals.
(a) Plot of the angular wave function, $A(\Theta,\Phi)$.

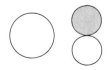

(b) Plot of the squared function, $A^2(\Theta,\Phi)$.

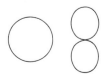

224

(c) Cross section of the squared total wave function, ψ^2, representing the electron density. Reprinted from [6-6] by permission of Harper & Row, Publishers, Inc. Copyright © 1981 by Thomas H. Lowry and Kathleen Schueller Richardson.

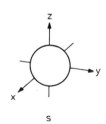

surface of this space is determined by the square of the angular function. It does not say anything, however, about the variation of the probability density within this surface. That information is contained in the radial distribution function. A way to illustrate the latter is shown for the $1s$ and $2p_z$ orbitals in Fig. 6-3c. This is a cross section of the electron density distribution. The varying amount of shading reflects the square of the radial function. Thus, this picture represents the squared total wave function, ψ^2. Rotating this picture around any axis for the $1s$ orbital and around the z axis for the $2p_z$ orbital would give the three-dimensional representation of the total wave function.

Whereas the square of the angular function has outstanding physical significance, the angular function itself contains valuable information regarding the symmetry properties of the wave function. These properties are lost in the squared angular function.

The well known shapes of the one electron orbitals are presented in Fig. 6-4; these are, in fact, representations of the

Figure 6-4.
Shapes of one-electron orbitals. They are representations of the angular wave function, $A(\Theta,\Phi)$.

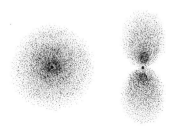

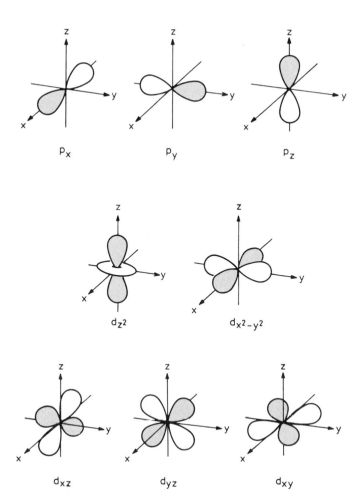

angular wave functions. Such representations are used commonly for illustrations because they describe accurately the symmetry properties of the wave function. In order to give the total wave function, however, they must be multiplied by an appropriate radial function. Another representation, shown in Fig. 6-5, is a three-dimensional computer drawing of the total function [6-9] including *both* the radial and the angular functions. These are still *not* real "pictures" of the orbitals, since they represent a cross section of the wave function in one plane only.

Figure 6-5.
Three-dimensional computer drawings of the total wave function, ψ. They show the values of ψ in a cross section. Reproduced with permission from [6-9]. Copyright (1973) Macmillan Publ. Co.
(a) $1s$ orbital.

(b) $2p_x$ orbital.

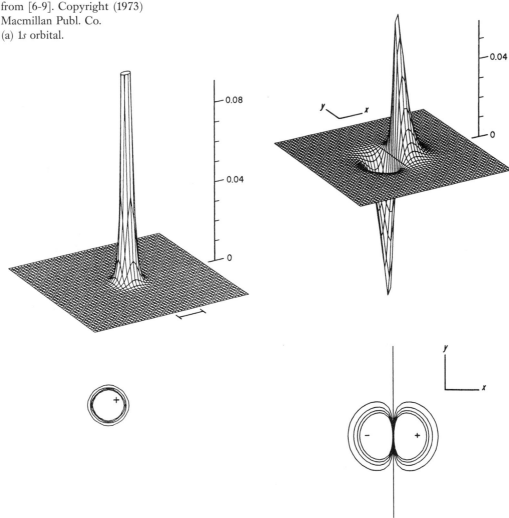

(c) $3d_{xy}$ orbital.

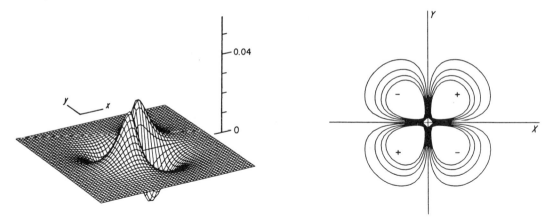

The vertical scale gives the value of ψ for each point in the xy plane. These diagrams show how the sign and magnitude of ψ varies in the xy plane, and they also help us to visualize the electronic wave function as a wave. On the other hand, they do not illustrate its symmetry properties as well as do the simple diagrams in Fig. 6-4.

As mentioned before, the symmetry properties of the one-electron wave function are shown by the simple plot of the angular wave function. But, what are the symmetry properties of an orbital and how can they be described? We can examine the behavior of an orbital under the different symmetry operations of a point group. This will be illustrated below via the inversion operation.

The s and d orbitals are transformed into themselves as the inversion operation is applied to them (Fig. 6-6). Both the magnitude and the "sign" of the wave function will remain the same under the inversion operation. These orbitals are said to

Figure 6-6.
The effect of inversion on the s and d orbitals. They are symmetric to this operation.

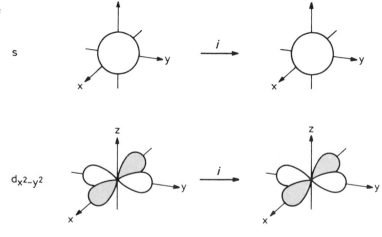

Figure 6-6. (continued)

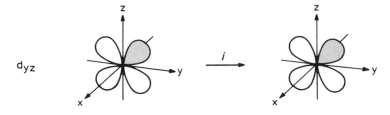

d_{yz}

Figure 6-7.
The inversion operation changes the sign of the p orbitals. They are antisymmetric to inversion.

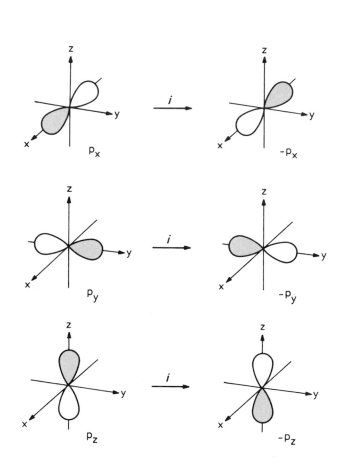

be *symmetric* with respect to inversion. The effect of the inversion operation on the p orbitals is demonstrated in Fig. 6-7. Whereas their magnitude does not change, their "sign" changes upon inversion. These orbitals are said to be *antisymmetric* with respect to inversion. In the character tables, this is indicated by $+1$ for symmetric and -1 for antisymmetric behavior under each symmetry operation. As mentioned in Chapter 4, the atomic orbitals always belong to the same irreducible representations of the given point group as do their subscripts (x, y, z, xy, x^2-y^2, etc.).

6.2 Many-Electron Atoms

There is interaction among all the electrons in a many-electron atom. Thus the wave function for even one electron in a many-electron system in principle will be different from the wave function for the one electron in the hydrogen atom. Since the electrons are mutually indistinguishable, it is not possible to describe rigorously the properties of a single electron in such a system. There is no exact solution to this problem, and approximate methods must be adopted.

In the most commonly utilized approximation, the many-electron wave functions are written in terms of products of one-electron wave functions similar to the solutions obtained for the hydrogen atom. These one-electron functions used to construct the many-electron wave function are called *atomic orbitals*. They are also called "hydrogen-like" orbitals since they are one-electron orbitals and also because their shape is similar to that of the hydrogen atom orbitals. Coulson referred to the atomic orbitals as "personal wave functions" [6-10] to emphasize that each electron is allocated to an individual orbital in this model.

At this point we can again appreciate the possibility of separating the total wave function into a radial and an angular wave function. The angular wave function does not depend on n and r, so it will be the same for every atom. This is why the "shapes" of atomic orbitals are always the same. Hence, symmetry operations can be applied to the orbitals of all atoms in the same way. The differences occur in the radial part of the wave function; the radial contribution depends on both n and r, and it determines the energy of the orbital which is, of course, different for different atoms.

Figure 6-8.
Radial distribution functions for the 1s, 2s, and 2p orbitals. Reproduced with permission from [6-3]. Copyright (1972) Pergamon Press.

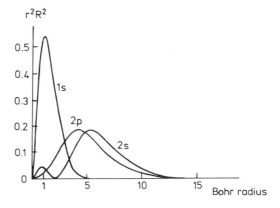

229

While the energy of a one-electron orbital depends only on *n*, in a many-electron atom the energy of the orbital is determined by both *n* and *l*. Thus, an electron in a 2*p* orbital has higher energy than an electron in a 2*s* orbital. This relationship is illustrated in Fig. 6-8 where the radial distribution functions of 1*s*, 2*s*, and 2*p* orbitals are displayed. The 2*s* orbital has a "bump" inside the 1*s* shell while the 2*p* orbital does not. This means that the probability of finding a 2*s* electron inside the 1*s* shell will be greater than that of finding a 2*p* electron. Consequently, the screening of the 2*p* electrons from the nuclear charge by the 1*s* electrons will be greater than will be the screening of the 2*s* electrons. As a result, the energy of the 2*s* orbital will be smaller than that of the 2*p* orbital.

The order of orbital energies in many-electron atoms is generally as follows:

$$1s < 2s < 2p < 3s < 3p < 4s \sim 3d < 4p < 5s < 4d < \ldots$$

There are some cases, however, when the order is changed somewhat. The 3*d* orbital, e.g., sometimes lies below the 4*s* orbital. A diagram which illustrates the order of orbital energies is shown in Fig. 6-9.

In addition to the three quantum numbers used to describe the one-electron wave function, the electron has also a fourth, the *spin quantum number*, m_s. It is related to the intrinsic angular momentum of the electron, called *spin*. This quantum number may assume the values of $+1/2$ or $-1/2$. Usually the sign of m_s is represented by arrows, (↑ and ↓), or by the Greek letters α and β. Thus for a many-electron atom the wave function of an orbital is expressed as

$$\psi_e = R(r) \cdot A(\Theta,\Phi) \cdot S(s) \qquad (6\text{-}4)$$

rather than as Eq. (6-3). However, the introduction of spin does not alter any of the properties discussed previously that relate to the shape and symmetry of the orbitals. The reason for this is that the spin function is independent of the spatial coordinates.

An important postulate in connection with the spin of the electron is called the *Pauli principle*. It states that if a system consists of identical particles with half integral spins, then all acceptable wave functions must be antisymmetric with respect to the exchange of the coordinates of any two particles. In our case, the particles are electrons, and the Pauli principle is formulated accordingly: No two electrons in an atom can have the same set of values for all four quantum numbers.

Figure 6-9.
The sequence of orbital energies.

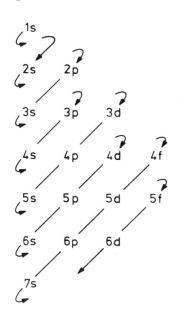

6.2.1 The Electronic Configuration of the Elements

The electronic configuration of an atom is called the atomic configuration and tells us how many electrons the atom has in its subshells. A *subshell* is a complete set of orbitals that have the same n and l. The first few elements in the Periodic Table build up their electronic configuration in the following way: Helium has two electrons with the same n, l, and m_l ($n = 1$, $l = 0$ and $m_l = 0$). The fourth quantum number must then be different, $m_s = +1/2$ for one electron and $m_s = -1/2$ for the other:

He   ↑↓
      $1s$

Continuing in this manner for the next three atoms:

Li   ↑↓   ↑
      $1s$ $2s$

Be   ↑↓   ↑↓
      $1s$ $2s$

B   ↑↓   ↑↓   ↑   __   __
      $1s$ $2s$ $2p$

For carbon, however, complication arises since different solutions can be written without violating the Pauli principle:

C   ↑↓   ↑↓   ↑↓   __   __
      $1s$ $2s$ $2p$
                               no unpaired spin

      ↑↓   ↑↓   ↑   ↑   __
      $1s$ $2s$ $2p$
                               two unpaired spins

      ↑↓   ↑↓   ↑   ↓   __
      $1s$ $2s$ $2p$
                               no unpaired spin

These three arrangements are equivalent electron configurations and are expressed as $1s^2 2s^2 2p^2$. However, they correspond to different electronic *states* wich have different energies. The state with the lowest energy is the ground state, and the other states are excited states.

"Hund's first rule" expresses that, for a given electronic configuration, the state with the greatest number of unpaired spins has the lowest energy. Thus, for carbon, the configuration with two unpaired spins will be the ground state. With the guidance of the Pauli principle and Hund's first rule, the buildup continues ("Aufbau principle"):

	1s	2s		2p	
N	↑↓	↑↓	↑ —	↑ —	↑ —
O	↑↓	↑↓	↑↓	↑ —	↑ —
F	↑↓	↑↓	↑↓	↑↓	↑ —
Ne	↑↓	↑↓	↑↓	↑↓	↑↓

An electron shell will be completely filled when the electron configuration of the inert gases is reached. It is customary to abbreviate the electron configuration of atoms by using the symbol of the preceding inert gas:

Li   [He] 2s
Be   [He] 2s^2
.
.
.
.
F   [He] 2$s^2 2p^5$
Ne   [He] 2$s^2 2p^6$ =   [Ne]
Na   [Ne] 3s
Mg   [Ne] 3s^2
.
.
.
.
Cl   [Ne] 3$s^2 3p^5$
Ar   [Ne] 3$s^2 3p^6$ =   [Ar]
K   [Ar] 4s
Ca   [Ar] 4s^2
.
.
.
.
Br   [Ar] 4$s^2 4p^5$

and so on.

There is a marked periodicity in these electron configurations which is the underlying idea of the Periodic Table. As the chemical properties of the atoms are determined by their electron configuration, the atoms with similar electron configurations will have similar chemical properties.

6.3 Molecules

6.3.1 Constructing Molecular Orbitals

In the discussion of the electronic structure of atoms, the Schrödinger equation could be reduced to one involving only the electrons. This was achieved by separating the electronic energy of the atom from the nuclear kinetic energy which is essentially determined by the translational motion of the atom.

Such a separation is exact for atoms. For molecules, only the translational motion of the whole system can be rigorously separated, while their kinetic energy includes all kinds of motion, vibration and rotation as well as translation. Here a two-step approximation can be introduced. First, as in the case of atoms, the translational motion of the molecule can be isolated. The remaining equation then describes the internal motion of the system. The second step is to apply the Born-Oppenheimer approximation. Since the relatively heavy nuclei move much more slowly than do electrons, the latter are assumed to move about a fixed nuclear arrangement. Accordingly, internal motion of the nuclei is also ignored in this approximation.

With these approximations the Hamiltonian of the molecular Schrödinger equation reduces to a form which depends only on the electron coordinates. Thus, the molecular wave function will be an electronic wave function.

As was emphasized before, a molecule is not simply a collection of its constituting atoms. Rather, it is a system of atomic nuclei and a common electron distribution. Nevertheless, in describing the electronic structure of molecules the most convenient way is to approximate the molecular electron distribution by the sum of atomic electron distributions. This approach is called the *linear combination of atomic orbitals* (LCAO) method. The orbitals produced by the LCAO procedure are called *molecular orbitals* (MOs). An important common property of the atomic and molecular orbitals is that both are one-electron wave functions. Combining a certain number of one-electron atomic orbitals, the same number of one-electron molecular orbitals are obtained. Finally, the total molecular wave function is the product or sum of products of the one-electron molecular orbitals. Thus, the final scheme is as follows:

One-electron atomic orbitals (AOs)

↓ LCAO

One-electron molecular orbitals (MOs)

↓ multiplication
(and summation)

Total molecular wave function

Figure 6-10.
Illustration by Ferenc Lantos.

Although both atomic orbitals and molecular orbitals are one-electron wave functions, the shape and symmetry of the molecular orbitals are different from those of the atomic orbitals of the isolated atom. The molecular orbitals extend over the entire molecule, and their spatial symmetry must conform to that of the molecular framework. Of course, the electron distribution is not uniform throughout the molecular orbital. In depicting these orbitals, usually only those portions are emphasized in which the electron density is substantial.

When constructing molecular orbitals from atomic orbitals, there may be a large number of possible linear combinations of the atomic orbitals. Many of these linear combinations, however, are unnecessary. Symmetry is instrumental as a criterion in choosing among them. The following statement is attributed to Michelangelo: "The sculpture is already there in the raw stone; the task of a good sculptor is merely to eliminate the unnecessary parts of the stone" (Fig. 6-10). In the LCAO procedure, the knowledge of symmetry eliminates the unnecessary linear combinations. All those linear combinations must be eliminated that do not belong to any irreducible representation of the molecular point group. The reverse of this statement constitutes the fundamental principle of forming the molecular orbitals: *Each possible molecular orbital must belong to an irreducible representation of the molecular point group.* Another, equally important rule for the construction of molecular orbitals is that only those atomic orbitals can form a molecular orbital which belong to the same irreducible representation of the molecular point group. This rule follows from the general theorem (see p. 220) about the value of an energy integral. This theorem can be restated for the special case of MO construction as follows: An energy integral will be nonzero only if the atomic orbitals used for the construction of molecular orbitals belong to the same irreducible representation of the molecular point group.

The atomic orbitals in an isolated atom possess spherical symmetry. When they are used for MO construction, however, their symmetry must be considered in the symmetry group of the particular molecule. When two atomic orbitals of the same symmetry form a molecular orbital, the symmetry of the molecular orbital will be the same as that of the component atomic orbitals.

In addition to complying with the symmetry rules, successful MO construction requires certain energy conditions. In order for two orbitals to interact appreciably, their energies cannot be too different.

The so-called overlap integral S_{ij} is a useful guide in constructing molecular orbitals. It is symbolized as

$$S_{ij} = \int \psi_i \cdot \psi_j \, d\tau \qquad (6\text{-}5)$$

where ψ_i and ψ_j are the two participating atomic orbitals. The physical meaning of S_{ij} is related to the measure of volume in which there is electron density contributed by both atoms i and j. The knowledge of the sign and magnitude of S_{ij} is especially instructive; they can be arrived at via the following considerations:

Positive overlap results from the combination of adjacent lobes that have the same sign. The electron density originating from both atoms will increase and concentrate in the region between the two nuclei. The resulting MO is a *bonding orbital*. Some typical bonding atomic orbital combinations are presented in Fig. 6-11. Two kinds of molecular orbitals are shown in

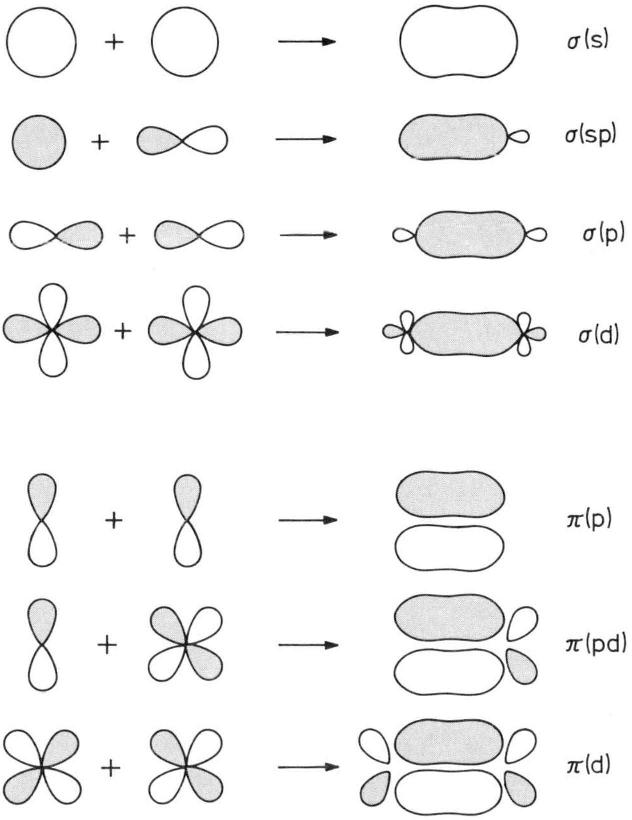

Figure 6-11.
Illustration of positive overlap between atomic orbitals. The result is a bonding orbital.
(a) σ orbitals.

(b) π orbitals.

235

this figure. A σ orbital is concentrated primarily along the internuclear axis. On the other hand, a π orbital has a nodal plane going through this axis, and its electron density is highest on either side of this nodal plane. Sigma orbitals are non-degenerate, while the π orbitals are always doubly degenerate.

Negative overlap occurs from the combination of adjacent lobes that have opposite sign. In such an instance, there will be no common electron density in the region between the two nuclei; instead, electron density will concentrate in the outside regions. Such an MO is an *antibonding orbital* and is illustrated in Fig. 6-12.

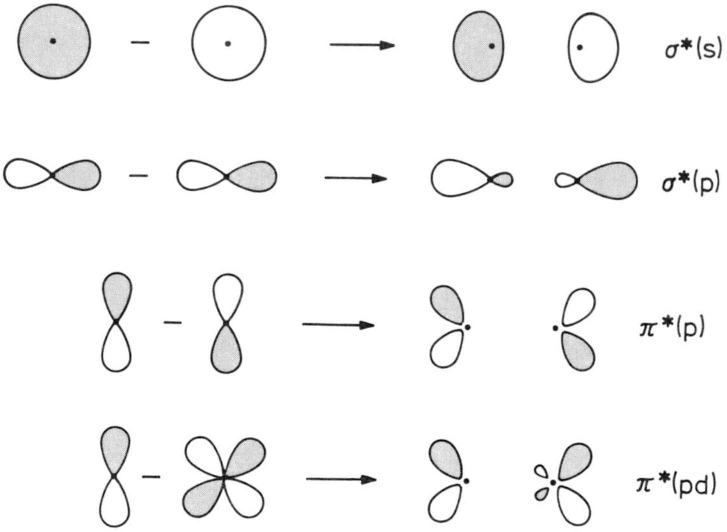

Figure 6-12.
Formation of antibonding orbitals by the combination of different lobes of atomic orbitals.
(a) σ antibonding orbitals.

(b) π antibonding orbitals.

Zero overlap means that there is no net interaction between the two atomic orbitals. They have both positive and negative overlaps that cancel each other. Some examples are shown in Fig. 6-13.

Figure 6-13.
Zero overlap between atomic orbitals. There is no net interaction.

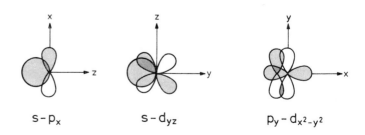

The energy changes in the formation of homonuclear and heteronuclear diatomic molecules are illustrated in Fig. 6-14.

Figure 6-14.
Energy changes during MO
formation.
(a) Homonuclear molecules.

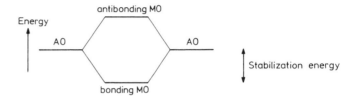

(b) Heteronuclear molecules.

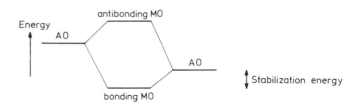

Figure 6-15.
Combination of the $2s$ and
$2p_x$ (or $2p_y$) atomic orbitals
does not result in a molecular
orbital.

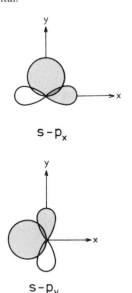

The energy of the bonding MO is smaller (larger negative value)
than is the energy of the interacting atomic orbitals. On the
other hand, the energy of the antibonding MO is larger than is
the energy of the interacting AOs. The largest energy changes
occur when the two participating AOs have equal energies. As
the energy difference between the participating AOs increases,
the stabilization of the bonding MO decreases. Molecular orbitals are not formed when the participating atomic orbitals
possess very different energies.

Thus, both symmetry and energy requirements must be fulfilled in order to form molecular orbitals. Energetically, the $2s$
and $2p$ atomic orbitals are sufficiently similar to form molecular
orbitals with each other. For symmetry reasons, however, the p_x
and p_y orbitals of one atom of a homonuclear diatomic molecule cannot combine with the $2s$ orbital of the other atom
because they belong to different irreducible representations
(see Fig. 6-15). On the other hand, $3d$ orbitals of first row transition metals often are prevented from forming molecular orbitals
with the ligand orbitals for energetic reasons despite their
matching symmetries. Quantum chemical calculations on transition metal dihydrides [6-11] support this suggestion.

Knowledge of the symmetry of the MOs is important for
practical reasons. The energy of the orbitals can be calculated
by costly quantum chemical calculations. The symmetry of the
molecular orbitals, on the other hand, can be deduced from the
molecular point group and using character tables, a process that
requires merely paper and pencil. Then, excluding all possible
solutions that are not allowed by symmetry, only the energies of
the remaining orbitals must be calculated.

We are, of course, concerned with the symmetry aspects of the MOs and their construction. As was discussed before, the degeneracy of AOs is determined by m_l. Thus, all p orbitals are three-fold degenerate, and all d orbitals are five-fold degenerate. The spherical symmetry of the atomic subshells, however, necessarily changes when the atoms enter the molecule, since the symmetry of molecules is nonspherical. The degeneracy of atomic orbitals will, accordingly, decrease; the extent of decrease will depend upon molecular symmetry.

Various methods (described in Chapter 4) can be used to determine the symmetry of atomic orbitals in the point group of a molecule, i.e., to determine the irreducible representation of the molecular point group to which the atomic orbitals belong. There are two possibilities depending on the position of the atoms in the molecule. For a central atom (like O in H_2O or N in NH_3), the coordinate system can always be chosen in a way such that the central atom lies at the intersection of all symmetry elements of the group. Consequently, each atomic orbital of this central atom will transform as one or another irreducible representation of the symmetry group. They will have the same symmetry properties as those basis functions in the third and fourth areas of the character table which are indicated in their subscripts. For all other atoms so-called "group orbitals" or "symmetry adapted linear combinations" (SALCs) must be formed from like orbitals. Several examples will illustrate how this is done.

First, however, consider the symmetry properties of the central atom orbitals. Take the C_{4v} point group as an example. Its character table is presented in Table 6-1. The p_z and d_{z^2}

Table 6-1. The C_{4v} Character Table

C_{4v}	E	$2C_4$	C_2	$2\sigma_v$	$2\sigma_d$		
A_1	1	1	1	1	1	z	$x^2+y^2,\ z^2$
A_2	1	1	1	−1	−1	R_z	
B_1	1	−1	1	1	−1		x^2-y^2
B_2	1	−1	1	−1	1		xy
E	2	0	−2	0	0	$(x, y)(R_x, R_y)$	(xz, yz)

atomic orbitals of the central atom will belong to the totally symmetric irreducible representation A_1; the $d_{x^2-y^2}$ orbital belongs to B_1 and d_{xy} to B_2. The symmetry properties of the (p_x, p_y) and (d_{xz}, d_{yz}) orbitals present a good opportunity for illustrating two-dimensional representations. Taking the three p orbitals as basis functions, the symmetry operations of the C_{4v}

point group are applied to them. This is shown in Fig. 6-16. The matrix representations are given here:

$$E = \begin{bmatrix} 1 & 0 & 0 \\ 0 & 1 & 0 \\ 0 & 0 & 1 \end{bmatrix}$$

$$C_4 = \begin{bmatrix} 0 & 1 & 0 \\ -1 & 0 & 0 \\ 0 & 0 & 1 \end{bmatrix} \quad C_4^{\,3} = \begin{bmatrix} 0 & -1 & 0 \\ 1 & 0 & 0 \\ 0 & 0 & 1 \end{bmatrix} \quad C_2 = \begin{bmatrix} -1 & 0 & 0 \\ 0 & -1 & 0 \\ 0 & 0 & 1 \end{bmatrix}$$

$$\sigma_v(xz) = \begin{bmatrix} 1 & 0 & 0 \\ 0 & -1 & 0 \\ 0 & 0 & 1 \end{bmatrix} \quad \sigma_v(yz) = \begin{bmatrix} -1 & 0 & 0 \\ 0 & 1 & 0 \\ 0 & 0 & 1 \end{bmatrix}$$

$$\sigma_d = \begin{bmatrix} 0 & 1 & 0 \\ 1 & 0 & 0 \\ 0 & 0 & 1 \end{bmatrix} \quad \sigma_{d'} = \begin{bmatrix} 0 & -1 & 0 \\ -1 & 0 & 0 \\ 0 & 0 & 1 \end{bmatrix}$$

All these matrices can be block-diagonalized into a 2x2 and a 1x1 matrix. The set of the 1x1 matrices corresponds to p_z and the set of the 2x2 size matrices corresponds to p_x and p_y. The representations are:

	E	$2C_4$	C_2	$2\sigma_v$	$2\sigma_d$	
p_z	1	1	1	1	1	A_1
(p_x, p_y)	2	0	-2	0	0	E

Notice that the operations C_4 and σ_d transform p_x into p_y, and *vice versa*. They cannot be separated from one another so they *together* belong to the two-dimensional representation E.

If two or more atomic orbitals are interrelated under a symmetry operation of the point group and, accordingly, they *together* belong to an irreducible representation, their energies will also be the same. In other words, these orbitals are *degenerate*. Such atomic orbitals are parenthesized in the character tables.

Figure 6-16.
The symmetry operations of
the C_{4v} point group applied to
the $2p$ orbitals.

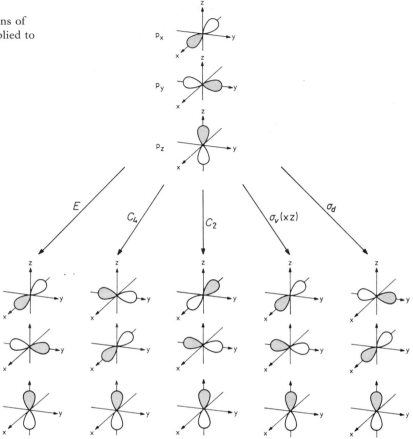

The direct connection between symmetry and degeneracy of the atomic orbitals is demonstrated here once again. The higher the symmetry of the molecule the greater will be the interrelation of the orbitals upon symmetry operations. Consequently, their energies become less and less distinguishable. The following example shows how the degeneracy of p orbitals decreases with decreasing symmetry:

free atom	spherical symmetry	(p_x, p_y, p_z)	three-fold degenerate
O_h point group	T_{1u}	(p_x, p_y, p_z)	three-fold degenerate
C_{4v} point group	A_1	p_z	non-degenerate
	E	(p_x, p_y)	two-fold degenerate
C_{2v} point group	A_1	p_z	
	B_1	p_x	non-degenerate
	B_2	p_y	

The degree of degeneracy of atomic orbitals always corresponds to the dimension of the irreducible representation to which these atomic orbitals belong. The same is true for molecular orbitals. Thus, knowing the symmetry of a molecule and looking at the character table, the maximum possible degeneracy of its molecular orbitals can be determined at once. The irreducible representation having the highest dimension will show this.

6.3.2 Electronic States

The concepts of orbitals and electronic configurations are useful descriptions. However, they are only models, and they employ approximations. The energy of an orbital has rigorous physical significance in systems that contain only a single electron. In many-electron systems, the energy of the orbitals loses its physical meaning, and only the energies of the (ground and excited) states is real. It is these states that are described by the total electronic wave functions. Electronic transitions, in fact, represent changes in the state of an atom or a molecule and not necessarily in the electronic configurations.

We shall not be concerned with the atomic states. The systematic way of determining them is given, for example, in [6-3] and [6-5]. Molecular states and the determination of their symmetries, however, will be briefly introduced [6-4].

First, let us consider the customary notations. Assume that a hypothetical ground state molecule of the C_{2v} point group has four electrons, two in an A_1 symmetry and two in a B_1 symmetry orbital. In short notation this can be written as $a_1^2 b_1^2$. An electron occupying an A_1 symmetry orbital is represented by a_1, the lower-case letter indicating that this is the symmetry of an *orbital* and not of an electronic state. If two electrons occupy an orbital, the notation is a_1^2. The symmetry of a state is represented by capital letters, just as are the irreducible representations.

The symmetry of the electronic states can be determined from the symmetry of the occupied orbitals. There are two different cases:

1. *States with fully occupied orbitals.* An electronic configuration in which all orbitals are completely filled possesses only one electronic state, and it will be totally symmetric. This can be seen for the case of non-degenerate orbitals. The wave function describing the electronic state can be written as the product of the one-electron orbitals. The symmetry of the product is given by the characters of the direct product representation. However, the product of any orbital with itself will always give the totally symmetric representation: no matter what characters it has, both $1 \cdot 1$ and $(-1) \cdot (-1)$ is 1, i.e., in each class of the point group the characters of the product will be 1. The same is

true for degenerate orbitals, although the procedure in this case is not as simple.

2. *States with partially occupied orbitals.* First of all, the completely filled orbitals are ignored for the reasons described above. The symmetry of the state will be given by the direct product of the partially filled orbitals.

Let us consider some examples for the above hypothetical molecule. The supposed ground state and the configurations of two different singly excited states are represented in Fig. 6-17.

Figure 6-17.
Different states of a molecule with C_{2v} symmetry.

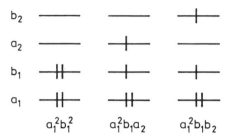

The ground state $a_1^2 b_1^2$ has only fully occupied orbitals, so its symmetry is A_1. The first excited state, $a_1^2 b_1 a_2$, has one fully occupied orbital, a_1^2, so this is not considered. The symmetry of this state will be given by the direct product $B_1 \cdot A_2$. Table 6-2 lists the direct products under the C_{2v} character table. The symmetry of the state is B_2. The other excited state in our example has the configuration $a_1^2 b_1 b_2$. The direct product is given in Table 6-2; the state symmetry is A_2. Since we are concerned only with the spatial symmetry properties, the electron spin and its role in determining the electronic states have been neglected in the above description.

Table 6-2. C_{2v} Character Table and Some Direct Product Representations

C_{2v}	E	C_2	$\sigma_v(xz)$	$\sigma_v'(yz)$		
A_1	1	1	1	1	z	x^2, y^2, z^2
A_2	1	1	-1	-1	R_z	xy
B_1	1	-1	1	-1	x, R_y	xz
B_2	1	-1	-1	1	y, R_x	yz
$B_1 \cdot A_2$	1	-1	-1	1	B_2	
$B_1 \cdot B_2$	1	1	-1	-1	A_2	

6.3.3 Examples for MO Construction

6.3.3.1 Homonuclear Diatomics

Hydrogen, H_2. There are two $1s$ hydrogen atomic orbitals available for bonding. The molecular point group is $D_{\infty h}$. This molecule does not have a central atom, so the symmetry operations of the point group are applied to both $1s$ orbitals, since they *together* form the basis for a representation of this point group. The $1s$ orbital of one hydrogen atom *alone* does not belong to any irreducible representation of the $D_{\infty h}$ point group. Several symmetry operations of this group transform one of the two $1s$ orbitals into the other rather than into itself (see Fig. 6-18a). Thus, they must be treated together; in this way they form a basis for a representation. All symmetry operations are indicated in Fig. 6-18b. The $D_{\infty h}$ character table is given in Table 5-3. The characters of this representation will be

$D_{\infty h}$	E	$2C_{\infty}^{\Phi}$	$\infty\sigma_v$	i	$2S_{\infty}^{\Phi}$	∞C_2
$2H(1s)$	2	2	2	0	0	0

Figure 6-18.
Some symmetry operations of the $D_{\infty h}$ point group applied to
(a) One $1s$ orbital in the hydrogen molecule.

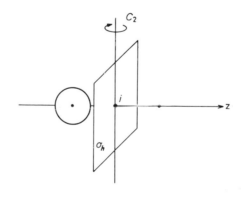

(b) The two $1s$ orbitals of the hydrogen molecule together.

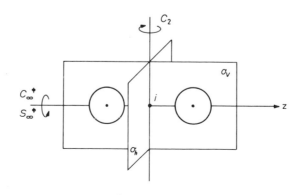

This is a reducible representation of the $D_{\infty h}$ point group which reduces to $\sigma_g + \sigma_u$. Two molecular orbitals must be generated, one with σ_g and the other with σ_u symmetry. The two possible combinations are the bonding and antibonding orbitals which can be formed from the two $1s$ atomic orbitals.

As there are altogether two electrons in the hydrogen molecule, they will occupy the bonding orbital, and none will go into the antibonding orbital.

$$\underline{} \quad \sigma_u \quad \text{antibonding}$$

$$\underline{+\!\!+} \quad \sigma_g \quad \text{bonding}$$

Hence, the molecule is stable.

Other homonuclear diatomic molecules. The principle utilized to construct molecular orbitals is the same as that for the hydrogen molecule. For helium, the MO picture is the same as for hydrogen except that here the additional two electrons occupy the antibonding σ_u orbital, and, therefore, the molecule is unstable.

In the series from lithium through neon, similar symmetry considerations apply, except that in these examples the second electron shell must be considered. The two $2s$ orbitals, as was found to be the case for the two $1s$ orbitals, form MOs that possess σ_g and σ_u symmetry. As regards the $2p$ orbitals, the two $2p_z$ orbitals lie along the molecular axis and belong to the same irreducible representation as the $2s$ orbitals. They also combine to give MOs that possess σ_g and σ_u symmetry.

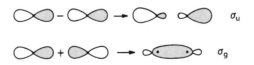

The $2s$ and $2p_z$ orbitals of the same atom belong to the same irreducible representation of the $D_{\infty h}$ point group. Their energies are also similar, so they cannot be separated completely. Another way of making linear combinations is first to combine the $2s$ and $2p_z$ orbitals of the same atom

and then combine the resulting orbitals into MOs.

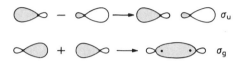

The result is essentially the same as before.

The $2p_x$ and $2p_y$ orbitals of the two atoms together form a representation that reduces to π_g and π_u. These correspond to two doubly degenerate π orbitals, one of which lies in the yz plane

and the other in the xz plane. The relative energies of these orbitals are known from energy calculations. In most cases the order is as follows:

$$1\sigma_g < 1\sigma_u < 2\sigma_g < 2\sigma_u < 3\sigma_g < 1\pi_u < 1\pi_g < 3\sigma_u,$$

while in some cases $1\pi_u < 3\sigma_g$.

6.3.3.2 Polyatomic Molecules

Before working out actual examples, let us recall what was said about the symmetry properties of atomic orbitals. If there is a central atom in the molecule, its atomic orbitals belong to some irreducible representation of the molecular point group. For the other atoms of these molecules, SALCs are formed from like orbitals. These new orbitals are then coupled with the central atom AOs to form MOs.

If the molecule does not have a central atom (e.g., C_6H_6), we begin with the second step, first forming different group orbitals and then combining them, if possible, into MOs. Examples will be given for both cases.

Water, H_2O. The molecular symmetry is C_{2v}. There are six atomic orbitals available for MO construction; two H $1s$, one oxygen $2s$ and three oxygen $2p$. They can combine to produce six MOs. As the molecule has a central atom, its AOs will belong to some of the irreducible representations of the C_{2v} point group by themselves. Group orbitals must be formed from the H $1s$ orbitals. The symmetry operations applied to them are shown in Fig. 6-19. The C_{2v} character table was given in Table 6-2. The reducible representation is:

Figure 6-19.
The C_{2v} symmetry operations applied to the two hydrogen $1s$ orbitals of water as basis functions.

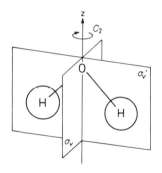

C_{2v}	E	C_2	$\sigma_v(xz)$	$\sigma_v'(yz)$
2 H($1s$)	2	0	0	2

This representation reduces to $A_1 + B_2$. The projection operator (see Chapter 4) is used to form these SALCs. Since we are interested only in symmetry aspects, numerical factors and normalization are omitted.

$$\hat{P}^{A_1}s_1 \approx 1 \cdot E \cdot s_1 + 1 \cdot C_2 \cdot s_1 + 1 \cdot \sigma \cdot s_1 + 1 \cdot \sigma' \cdot s_1 =$$
$$= s_1 + s_2 + s_2 + s_1 = 2s_1 + 2s_2 \approx s_1 + s_2$$

$$\hat{P}^{B_2}s_1 \approx 1 \cdot E \cdot s_1 + (-1) \cdot C_2 \cdot s_1 + (-1) \cdot \sigma \cdot s_1 + 1 \cdot \sigma' \cdot s_1 =$$
$$= s_1 - s_2 - s_2 + s_1 = 2s_1 - 2s_2 \approx s_1 - s_2$$

Thus, the two hydrogen group orbitals (φ_1 and φ_2) will have the forms:

$\varphi_2 = s_1 - s_2$:

$\varphi_1 = s_1 + s_2$:

The available AOs are summarized according to their symmetry properties in Table 6-3. Since only orbitals of the same symmetry can overlap, two combinations are possible: one has A_1 symmetry and the other has B_2 symmetry. The remaining two orbitals of oxygen (one with A_1 and the other with B_1 symmetry) will be non-bonding in the water molecule.

Table 6-3. The Atomic Orbitals of Water Grouped According to Their Symmetry Properties

	O Orbitals	H Group Orbitals
A_1	$2s, 2p_z$	φ_1
A_2		
B_1	$2p_x$	
B_2	$2p_y$	φ_2

If we choose the oxygen $2s$ orbital for bonding and leave the $2p_z$ orbital antibonding (from the symmetry point of view the opposite choice or a mixed orbital would do just as well), the MOs of the water molecule can be constructed as shown in Fig. 6-20. These MOs are compared with the calculated contour diagrams of the water molecular orbitals in Fig. 6-21 [6-12].

Figure 6-20.
Construction of the molecular orbitals of water.

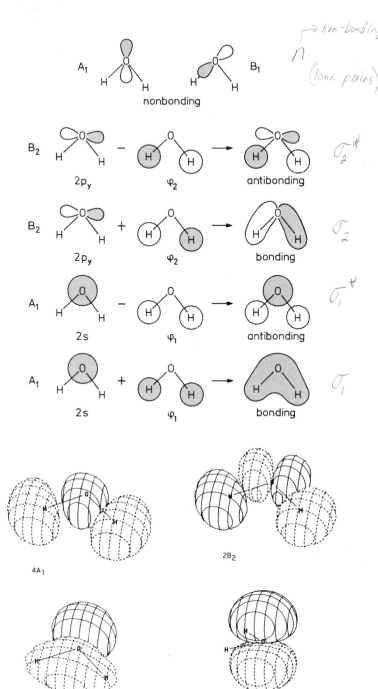

Figure 6-21.
Contour diagrams of the molecular orbitals of water. Reproduced with permission from [6-12]. Copyright (1973) Academic Press.

247

The construction of the molecular orbitals of the water molecule can also be represented by a qualitative MO diagram (see Fig. 6-22). The relative energies of the orbitals are also indicated in Fig. 6-22. What information can be deduced from such a diagram? First, there are two bonding orbitals occupied by four electrons; these correspond to the two O-H bonds of water. There are two non-bonding orbitals also occupied; these are the two lone pairs of oxygen. And, finally, there are two antibonding orbitals that are empty, so there is a net energy gain in the formation of H_2O and the molecule is stable.

Figure 6-22.
Qualitative MO diagram for water.

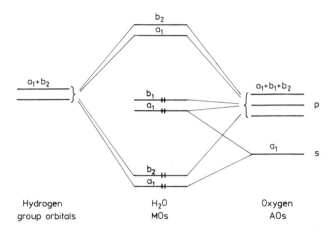

Hydrogen group orbitals — H_2O MOs — Oxygen AOs

Ammonia, NH_3. This example is given primarily to illustrate the construction of degenerate molecular orbitals. The symmetry of the molecule is C_{3v}. There are seven atomic orbitals available for bonding: three H $1s$, one N $2s$ and three N $2p$ AOs; hence, seven MOs must be formed. Since the nitrogen atom is a central atom, the coordinate axes can be chosen so that its AOs lie on all symmetry elements of the C_{3v} point group. The pertinent character table is given in Table 6-4. The N $2s$ and $2p_z$

Table 6-4. The C_{3v} Character Table and the Reducible Representation of the Hydrogen Group Orbitals of Ammonia

C_{3v}	E	$2C_3$	$3\sigma_v$		
A_1	1	1	1	z	x^2+y^2, z^2
A_2	1	1	-1	R_z	
E	2	-1	0	$(x, y)(R_x, R_y)$	$(x^2-y^2, xy)(xz, yz)$
3 H($1s$)	3	0	1		

Figure 6-23.
The C_{3v} symmetry operations applied to the three hydrogen atom $1s$ orbitals of ammonia as basis functions.

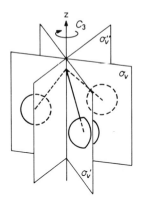

orbitals will have A_1 symmetry, and the $2p_x$ and $2p_y$ orbitals together belong to the E irreducible representation. Group orbitals must be formed from the three hydrogen $1s$ orbitals. The symmetry elements of the C_{3v} point group applied to these orbitals are shown in Fig. 6-23; their representation is given in Table 6-4.

This representation can now be reduced by using the reduction formula introduced in Chapter 4:

$$a_{A_1} = (1/6)\ (1 \cdot 3 \cdot 1 + 2 \cdot 0 \cdot 1 + 3 \cdot 1 \cdot 1) = 1$$
$$a_{A_2} = (1/6)\ (1 \cdot 3 \cdot 1 + 2 \cdot 0 \cdot 1 + 3 \cdot 1 \cdot (-1)) = 0$$
$$a_E = (1/6)\ (1 \cdot 3 \cdot 2 + 2 \cdot 0 \cdot (-1) + 3 \cdot 1 \cdot 0) = 1$$

Thus the representation reduces to $A_1 + E$. Next, let us use the projection operator to generate the form of these SALCs:

$$\hat{P}^{A_1} s_1 \approx 1 \cdot E \cdot s_1 + 1 \cdot C_3 \cdot s_1 + 1 \cdot C_3^2 \cdot s_1 + 1 \cdot \sigma \cdot s_1 +$$
$$+ 1 \cdot \sigma' \cdot s_1 + 1 \cdot \sigma'' \cdot s_1 =$$
$$= s_1 + s_2 + s_3 + s_1 + s_2 + s_3 =$$
$$= 2(s_1 + s_2 + s_3) \approx s_1 + s_2 + s_3$$

The same procedure is illustrated pictorially in Fig. 6-24 after [6-5].

Figure 6-24.
Generation of the A_1 symmetry orbital of the 3H group orbitals of ammonia.

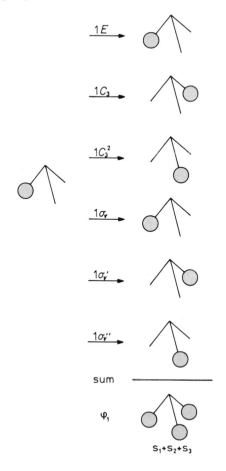

$1E$

$1C_3$

$1C_3^2$

$1\sigma_v$

$1\sigma_v'$

$1\sigma_v''$

sum

φ_1

$S_1 + S_2 + S_3$

Before constructing the E symmetry group orbitals, a time-saving simplification can be introduced [6-5]. This simplification follows from the fact that the rotational subgroup C_n in itself contains all the information needed to construct the SALCs in a molecule that possesses a principal axis C_n. The rotational subgroup of C_{3v} is C_3, and its character table is given in Table 6-5. If we perform the three symmetry operations of the C_3 point group and check the generation of the A_1 symmetry SALC of NH_3 (Fig. 6-24), we see that the application of these three operations suffices to define the form of this orbital.

Table 6-5. The C_3 Character Table

C_3	$E \quad C_3 \quad C_3{}^2$		$\varepsilon = \exp(2\pi i/3)$
A_1	$1 \quad 1 \quad 1$	z, R_z	x^2+y^2, z^2
E	$\begin{Bmatrix} 1 & \varepsilon & \varepsilon^* \\ 1 & \varepsilon^* & \varepsilon \end{Bmatrix}$	$(x, y)(R_x, R_y)$	$(x^2-y^2, xy)(yz, xz)$

The C_3 character table contains imaginary characters for the E representation. They can be eliminated by following the procedure used in [6-5]. First, the irreducible representation E is written as:

$$E = \begin{Bmatrix} 1 & \varepsilon & \varepsilon^* \\ 1 & \varepsilon^* & \varepsilon \end{Bmatrix} =$$

$$= \begin{Bmatrix} 1 & \cos 2\pi/3 + i \sin 2\pi/3 & \cos 2\pi/3 - i \sin 2\pi/3 \\ 1 & \cos 2\pi/3 - i \sin 2\pi/3 & \cos 2\pi/3 + i \sin 2\pi/3 \end{Bmatrix} =$$

$$= \begin{Bmatrix} 1 & \cos 120° + i \sin 120° & \cos 120° - i \sin 120° \\ 1 & \cos 120° - i \sin 120° & \cos 120° + i \sin 120° \end{Bmatrix} =$$

$$= \begin{Bmatrix} 1 & -1/2 + i \sqrt{3}/2 & -1/2 - i \sqrt{3}/2 \\ 1 & -1/2 - i \sqrt{3}/2 & -1/2 + i \sqrt{3}/2 \end{Bmatrix} \qquad \begin{matrix} \text{(a)} \\ \text{(b)} \end{matrix}$$

When the last two lines are summed the imaginary characters vanish:

add (a) and (b):

$$2 \quad -1 \quad -1.$$

The imaginary characters can also be eliminated if (b) is subtracted from (a) and the result is divided by $i\sqrt{3}$:

subtract (b) from (a): $\quad\quad 0 \quad i\sqrt{3} \quad -i\sqrt{3}$ (c)

divide (c) by $i\sqrt{3}$: $\quad\quad 0 \quad 1 \quad -1.$

This "manipulation" of characters is allowed since it simply produces a new linear combination of the original representation. The C_3 character table with real characters is given in Table 6-6. When the projection operator is applied to one of the $1s$ orbitals of the hydrogen group orbitals with the two E representations, the two E symmetry doubly degenerate SALCs result:

$$\hat{P}^{E^1} s_1 \approx 2 \cdot E \cdot s_1 + (-1) \cdot C_3 \cdot s_1 + (-1) \cdot C_3{}^2 \cdot s_1 =$$
$$= 2 s_1 - s_2 - s_3$$
$$\hat{P}^{E^2} s_1 \approx 0 \cdot E \cdot s_1 + 1 \cdot C_3 \cdot s_1 + (-1) \cdot C_3{}^2 \cdot s_1 =$$
$$= s_2 - s_3$$

Table 6-6. The C_3 Character Table with Real Characters

C_3	E	C_3	$C_3{}^2$
A	1	1	1
E	$\begin{cases} 2 \\ 0 \end{cases}$	$\begin{matrix} -1 \\ 1 \end{matrix}$	$\begin{matrix} -1 \\ -1 \end{matrix}$

Fig. 6-25 illustrates the same procedure pictorially.

Figure 6-25.
Projection of the two E symmetry group orbitals of the three H $1s$ orbitals in ammonia.

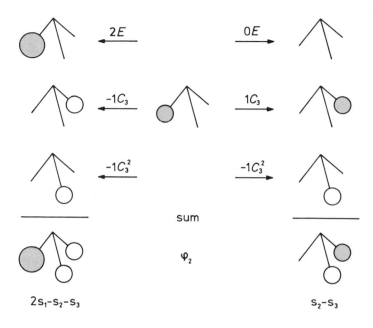

The next step is the MO construction. The orbitals used for this purpose are summarized in Table 6-7. An A_1 and a doubly degenerate E symmetry combination is possible here, and there will be a non-bonding orbital with A_1 symmetry left on nitrogen. Fig. 6-26 illustrates the building of MOs. Again, they can be compared with the calculated contour diagrams of the

Table 6-7. The Atomic Orbitals of Ammonia Sorted According to Their Symmetry Properties

	N Orbitals	H Group Orbitals
A_1	s, p_z	φ_1
E	(p_x, p_y)	φ_2

Figure 6-26.
Construction of molecular
orbitals for ammonia.

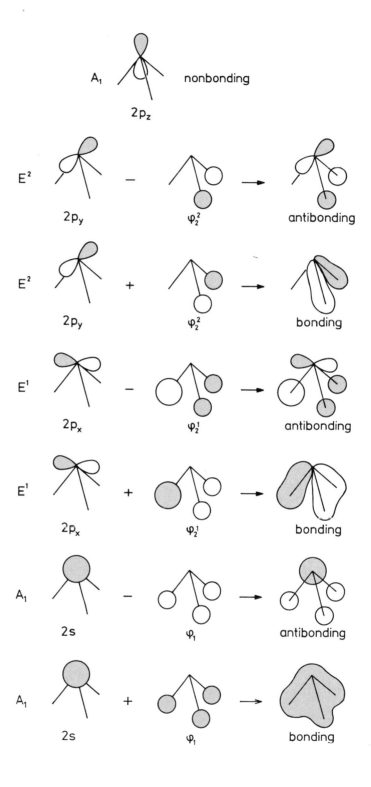

Figure 6-27.
Contour diagrams of the MOs
of ammonia. Reproduced with
permission from [6-12]. Copyright (1973) Academic Press.

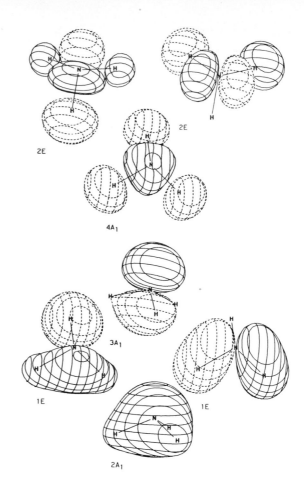

ammonia molecular orbitals in Fig. 6-27. The qualitative MO
diagram is given in Fig. 6-28. The following conclusions can be
drawn: 1. There are three bonding orbitals occupied by electrons; these correspond to the three N-H bonds. 2. There is
a non-bonding orbital also occupied by electrons; this
corresponds to the lone electron pair. 3. The three antibonding
orbitals are unoccupied, so the MO construction is energetically
favorable, and the molecule is stable.

Figure 6-28.
Qualitative MO diagram for ammonia.

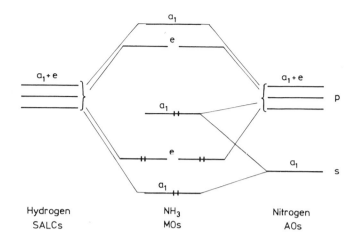

Benzene, C_6H_6. The molecular symmetry is D_{6h}. There are 30 AOs that can be used for MO construction; six H $1s$, six C $2s$ and 18 C $2p$ orbitals. Since this molecule does not contain a central atom, each AO must be grouped into SALCs in such a way that they can transform according to the symmetry operations of the D_{6h} point group. It is straightforward to combine like orbitals, for example, the hydrogen $1s$ orbitals, the carbon $2s$ orbitals, and so on. The following combinations will be used here:

$$\Phi_1(6\ H\ 1s), \quad \Phi_2(6\ C\ 2s), \quad \Phi_3(6\ C\ 2p_x,2p_y) \text{ and } \Phi_4(6\ C\ 2p_z).$$

The next step is to determine how these group orbitals transform in the D_{6h} point group. The D_{6h} character table is given in Table 6-8. Since most of the AOs in the suggested group orbitals

Table 6-8. The D_{6h} Character Table

D_{6h}	E	$2C_6$	$2C_3$	C_2	$3C_2'$	$3C_2''$	i	$2S_3$	$2S_6$	σ_h	$3\sigma_d$	$3\sigma_v$		
A_{1g}	1	1	1	1	1	1	1	1	1	1	1	1		$x^2+y^2,\ z^2$
A_{2g}	1	1	1	1	-1	-1	1	1	1	1	-1	-1	R_z	
B_{1g}	1	-1	1	-1	1	-1	1	-1	1	-1	1	-1		
B_{2g}	1	-1	1	-1	-1	1	1	-1	1	-1	-1	1		
E_{1g}	2	1	-1	-2	0	0	2	1	-1	-2	0	0	$(R_x,\ R_y)$	$(xz,\ yz)$
E_{2g}	2	-1	-1	2	0	0	2	-1	-1	2	0	0		$(x^2-y^2,\ xy)$
A_{1u}	1	1	1	1	1	1	-1	-1	-1	-1	-1	-1		
A_{2u}	1	1	1	1	-1	-1	-1	-1	-1	-1	1	1	z	
B_{1u}	1	-1	1	-1	1	-1	-1	1	-1	1	-1	1		
B_{2u}	1	-1	1	-1	-1	1	-1	1	-1	1	1	-1		
E_{1u}	2	1	-1	-2	0	0	-2	-1	1	2	0	0	$(x,\ y)$	
E_{2u}	2	-1	-1	2	0	0	-2	1	1	-2	0	0		

255

are transformed into another AO by most of the symmetry operations, the representations will be quite simple, though still reducible:

Γ_{Φ_1}	6	0	0	0	2	0	0	0	0	6	0	2
Γ_{Φ_2}	6	0	0	0	2	0	0	0	0	6	0	2
Γ_{Φ_3}	12	0	0	0	0	0	0	0	0	12	0	0
Γ_{Φ_4}	6	0	0	0	-2	0	0	0	0	-6	0	2

These representations now can be reduced by applying the reduction formula:

First Φ_1:

$$
\begin{aligned}
a_{A_{1g}} &= (1/24)\,(1 \cdot 6 \cdot 1 + 2 \cdot 0 \cdot 1 + 2 \cdot 0 \cdot 1 + 1 \cdot 0 \cdot 1 + \\
&\quad + 3 \cdot 2 \cdot 1 + 3 \cdot 0 \cdot 1 + 1 \cdot 0 \cdot 1 + 2 \cdot 0 \cdot 1 + 2 \cdot 0 \cdot 1 + \\
&\quad + 1 \cdot 6 \cdot 1 + 3 \cdot 0 \cdot 1 + 3 \cdot 2 \cdot 1) = \\
&= (1/24)\,(6 + 6 + 6 + 6) = 24/24 = 1 \\
a_{A_{2g}} &= (1/24)\,(6 - 6 + 6 - 6) = 0 \\
a_{B_{1g}} &= (1/24)\,(6 + 6 - 6 - 6) = 0 \\
a_{B_{2g}} &= (1/24)\,(6 - 6 - 6 + 6) = 0 \\
a_{E_{1g}} &= (1/24)\,(12 - 12) = 0 \\
a_{E_{2g}} &= (1/24)\,(12 + 12) = 1 \\
a_{A_{1u}} &= (1/24)\,(6 + 6 - 6 - 6) = 0 \\
a_{A_{2u}} &= (1/24)\,(6 - 6 - 6 + 6) = 0 \\
a_{B_{1u}} &= (1/24)\,(6 + 6 + 6 + 6) = 1 \\
a_{B_{2u}} &= (1/24)\,(6 - 6 + 6 - 6) = 0 \\
a_{E_{1u}} &= (1/24)\,(12 + 12) = 1 \\
a_{E_{2u}} &= (1/24)\,(12 - 12) = 0
\end{aligned}
$$

Thus, the first representation reduces to the following irreducible representations:

$$
\Gamma_{\Phi_1} = A_{1g} + E_{2g} + B_{1u} + E_{1u}
$$

Without giving details of the other three reductions, the results are:

$$
\begin{aligned}
\Gamma_{\Phi_2} &= A_{1g} + E_{2g} + B_{1u} + E_{1u} \\
\Gamma_{\Phi_3} &= A_{1g} + A_{2g} + 2\,E_{2g} + B_{1u} + B_{2u} + 2\,E_{1u} \\
\Gamma_{\Phi_4} &= B_{2g} + E_{1g} + A_{2u} + E_{2u}
\end{aligned}
$$

Before using the projection operator to generate these SALCs, a simplification which is similar to the one used in the case of ammonia is introduced. Instead of the D_{6h} point group, its rotational subgroup, C_6, will be used. The C_6 character table is given in Table 6-9. Again, imaginary characters appear with

Table 6-9. The C_6 Character Table

C_6	E	C_6	C_3	C_2	$C_3{}^2$	$C_6{}^5$		$\varepsilon = \exp(2\pi i/6)$
A	1	1	1	1	1	1	z, R_z	x^2+y^2, z^2
B	1	−1	1	−1	1	−1		
E_1	$\begin{cases} 1 \\ 1 \end{cases}$	$\begin{matrix}\varepsilon \\ \varepsilon^*\end{matrix}$	$\begin{matrix}-\varepsilon^* \\ -\varepsilon\end{matrix}$	$\begin{matrix}-1 \\ -1\end{matrix}$	$\begin{matrix}-\varepsilon \\ -\varepsilon^*\end{matrix}$	$\begin{matrix}\varepsilon^* \\ \varepsilon\end{matrix}$	(x, y) (R_x, R_y)	(xz, yz)
E_2	$\begin{cases} 1 \\ 1 \end{cases}$	$\begin{matrix}-\varepsilon^* \\ -\varepsilon\end{matrix}$	$\begin{matrix}-\varepsilon \\ -\varepsilon^*\end{matrix}$	$\begin{matrix}1 \\ 1\end{matrix}$	$\begin{matrix}-\varepsilon^* \\ -\varepsilon\end{matrix}$	$\begin{matrix}-\varepsilon \\ -\varepsilon^*\end{matrix}$		(x^2-y^2, xy)

the E representations. They can be turned into real representations by using the same procedure detailed earlier. The values of ε and ε^* are given by:

$$\varepsilon = \cos 2\pi/6 + i \sin 2\pi/6 = \cos 60° + i \sin 60° =$$
$$= 1/2 + i \sqrt{3}/2$$
$$\varepsilon^* = \cos 2\pi/6 - i \sin 2\pi/6 = \cos 60° - i \sin 60° =$$
$$= 1/2 - i \sqrt{3}/2$$

Substituting these expressions into the E_1 representation, we obtain:

$$E_1 = \begin{cases} 1 & 1/2+i\sqrt{3}/2 & -1/2+i\sqrt{3}/2 & -1 & -1/2-i\sqrt{3}/2 & 1/2-i\sqrt{3}/2 \\ 1 & 1/2-i\sqrt{3}/2 & -1/2-i\sqrt{3}/2 & -1 & -1/2+i\sqrt{3}/2 & 1/2+i\sqrt{3}/2 \end{cases}$$
(a)
(b)

Making new linear combinations by adding (a) and (b), we have:

$$2 \qquad 1 \qquad -1 \qquad -2 \qquad -1 \qquad 1.$$

When (b) is subtracted from (a), we obtain:

$$0 \qquad i\sqrt{3} \qquad i\sqrt{3} \qquad 0 \qquad i\sqrt{3} \qquad -i\sqrt{3}. \quad \text{(c)}$$

Finally, dividing (c) by $i\sqrt{3}$, we arrive at:

$$0 \qquad 1 \qquad 1 \qquad 0 \qquad -1 \qquad -1.$$

The E_2 representation can be changed in the same way. The C_6 character table which contains these real characters is shown in Table 6-10.

Table 6-10. The C_6 Character Table with Real Characters

C_6	E	C_6	C_3	C_2	$C_3{}^2$	$C_6{}^5$
A	1	1	1	1	1	1
B	1	−1	1	−1	1	−1
E_1	2	1	−1	−2	−1	1
	0	1	1	0	−1	−1
E_2	2	−1	−1	2	−1	−1
	0	1	−1	0	1	−1

Benzene consists of 30 MOs; only some representative benzene MOs will be constructed and shown here. It may be a good exercise for the reader to construct the remaining MOs of benzene by following the procedure demonstrated here. The SALCs are sorted according to their symmetry properties in Table 6-11. Inspection of this table reveals that the first three

Table 6-11. The Symmetry of the Different Group Orbitals of Benzene

	Φ_1 H Group Orbital	Φ_2 C 2s Group Orbital	Φ_3 C 2p_x, 2p_y Group Orbital	Φ_4 C 2p_z Group Orbital
A_{1g}	+	+	+	
A_{2g}			+	
B_{1g}				
B_{2g}				+
E_{1g}				+
E_{2g}	+	+	++	
A_{1u}				
A_{2u}				+
B_{1u}	+	+	+	
B_{2u}			+	
E_{1u}	+	+	++	
E_{2u}				+

group orbitals have common irreducible representations, so they can be mixed with each other. They consist of 24 AOs; thus, 24 MOs will be formed. Since each bonding MO has its antibonding counterpart, there will be 12 bonding and 12 anti-bonding molecular orbitals. The former will be the σ bonding orbitals of benzene, since there are six C-C and six C-H bonds. The fourth group orbital does not belong to any irreducible representation common to the other three, so it will not be mixed with them. This representation corresponds to the π orbitals of benzene by itself.

Let us now construct some of the σ orbitals of benzene, viz. MOs that possess A_{1g}, B_{1u}, B_{2u}, and E_{1u} symmetry. The totally symmetric representation, A_{1g}, appears three times, once in Φ_1, in Φ_2, and in Φ_3. Two A_{1g} representations can be combined into an MO, and the third one can represent an MO by itself. These three SALCs can be generated by using the projection operator pictorially as shown in Fig. 6-29. The forms of these

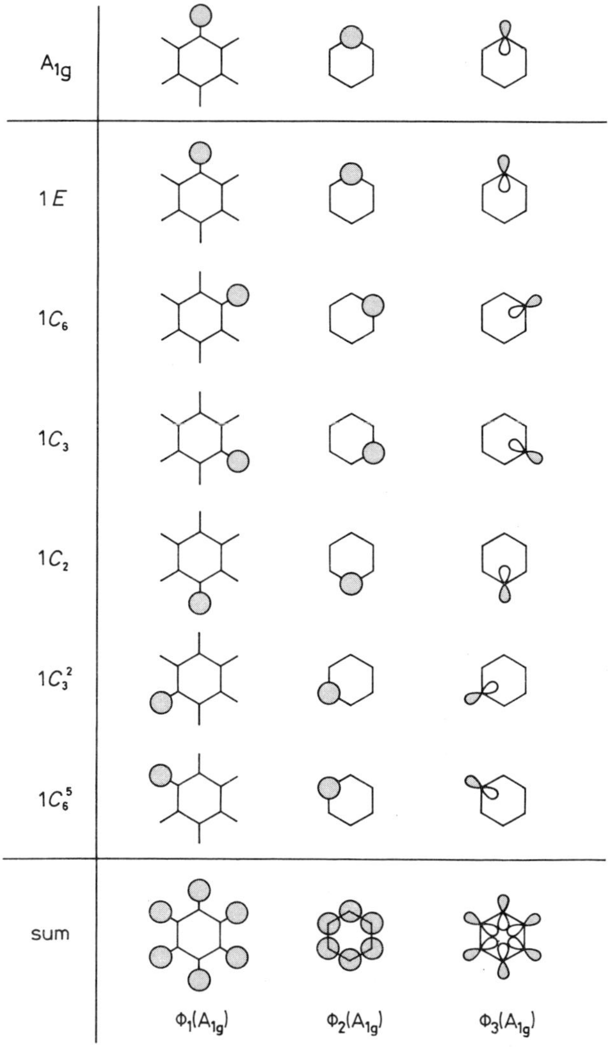

Figure 6-29.
Generation of the A_{1g} symmetry group orbitals of benzene.

Figure 6-30.
Contour diagrams of some σ and π bonding molecular orbitals of benzene. Reproduced with permission from [6-12]. Copyright (1973) Academic Press.

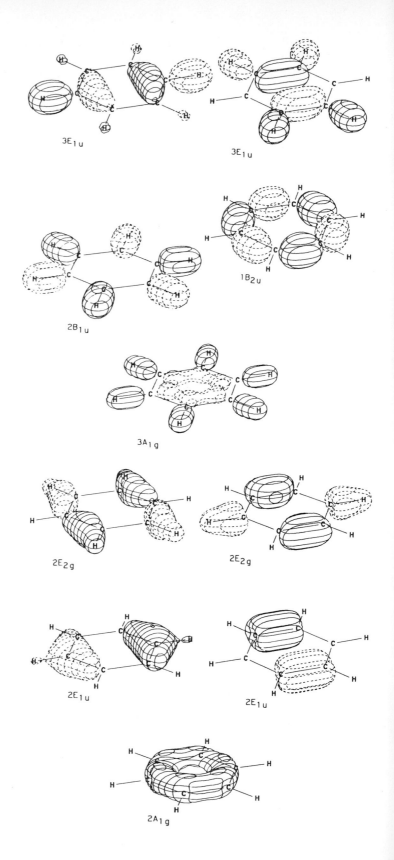

group orbitals are such that $\Phi_2(A_{1g})$ can be taken as an MO by itself (C-C σ bond, cf. also the corresponding orbital, $2\ A_{1g}$, in the contour diagram in Fig. 6-30), and the other two group

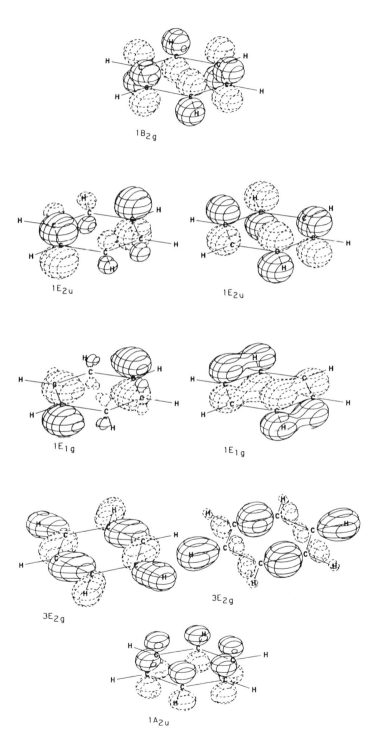

$1B_{2g}$

$1E_{2u}$

$1E_{2u}$

$1E_{1g}$

$1E_{1g}$

$3E_{2g}$

$3E_{2g}$

$1A_{2u}$

orbitals can be combined into molecular orbitals as shown in Fig. 6-31. The contour diagram of the bonding MO is depicted by the 3 A_{1g} orbital in Fig. 6-30.

Figure 6-31.
Bonding and antibonding combination of A_{1g} symmetry group orbitals of benzene.

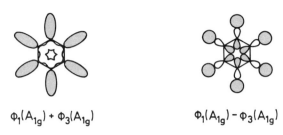

$$\Phi_1(A_{1g}) + \Phi_3(A_{1g}) \qquad\qquad \Phi_1(A_{1g}) - \Phi_3(A_{1g})$$

The next MO will be of B_{1u} symmetry. This irreducible representation also appears in Φ_1, Φ_2, and Φ_3. Take this time the corresponding Φ_1 and Φ_2 group orbitals and combine them into molecular orbitals:

$$\hat{P}^{B_{1u}} s_1 \approx 1 \cdot E \cdot s_1 + (-1) \cdot C_6 \cdot s_1 + 1 \cdot C_3 \cdot s_1 + (-1) \cdot C_2 \cdot s_1 +$$
$$+ 1 \cdot C_3^{\,2} \cdot s_1 + (-1) \cdot C_6^{\,5} \cdot s_1 =$$
$$= s_1 - s_2 + s_3 - s_4 + s_5 - s_6,$$

or pictorially:

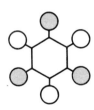

The B_{1u} symmetry SALC of Φ_2, i.e. the group orbital of the six C $2s$ AOs, will have a similar form:

The combination of these Φ_1 and Φ_2 SALCs affords the bonding and antibonding combinations shown in Fig. 6-32. The contour diagram corresponding to the bonding MO is the 2 B_{1u} orbital in Fig. 6-30.

Figure 6-32.
Bonding and antibonding
combination of B_{1u} symmetry
group orbitals of benzene.

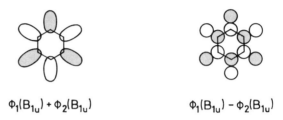

$$\Phi_1(B_{1u}) + \Phi_2(B_{1u}) \qquad\qquad \Phi_1(B_{1u}) - \Phi_2(B_{1u})$$

Since there is only one B_{2u} symmetry orbital among the SALCs, the one in Φ_3, it will be a MO by itself. Let us generate this MO:

$$\hat{P}^{B_{2u}} p_y(C_1) \approx 1 \cdot E \cdot p_{y1} + (-1) \cdot C_6 \cdot p_{y1} + 1 \cdot C_3 \cdot p_{y1} +$$
$$+ (-1) \cdot C_2 \cdot p_{y1} + 1 \cdot C_3^{\,2} \cdot p_{y1} + (-1) \cdot C_6^{\,5} \cdot p_{y1} =$$
$$= p_{y1} - p_{y2} + p_{y3} - p_{y4} + p_{y5} - p_{y6}.$$

This group orbital has the following shape:

Compare the above orbital with $1\,B_{2u}$ (Fig. 3-30).

Of the four doubly degenerate E_{1u} symmetry SALCs, two will be considered here; namely, that of Φ_1 and one of the two E_{1u} symmetry SALCs of Φ_3. The generation of both of the E_{1u} symmetry degenerate orbitals of Φ_1 is shown in Fig. 6-33. The left and right columns each refer to one of the orbitals. The application of the other method for the p_x SALC of Φ_3 leads to the following:

$$\hat{P}^{E_{1u}^1} p_x(C_1) \approx 2 \cdot E \cdot p_{x1} + 1 \cdot C_6 \cdot p_{x1} + (-1) \cdot C_3 \cdot p_{x1} +$$
$$+ (-2) \cdot C_2 \cdot p_{x1} + (-1) \cdot C_3^{\,2} \cdot p_{x1} + 1 \cdot C_6^{\,5} \cdot p_{x1} =$$
$$= 2\,p_{x1} + p_{x2} - p_{x3} - 2p_{x4} - p_{x5} + p_{x6}$$

$$\hat{P}^{E_{1u}^2} p_x(C_1) \approx 0 \cdot E \cdot p_{x1} + 1 \cdot C_6 \cdot p_{x1} + 1 \cdot C_3 \cdot p_{x1} + 0 \cdot C_2 \cdot p_{x1} +$$
$$+ (-1) \cdot C_3^{\,2} \cdot p_{x1} + (-1) \cdot C_6^{\,5} \cdot p_{x1} =$$
$$= p_{x2} + p_{x3} - p_{x4} - p_{x5}.$$

Figure 6-33.
Generation of the two E_{1u} symmetry group orbitals of the hydrogen atomic orbitals in benzene.

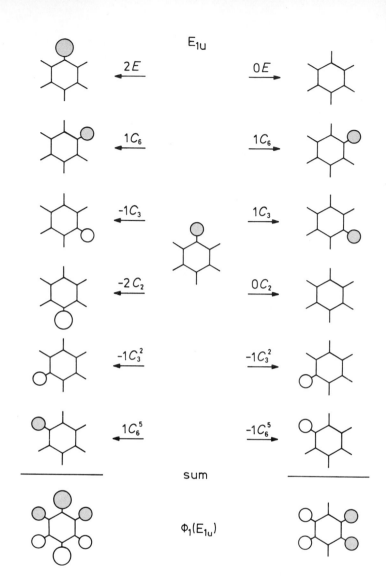

E_{1u}

$\Phi_1(E_{1u})$

The form of the E_{1u} symmetry SALC of Φ_3 is:

The combination of the Φ_1 and Φ_3 doubly degenerate group orbitals affords the bonding and antibonding MOs shown in Fig. 6-34. The bonding combination can be compared to 3 E_{1u} in Fig. 6-30.

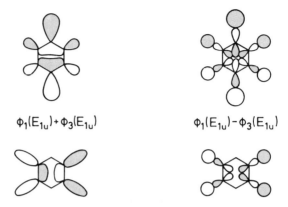

$$\phi_1(E_{1u}) + \phi_3(E_{1u}) \qquad\qquad \phi_1(E_{1u}) - \phi_3(E_{1u})$$

The π orbitals of benzene will be the two doubly degenerate and the two non-degenerate combinations of the Φ_4 group orbital itself. All of these are shown below.

A_{2u} symmetry orbital: this corresponds to the totally symmetric representation in the rotational subgroup C_6; so, even without using the projection operator, its form can be given by:

$$\Phi_4(A_{2u}) = p_{z1} + p_{z2} + p_{z3} + p_{z4} + p_{z5} + p_{z6}$$

B_{2g} symmetry orbital: using the projection operator, we obtain:

$$\hat{P}^{B_{2g}} p_z(C_1) \approx 1 \cdot E \cdot p_{z1} + (-1) \cdot C_6 \cdot p_{z1} + 1 \cdot C_3 \cdot p_{z1} +$$
$$+ (-1) \cdot C_2 \cdot p_{z1} + 1 \cdot C_3^2 \cdot p_{z1} + (-1) \cdot C_6^5 \cdot p_{z1} =$$
$$= p_{z1} - p_{z2} + p_{z3} - p_{z4} + p_{z5} - p_{z6}$$

The two E_{1g} symmetry SALCs are constructed in Fig. 6-35.

Figure 6-35.
The two E_{1g} symmetry group orbitals formed from the carbon $2p_z$ orbitals in benzene.

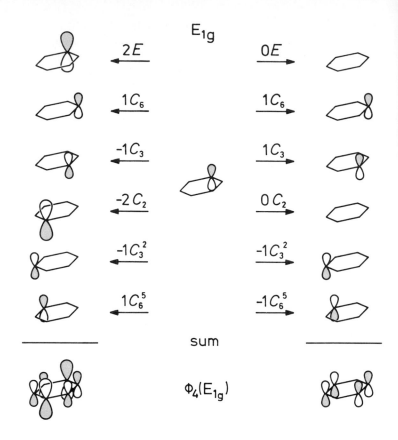

E_{1g}

sum

$\phi_4(E_{1g})$

Finally, the two E_{2u} symmetry orbitals are expressed as follows:

$$\hat{P}^{E_{2u}^1} p_z(C_1) \approx 2 \cdot E \cdot p_{z1} + (-1) \cdot C_6 \cdot p_{z1} + (-1) \cdot C_3 \cdot p_{z1} +$$
$$+ 2 \cdot C_2 \cdot p_{z1} + (-1) \cdot C_3^2 \cdot p_{z1} + (-1) \cdot C_6^5 \cdot p_{z1} =$$
$$= 2\, p_{z1} - p_{z2} - p_{z3} + 2p_{z4} - p_{z5} - p_{z6}$$
$$\hat{P}^{E_{2u}^2} p_z(C_1) \approx 0 \cdot E \cdot p_{z1} + 1 \cdot C_6 \cdot p_{z1} + (-1) \cdot C_3 \cdot p_{z1} +$$
$$+ 0 \cdot C_2 \cdot p_{z1} + 1 \cdot C_3^2 \cdot p_{z1} + (-1) \cdot C_6^5 \cdot p_{z1} =$$
$$= p_{z2} - p_{z3} + p_{z5} - p_{z6}.$$

Their forms are:

Compare these SALCs to the contour diagram of the 1 E_{2u} orbital (Fig. 6-30). Fig. 6-36 shows the relative energies of the benzene π orbitals.

Figure 6-36.
Relative energies of the benzene π orbitals.

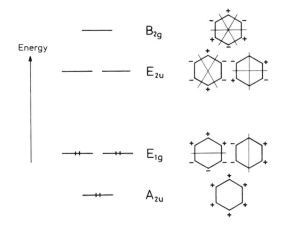

6.3.3.3 Short Summary of MO Construction

The steps of MO construction can now be summarized as follows:

1. Identify the symmetry of the molecule.

2. List all atomic orbitals that are intended to be used for MO construction.

3. See whether or not the molecule has a central atom. If it does, then look up in the character table the irreducible representations to which its atomic orbitals belong. If there is no central atom in the molecule, proceed to the next step.

4. Construct group orbitals (SALCs) from the atomic orbitals of like atoms.

5. Use these orbitals as bases for representations of the point group.

6. Reduce these representations to their irreducible components.

7. Apply the projection operator to the AOs for each of these irreducible representations to obtain the forms of the SALCs.

8. These SALCs will either be MOs by themselves, or they can be combined with other SALCs or central atom orbitals of the same symmetry. Each of these combinations will give one bonding and one antibonding MO of the same symmetry.

267

9. Normalization has been ignored throughout our discussion. However, the SALCs must be properly normalized in all calculations [6-4]. This may be done at the end of the SALC construction, i.e. after step 7 in our list.

6.4 Influence of Environmental Symmetry

Symmetry has a major role in two widely used and successful methods of chemistry, viz., the crystal field and ligand field theories of coordination compounds. This topic has been thoroughly covered in existing textbooks and monographs on coordination chemistry. Therefore, it is mentioned here only in passing. Bethe has shown that the degenerate electronic state of a cation is split by a crystal field into nonequivalent states [6-13]. The change is determined entirely by the symmetry of the crystal lattice.

Bethe's original work was concerned with ionic crystals, but his concept has more general applications. When an atom or an ion enters a ligand environment, the symmetry of the ligand arrangement will influence the electron density distribution of that atom or ion. The original spherical symmetry of the atomic orbitals will be lost, and the symmetry of the ligand environment will be adopted. As a consequence of the usual decrease of symmetry, the degree of degeneracy of the orbitals decreases.

The s electrons are already nondegenerate in the free atom, so their degeneracy does not change. They will always belong to the totally symmetric irreducible representation of the symmetry group. The p orbitals, however, are threefold degenerate, and the d orbitals are five-fold degenerate. To determine their splitting in a certain point group, we must use them, in principle, as bases for a representation of the group. In practice, we can find in the character table of the point group the irreducible representations to which the orbitals belong. An orbital always belongs to the same irreducible representation as do its subscripts. Some orbital splittings that accompany the decrease in environmental symmetry are shown in Table 6-12.

Table 6-12. Splitting of Atomic Orbitals in Different Symmetry Environment

	s	p	d
O_h	a_{1g}	t_{1u}	$e_g + t_{2g}$
T_d	a_1	t_2	$e + t_2$
$D_{\infty h}$	σ_g	$\sigma_u + \pi_u$	$\sigma_g + \pi_g + \Delta_g$
D_{4d}	a_1	$b_2 + e_1$	$a_1 + e_2 + e_3$
D_{4h}	a_{1g}	$a_{2u} + e_u$	$a_{1g} + b_{1g} + b_{2g} + e_g$
C_{4v}	a_1	$a_1 + e$	$a_1 + b_1 + b_2 + e$
C_{2v}	a_1	$a_1 + b_1 + b_2$	$2\,a_1 + a_2 + b_1 + b_2$

Figure 6-37.
The orientation of the different symmetry d orbitals in an octahedral environment.

$d_{x^2-y^2}$

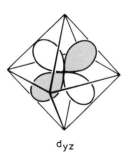

d_{yz}

As environmental symmetry decreases, the orbitals will become split to an increasing extent. At the C_{2v} point group, e.g., all atomic orbitals become split into non-degenerate levels. This is not surprising, since the C_{2v} character table contains only one-dimensional irreducible representations. This result shows at once that there are no degenerate energy levels in this point group. This has been stressed in Chapter 4 in the discussion of irreducible representations.

The symmetry of the ligand environment gives an important but limited amount of information about orbital splitting. Both the octahedral and cubic ligand arrangements, for example, belong to the O_h point group, and we can tell that the d orbitals of the central atom will split into a doubly degenerate and a triply degenerate pair. But nothing is revealed about the relative energies of these two sets of degenerate orbitals.

The problem of relative energies is, however, dealt with by crystal field theory. This theory examines the repulsive interaction between the ligands and the central atom orbitals. Consider first an octahedral molecule (Fig. 6-37), and compare the positions of one e_g (e.g. $d_{x^2-y^2}$) and one t_{2g} (e.g. d_{yz}) orbital. The others need not be considered, as they are degenerate with one of the e_g or t_{2g} orbitals and, thus, they have the same energy. The lobes of the $d_{x^2-y^2}$ orbital point towards the ligands. The resulting electrostatic repulsion will destabilize these orbitals, and their energies will increase accordingly. The d_{yz} orbital, on the other hand, points in directions between the ligands. This is an energetically more favorable position; hence, the energy of these orbitals will decrease.

Examine now the cubic arrangement in Fig. 6-38. It can be

Figure 6-38.
The orientation of the
different symmetry d orbitals
in a cubic environment.

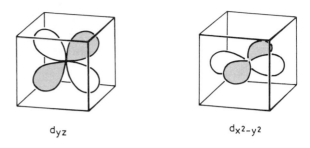

d_{yz} $d_{x^2-y^2}$

seen that the d_{yz} orbital is in a more unfavorable situation relative to the ligands than is the $d_{x^2-y^2}$ orbital, so their relative energies will be reversed (see Fig. 6-39). Some other typical orbital

Figure 6-39.
Relative energies of the d
orbitals in octahedral and
cubic ligand environment.

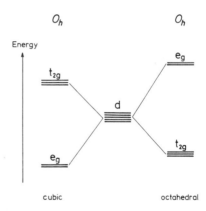

splittings and the corresponding changes in the relative energies are shown in Fig. 6-40.

Figure 6-40.
d orbital splittings in different
ligand environment.

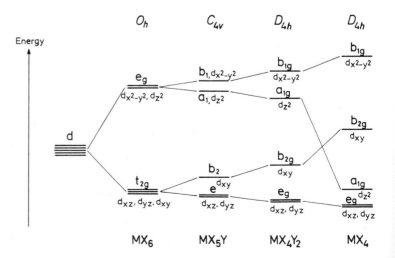

Prediction of structural changes. Crystal field theory is frequently applied to account for and even predict structural and chemical changes. A well-known example is the variation of first row transition metal ionic radii in an octahedral environment [6-14] as illustrated in Fig. 6-41, curve A. The dashed line

Figure 6-41.
The variation of octahedral ionic radii according to [6-14] (Curve A); and of the bond lengths of transition metal dichlorides according to [6-15] (Curve B), and dibromides (Curve C).

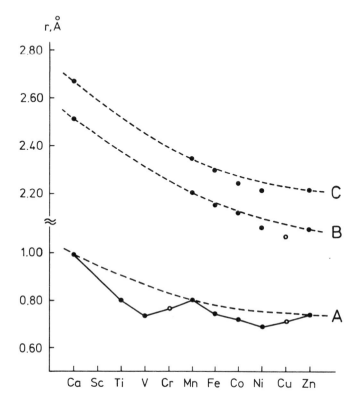

connects the points for Ca–Mn–Zn, i.e., atoms with spherically symmetric distribution of d electrons. Since the shielding of one d electron by another is imperfect, a contraction in the ionic radius is expected along this series. This in itself would account only for a steady decrease in the radii, whereas the ionic radii of all the other atoms are smaller than interpolation from the Ca–Mn–Zn curve would suggest. As is well known, the non-uniform distribution of d electrons around the nuclei is the origin of this phenomenon. In the octahedral environment the d orbitals split into orbitals with t_{2g} and e_g symmetry. The electrons added gradually occupy t_{2g} orbitals in Sc^{2+}, Ti^{2+}, and V^{2+} as well as in Fe^{2+}, Co^{2+}, and Ni^{2+}, if only high spin configurations are considered. Since these orbitals are not oriented towards the ligands, the degree of shielding between the ligands and the positively charged atomic cores decreases along with the ionic radius. The fourth electron in Cr^{2+} as well as the ninth electron in Cu^{2+} occupy e_g symmetry orbitals. The degree of shielding thus somewhat increases and, accordingly, there is a smaller relative decrease in the ionic radii.

The bond length variation among the first row transition metal dihalides has been interpreted by similar symmetry arguments [6-15]. Curve B in Fig. 6-41 presents the available experimental data on the bond lengths of first row transition metal dichlorides in the vapor phase [6-16]. The value for $CuCl_2$ originates from quantum chemical calculation [6-17]. Curve C shows the bond length variation of the corresponding dibromides. The dashed lines again connect the points for atoms that possess spherically symmetric electron distributions.

The observed decrease in the bond lengths with increasing atomic number is even more pronounced than what was observed for the ionic radii. This difference between the slopes of curves may originate from differences in coordination numbers. The coordination number is smaller in the dihalides than in the octahedral complexes. The van der Waals repulsion of the ligands may counter the attraction by the central atom in the octahedral environment and may partially compensate for the imperfect shielding. On the other hand, van der Waals repulsion of the ligands probably has a negligibly small influence on the metal-halogen bond length in the dihalides.

Even more interesting is the extent of deviation of the experimental data from the smooth curves. Here, the explanation may be sought by considering molecular symmetry, i.e., the symmetry of the "ligand field". Assuming that these molecules are linear, the splitting of the d orbitals will be different from that in the case of octahedral environment. Fig. 6-42 depicts the two

Figure 6-42.
d orbital splittings in octahedral (O_h), linear $(D_{\infty h})$, and in C_{2v} symmetry ligand environment.

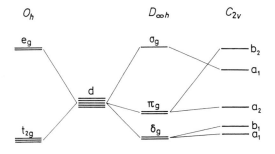

different orbital splittings. In the $D_{\infty h}$ symmetry dihalides, only the d_{z^2} orbital is oriented towards the ligands. Since this is energetically the least favorable orbital, it will be occupied only by the fifth and the tenth electrons. Thus, the least shielding occurs with electrons four and nine. Accordingly, the largest deviations from the Ca – Mn – Zn line could be anticipated for the bond lengths in $CrCl_2$ and $CuCl_2$.

These arguments [6-15] are valid only if all the first row transition metal dihalides are linear. According to various experimental evidences [6-16, 6-18 to 6-21] this assumption is correct for the dihalides of manganese, iron, cobalt, and nickel. *Ab initio* calculations on $CuCl_2$ suggest that this dihalide also has linear geometry [6-17]. Linearity has been supposed for a long time for the early transition metal dihalides as well [6-20, 6-21]. New experimental evidence, however, seems to indicate, that VCl_2 and $CrCl_2$ are highly bent molecules [6-22]. In this case, their symmetry is C_{2v}, and the d orbital splitting is different from that in a linear environment (Fig. 6-42). The relative energies of the orbitals cannot be predicted by qualitative reasoning in this non-linear environment. The order given here for the bent molecule originates from results of quantum chemical calculations [6-23]. The prediction for the bond length variation for the linear arrangements, of course, cannot be valid for VCl_2 and $CrCl_2$ as they are bent.

6.5 Jahn-Teller Effect

"Somewhat paradoxically, symmetry is seen to play an important role in the understanding of the ... Jahn-Teller effect, the very nature of which is symmetry destruction" [6-24]. According to the original formulation of the Jahn-Teller effect [6-25], a non-linear symmetrical nuclear configuration in a degenerate electronic state is unstable and gets distorted, thereby removing the electronic degeneracy until a non-degenerate ground state is achieved. This formulation indicates the strong relevance of this effect to orbital splitting and generally to the relationship of symmetry and electronic structure discussed in previous sections.

Only molecules with partially filled orbitals display Jahn-Teller distortion. As was shown in Section 6.3.2, the electronic ground state of molecules with completely filled orbitals is always totally symmetric, and thus cannot be degenerate. Since transition metals have partially filled d or f orbitals, their compounds may be Jahn-Teller systems.

Let us consider an example from among the much studied cupric compounds (cf. [6-14]). Suppose that the Cu^{2+} ion with its d^9 electronic configuration is surrounded by six ligands in an octahedral arrangement. We have already seen (Table 6-12 and Fig. 6-40) that in an octahedral environment d orbitals split into a triply (t_{2g}) and a doubly (e_g) degenerate level. For Cu^{2+} the only possible electronic configuration is $t_{2g}^6 e_g^3$.

Suppose now that of the two e_g orbitals, d_{z^2} is doubly, while $d_{x^2-y^2}$ is only singly occupied. Thus the two ligands along the z axis are better screened from the electrostatic attraction of the central ion, and will move farther away from it, than the four ligands in the xy plane. The opposite happens if the unpaired electron occupies the d_{z^2} orbital. In both cases the octahedral arrangement undergoes tetragonal distortion along the z axis, in the former by elongation, while in the latter, by compression. The original O_h symmetry reduces to D_{4h}. The splitting of d orbitals in both environments is given in Table 6-12 and is also shown here:

O_h	$\rightarrow$	D_{4h}		
e_g $(d_{x^2-y^2},\ d_{z^2})$	$\rightarrow$	a_{1g} (d_{z^2})	$+$	b_{1g} $(d_{x^2-y^2})$
t_{2g} $(d_{xz},\ d_{yz},\ d_{xy})$	$\rightarrow$	e_g $(d_{xz},\ d_{yz})$	$+$	b_{2g} (d_{xy})

Fig. 6-43 illustrates the tetragonal elongation and compression for an octahedron. For the Cu^{2+} ion the relative energies of the d_{z^2} and d_{x-y^2} orbitals depend on the location of the unpaired electron.

Figure 6-43.
Tetragonal distortions of the regular octahedral arrangement around a d^9 ion.

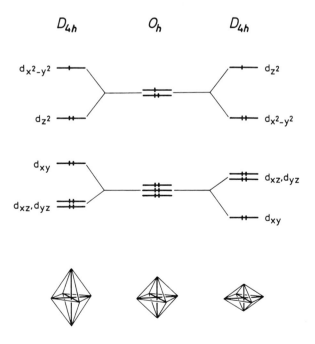

274

Consider now a qualitative picture of the splitting of the t_{2g} orbitals. If the ligands are somewhat further away along the z axis, their interaction with the d_{xz} and d_{yz} orbitals will decrease, and so will their energy, compared with that of the d_{xy} orbital. This is illustrated by the left-hand side of Fig. 6-43. Tetragonal compression can be accounted for by similar reasoning (cf. right-hand side of Fig. 6-43).

The splitting of the d orbitals in Fig. 6-43 shows the validity of the "center of gravity rule". One of the e_g orbitals goes up in energy as much as the other goes down. From among the t_{2g} orbitals the doubly degenerate pair goes up (or down) in energy half as much as the non-degenerate orbital goes down (or up). Thus, for the Cu(II) compounds the splitting of the fully occupied t_{2g} orbitals does not bring about a net energy change. The same is true for all other symmetrically occupied degenerate orbitals, such as t_{2g}^3, e_g^4, or e_g^2. On the other hand, the occupancy of the e_g orbitals of Cu^{2+} is unsymmetrical, since two electrons go down and only one goes up in energy, and here there is a net energy gain in the tetragonal distortion. This energy gain is the Jahn-Teller stabilization energy.

The above example referred to an octahedral configuration. Other highly symmetrical systems, for example tetrahedral arrangements, can also display this effect. For general discussion, see, e.g. [6-26 – 6-28].

The Jahn-Teller effect enhances the structural diversity of Cu(II) compounds [6-29]. Most of the octahedral complexes of Cu^{2+}, for example, show elongated tetragonally distorted geometry. Crystalline cupric fluoride and cupric chloride both have four shorter and two longer copper-halogen interatomic distances, 1.93 vs. 2.27 Å and 2.30 vs. 2.95 Å, respectively [6-29].

The square planar arrangement can be regarded as a limiting case of the elongated octahedral configuration. The four oxygen atoms are at 1.96 Å from the copper atom in a square configuration in crystalline cupric oxide, whereas the next nearest neighbors, two other oxygen atoms, are at 2.78 Å. The ratio of the two distances is much larger than in the usual distorted octahedral configuration [6-29].

Tetragonal compression around the central Cu^{2+} ion is much rarer. K_2CuF_4 is an example with two shorter and four longer Cu-F distances, viz. 1.95 vs. 2.08 Å [6-29].

There are also numerous cases when experimental investigation fails to provide evidence for Jahn-Teller distortion. For example, several chelate compounds of Cu(II), as well as some compounds containing the $[Cu(NO_2)_6]^{4-}$ ion show no detectable distortion from the regular octahedral structure (see [6-30] and references therein).

Bersuker [6-27, 6-28] has shown the need for a more sophisticated approach to account for such phenomena. We attempt to convey at least the flavor of his ideas here. Jahn-Teller distortions are of a *dynamic* nature in systems under no external influence. This means that there may be many minimum-energy distorted structures in such systems. Whether an experiment will or will not detect such a dynamic Jahn-Teller effect, depends on the relationship between the time scale of the physical measurement used for the investigation, and the mean lifetime of the distorted configurations. If the time of the measurement is longer than the mean life-time of the distorted configurations, only an average structure will be detected corresponding to the undistorted high-symmetry configuration. Since different physical techniques have different time scales, one technique may detect a distortion which appears to be undetected by another.

The *static* Jahn-Teller effect can be observed only in the presence of an external influence. Bersuker [6-28] stresses this point as the opposite statement is found often in the literature. According to the statement criticized, the effect is not to be expected in systems where low-symmetry perturbations remove electronic degeneracy. According to Bersuker, it is exactly the low-symmetry perturbations that make the Jahn-Teller distortions static and thus observable. Such a low-symmetry perturbation can be the substitution of one ligand by another. In this case one of the previously equivalent minimum-energy structures, or a new one, will become energetically more favored than the others.

The so-called *cooperative* Jahn-Teller effect is another occurrence of the static distortions. Here interaction, i.e., cooperation between different crystal centers make the phenomenon observable. Without interaction, the nuclear motion around each center would be independent and of a dynamic character.

Lattice vibrations tend to destroy the correlation among Jahn-Teller centers. Thus with increasing temperature, at a certain point, these centers may become independent of each other, and their static Jahn-Teller effects convert to dynamic ones. At this point the crystal as a whole becomes more symmetric. This temperature-dependent static ⇄ dynamic transition is called the *Jahn-Teller phase transition*. Below the phase transition temperature the cooperative Jahn-Teller effect governs the situation providing static distortion; the overall structure of the crystal is of a lesser symmetry. Above this temperature the cooperation breaks down, the Jahn-Teller distortion becomes dynamic and the crystal itself becomes more symmetric.

The temperature of the Jahn-Teller phase transition is very high for CuF_2, $CuCl_2$, and K_2CuF_4 among the examples mentioned above [6-28]. Therefore, at room temperature their crystal structures display distortions. Other compounds have symmetric crystal structures at room temperature as their Jahn-Teller phase transition occurs at lower temperatures. Cupric chelate compounds and $[Cu(NO_2)_6]^{4-}$ compounds, such as $K_2PbCu(NO_2)_6$ and $Tl_2PbCu(NO_2)_6$, can be mentioned as examples [6-30]. Further cooling, however, may make also these structures distorted [6-30].

6.6 Quantum Chemical Calculations

The results of quantum chemical calculations usually yield the wave functions and the energies of a system. Numerous integrals must be evaluated even for the simplest molecules. Their number can be conveniently reduced, however, by applying the theorem according to which an energy integral, $\int \psi_i \hat{H} \psi_j d\tau$ is nonzero only if ψ_i and ψ_j belong to the *same* irreducible representation of the molecular point group.

Many chemical and physical properties of the molecules can be calculated, including conformational properties, barrier to internal rotation, relative stabilities of various isomers as well as different electronic states. The electron spectroscopic and vibrational spectroscopic constants and other parameters can also be determined. We will now consider just one of the many characteristics of the molecule, viz. its equilibrium geometry. Today, the state-of-the-art calculations of molecular geometry involving relatively light atoms are as reliable as the results of the best experiments. Another thing to bear in mind is that while calculations provide information relating to the equilibrium geometry, various experiments yield some effective geometries for the molecule which are averaged over intramolecular vibrations. Depending on the magnitude of these vibrations and their structural influence, the equilibrium and average structures may differ to various extents. The results of calculations are less reliable for molecules involving heavier atoms, for example, transition metals. Even less sophisticated calculations may be instructive if structural differences rather than absolute values of the structural parameters are sought. Important systematic errors usually cancel in the determination of structural differences in calculations as well as in experiments. The importance of small structural differences in understanding various effects in series of substances is becoming recognized increasingly [6-31]. Quantum chemical calculations, of course, are the exclusive source of information for systems that are not amenable to experimental study. Such systems may include unstable or even unknown species and transition states. Quantum chemical calculations have proved to be complementary with experiments or can even be their alternatives. The situa-

277

tion is evolving and a host of problems must be resolved yet for individual systems. A case in point involves calculations of the geometrical parameters of transition metal compounds.

It is obvious that the larger the size and number of the atoms in a molecule, the more complicated and time-consuming the calculation of the molecular energy and geometry will be. Since the heavier elements have many electrons, a large number of one-electron wave functions must be used to approximate their electronic wave function. The problem of electron correlation also becomes increasingly important. There are many ways to construct sets of basis functions for these molecules. It is, however, important to choose the right sets, since the reliability of the results depends strongly on them [6-32]. For the main group elements, especially for molecules composed of lighter atoms, systematic studies have been carried out; consequently, a number of generally applicable methods and basis sets have been developed.

For transition metal compounds another difficulty arises due to the open shell electronic structure of the transition metal atom. Yet another problem is their molecular size, especially that of organometallic complexes. Even with today's computers they are often too large for routine calculations. Therefore, no systematic investigations have yet been carried out on these systems, and generally acceptable calculation procedures and optimal basis sets have not yet been developed. New development is anticipated in this area in the near future.

Bond lengths of transition metal compounds. It has been observed recently, that some bond lengths calculated by quantum chemical methods for transition metal compounds are not compatible with the experimental data [6-23, 6-33]. Since the calculated geometry corresponds to the equilibrium structure, the calculated bond lengths are expected to be smaller than the corresponding experimental values. The latter are usually measured at higher temperatures where the enhanced stretching vibrations result somewhat elongated bonds. In case of $MnCl_2$ the experimental bond length is: 2.205(5) Å (from electron diffraction [6-16]), compared to the calculated values of 2.267-2.294 Å, obtained with high-quality basis sets [6-23]. Interestingly, minimal basis sets yielded 2.149 and 2.208 Å in better agreement with the experiment. High-quality theoretical calculation [6-33] of the geometry of ferrocene, $(C_5H_5)_2Fe$, yielded a 1.89 Å metal-ring distance while the experimental results are 1.66 Å (from electron diffraction [6-34]), 1.64 Å (from neutron diffraction [6-35]), and 1.65 Å (from X-ray diffraction [6-36]). Although it is difficult to account for this discrepancy, one possibility may lie in the choice of basis sets, i.e., the atomic orbitals used for the MO construction [6-33].

In another study the distance in Cu_2 was found to be highly sensitive to the choice of basis sets [6-37]. The Cu-Cu distance varied from 2.248 Å to 2.40 Å; the experimental bond distance is 2.2195 Å [6-38]. Strangely enough, another theoretical investigation, which employed a much less sophisticated method that is generally not applicable to geometry determinations, yielded a Cu-Cu distance of 2.17 Å [6-39]. This result is in better agreement with the experiment.

Discrepancies do not always occur; examples where good agreement between calculated and experimental bond lengths is obtained are also known. For example, in ScF_3, the calculated Sc-F bond distance of 1.88 Å [6-40] compares well with the experimental value of 1.926(5) Å (electron diffraction [6-41]).

It will probably be some time before enough theoretical calculations will have been performed on transition metal compounds to permit full understanding the reasons that underly these failures and successes. However, further theoretical studies will eventually lead to a consistent and reliable approach. The ever-increasing capacity of computers is bringing investigators even closer to the attainment of these goals. Until the goals will be reached continued critical evaluation of computational results and the concerted application of theory and experiment are especially important.

References

[6-1]   M. W. Hanna, *Quantum Mechanics in Chemistry,* Second Edition, W. A. Benjamin, Inc., New York, Amsterdam, 1969.

[6-2]   J. N. Murrel, S. F. A. Kettle, and J. M. Tedder, *Valence Theory,* Second Edition, John Wiley & Sons, New York, 1970.

[6-3]   D. V. George, *Principles of Quantum Chemistry,* Pergamon Press, Inc., New York, 1972.

[6-4]   F. A. Cotton, *Chemical Applications of Group Theory,* Second Edition, Wiley-Interscience, New York, 1971.

[6-5]   D. C. Harris and M. D. Bertolucci, *Symmetry and Spectroscopy, An Introduction to Vibrational and Electronic Spectroscopy,* Oxford University Press, New York, 1978.

[6-6]   T. H. Lowry and K. S. Richardson, *Mechanism and Theory in Organic Chemistry,* Second Edition, Harper & Row, Publishers, New York, 1981.

[6-7]   M. Orchin and H. H. Jaffe, *Symmetry, Orbitals, and Spectra (S.O.S.),* Wiley-Interscience, New York, 1971.

[6-8]   E. P. Wigner, *Group Theory and its Application to the Quantum Mechanics of Atomic Spectra,* Academic Press, New York, 1959.

[6-9]   A. Streitwieser, Jr. and P. H. Owens, *Orbital and Electron Density Diagrams, An Application of Computer Graphics,* The Macmillan Company, New York, 1973.

[6-10]   C. A. Coulson, *The Shape and Structure of Molecules,* Clarendon Press, Oxford, 1973.

[6-11]   J. Demuynck and H. F. Schaefer III. J. Chem. Phys. **72**, 311 (1980).

[6-12]   W. L. Jorgensen and L. Salem, *The Organic Chemist's Book of Orbitals,* Academic Press, New York, 1973.

[6-13]   H. Bethe, Ann. Phys. **3**, 133 (1929).

[6-14]   F. A. Cotton and G. Wilkinson, *Advanced Inorganic Chemistry,* Third Edition, Interscience, New York, 1972.

[6-15]   M. Hargittai, Inorg. Chim. Acta Lett. **53**, L111 (1981).

[6-16]   M. Hargittai, O. V. Dorofeeva, and J. Tremmel, Inorg. Chem., **24**, 245 (1985); I. Hargittai and J. Tremmel, Coord. Chem. Rev., **18**, 257 (1976); and references therein.

[6-17]   D. W. Smith, Inorg. Chim. Acta **22**, 107 (1977).

[6-18]   G. E. Leroi, T. C. James, J. T. Hougen, and W. Klemperer, J. Chem. Phys. **36**, 2879 (1962).

[6-19]   K. R. Thompson and K. D. Carlson, J. Chem. Phys. **49**, 4379 (1968).

[6-20]   J. W. Hastie, R. H. Hauge, and J. L. Margrave, High Temp. Science **3**, 257 (1971).

[6-21]   M. C. Drake and G. M. Rosenblatt, J. Electrochem. Soc. **126**, 1387 (1979).

[6-22]   M. Hargittai, O. V. Dorofeeva, and J. Tremmel, Inorg. Chem. **24**, 3963 (1985).

[6-23]   M. Hargittai and A. Rossi, Inorg. Chem. **24**, 4758 (1985).

[6-24]   A. Ceulemans, D. Beyens, and L. G. Vanquickenborne, J. Am. Chem. Soc. **106**, 5824 (1984).

[6-25]   H. A. Jahn and E. Teller, Proc. Roy. Soc. **A161**, 220 (1937).

[6-26]   R. Englman, *The Jahn-Teller Effect in Molecules and Crystals,* Wiley, New York, 1972.

[6-27]   I. B. Bersuker, *The Jahn-Teller Effect and Vibronic Interactions in Modern Chemistry,* Plenum Press, New York, 1984.

[6-28]   I. B. Bersuker, Coord. Chem. Rew. **14**, 357 (1957).

[6-29]   A. F. Wells, *Structural Inorganic Chemistry,* Fourth Edition, Clarendon Press, Oxford, 1975.

[6-30]   J. E. Huheey, *Inorganic Chemistry, Principles of Structure and Reactivity,* Third Edition, Harper & Row, Publishers, New York, 1983.

[6-31]   I. Hargittai and M. Hargittai, in *Molecular Structure and Energetics,* Vol. 2, Chapter 1, J. F. Liebman and A. Greenberg, eds., VCH Publishers, Dearfield Beach, FL, 1986.

[6-32]   H. F. Schafer III. J. Mol. Struct. **76**, 117 (1981).

[6-33]   H. P. Luthi, J. H. Ammeter, J. Almløf, and K. Faegri, Jr., J. Chem. Phys. **77**, 2002 (1982).

[6-34]   A. Haaland and J. E. Nilsson, Acta Chem. Scand. **22**, 2653 (1968).

[6-35]   F. Takusagawa and T. F. Koetzle, Acta Crystallogr. **B35**, 1074 (1979).

[6-36]   P. Seiler and J. D. Dunitz, Acta Crystallogr. **B35**, 1068; 2020 (1979).

[6-37]   M. Pelissier, J. Chem. Phys. **75**, 775 (1981).

[6-38]   J. Lochet, J. Phys. B **11**, 55 (1978).

[6-39]   G. A. Ozin, H. Huber, D. McIntoch, S. Mitchell, J. G. Norman, Jr., and L. Noodleman, J. Am. Chem. Soc. **101**, 3504 (1979).

[6-40]   J. H. Yates and R. M. Pitzer, J. Chem. Phys. **70**, 4049 (1979).

[6-41]   N. I. Giricheva, E. Z. Zasorin, G. V. Girichev, K. S. Krasnov, and V. P. Spiridonov, Zh. Strukt. Khim. **17**, 797 (1976) [Russ. J. Struct. Chem., **17**, 686 (1976)].

7 Chemical Reactions

The chemical reaction is the "most chemical" event. Our encounter with the role of symmetry in chemistry would certainly be one-sided without looking at chemical reactions. In fact, it is perhaps the most flourishing, booming area today of all chemistry-related applications of the symmetry concept. For this very reason, we shall present only a short survey and refer to the vast recent literature (see e.g. [7-1 through 7-11]). Our discussion fully relies on these papers and monographs.

The first application of symmetry considerations to chemical reactions can be attributed to Wigner and Witmer [7-12]. The Wigner-Witmer rules are concerned with the conservation of spin and orbital angular momentum in the reaction of diatomic molecules. Although symmetry is not explicitly mentioned, it is present implicitly in the principle of conservation of orbital angular momentum. The real breakthrough in recognizing the role that symmetry plays in determining the course of chemical reactions has occurred only recently, mainly through the activities of Woodward and Hoffmann, Fukui, Pearson, and others.

The main idea in their work is that symmetry phenomena may play as important a role in chemical reactions as they do in the construction of molecular orbitals or in molecular spectroscopy. It is even possible to make certain symmetry based "selection rules" for the "allowedness" and "forbiddenness" of a chemical reaction, just as is done for spectroscopic transitions.

Before describing these rules, however, we would like to mention some limitations. Symmetry rules can usually be applied to comparatively simple reactions, the so-called *concerted reactions*. In a concerted reaction all relevant changes occur simultaneously; the transformation of reactants into products happens in one step with no intermediates.

At first sight it would seem logical that symmetry rules can be applied only to symmetrical molecules. However, even nonsymmetric reactants can be "simplified" to related symmetrical parent molecules. As Woodward and Hoffmann put it, they can be "reduced to their highest inherent symmetry" [7-3]. This is, in fact, a necessary criterion if symmetry principles are to be applied.

What does this mean? For example, propylene, $H_2C = CHCH_3$, must be treated as its "parent molecule", ethylene. The reason is that it is the double bond of propylene which changes during the reaction, and it nearly possesses the symmetry of ethylene. Salem calls this feature "pseudosymmetry" [7-7].

The statement: a chemical reaction is "symmetry allowed" or "symmetry forbidden", should not be taken literally. When a reaction is symmetry allowed, it means that it has a low activation energy. This makes it possible for the given reaction to occur, though it does not mean that it always will. There are other factors which can impose a substantial activation barrier. Such factors may be steric repulsions, difficulties in approach and unfavorable relative energies of orbitals. Similarly, "symmetry forbidden" means that the reaction, as a concerted one, would have a high activation barrier. However, various factors may make the reaction still possible; for example, it may happen as a stepwise reaction through intermediates. In this case, of course, it is no longer a concerted reaction.

Most of the symmetry rules explaining and predicting chemical reactions deal with changes in the electronic structure. However, a chemical reaction is more than just that. Breakage of bonds and formation of new ones are also accompanied by nuclear rearrangements and changes in the vibrational behavior of the molecule. (Molecular translation and rotation as a whole can be ignored.)

As has been shown previously, both the vibrational motion and the electronic structure of the molecules strongly depend on symmetry. This dependence can be fully utilized when discussing chemical reactions.

Describing the structures of both reactant and product molecules with the help of symmetry would not add anything new to our previous discussion. What is new and important is that certain symmetry rules can be applied to the transition state in between the reactants and products. This is indeed the topic of the present Chapter.

7.1 Potential Energy Surface

The potential energy surface is the cornerstone of all theoretical studies of reaction mechanisms [7-7]. The topography of a potential energy surface contains all possible information about a chemical reaction. However, how this potential energy surface can be depicted is another matter.

The total energy of a molecule consists of the potential and the kinetic energy of both the nuclei and the electrons. The coulombic energy of the nuclei and the electronic energy together represent the whole potential energy under whose influence the nuclei carry out their vibrations. Since the energies of the (ground and various excited) electronic states are different, each state has its own potential energy surface. We are usually interested in the lowest energy potential surface which corresponds to the ground state of the molecule. An N atomic molecule has $3N-6$ internal degrees of freedom (a linear molecule has $3N-5$). The potential energy for such a molecule can be represented by a $3N-6$ dimensional hypersurface in a $3N-5$ dimensional space. Clearly the actual representation of this surface is impossible in our limited dimensions.

There are ways, however, to plot parts of the potential energy hypersurface. For example, the energy is plotted with respect to the change of two coordinates during a reaction in both Figs. 7-1a and b. Such drawings help to visualize the real potential energy surface. It is like a rough topographic map with mountains of different heights, long valleys of different depths, mountain paths and holes. Since energy increases along the vertical coordinate, the mountains correspond to energy barriers and the holes and valleys to different energy minima.

Studying reaction mechanisms means essentially finding the most economical way to go from one valley to another. Two adjacent valleys are connected by a mountain path: this is the road that the reactant molecules must follow if they want to reach the valley on the other side, which will correspond to the product(s). The top of the pass is called the *saddle point* or *col*. The configuration of nuclei at this point is sometimes called a *transition state*, sometimes an *activated complex*, in other cases a *transition structure*, and yet in other cases, a *supermolecule*. Transition state is the most commonly used term.

Figure 7-1.
Three-dimensional potential energy surfaces.
(a) Energy hypersurface for FSSF $\rightleftarrows$ SSF$_2$ isomerization (detail). Reproduced with permission from Solouki and Bock [7-13(a)]. Copyright (1977) American Chemical Society.

(b) Rotation-inversion surface of $^-$CH$_2$OH (detail). Reproduced with permission from Wolfe et al. [7-13(b)]. Copyright (1975) American Chemical Society.

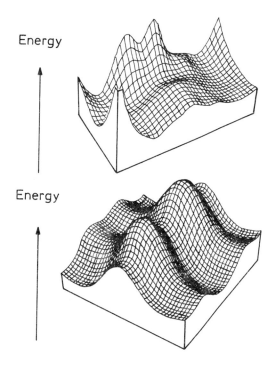

Energy

Energy

285

7.1.1 Reaction Coordinate

Where does symmetry come into the picture? It happens through the movement of the nuclei along the potential energy surface. As discussed in detail in Chapter 5, all possible internuclear motions of a molecule can be resolved into sets of special motions corresponding to the normal modes of the molecule. These normal modes already have a symmetry label since they belong to one of the irreducible representations of the molecular point group. The changing nuclear positions during the course of a reaction are collectively described by the term *reaction coordinate*. It is usually a good approximation, however, to assume that the chemical reaction is dominated by one of the normal modes of vibration and that the other modes remain essentially unchanged during the reaction. In such a case, this one vibrational mode is the reaction coordinate. By selecting this coordinate, we may cut a slice through the potential energy hypersurface along this particular motion. Such a slice is shown in Fig. 7-2. This is another way to represent the potential energy. The curve shows the reaction path along the reaction coordinate. Points a and c are energy minima, corresponding to the initial and final stages of the reaction. Point b is the saddle point corresponding to the transition structure and the energy barrier. This diagram has several important features.

Figure 7-2.
Potential energy curve plotted against the reaction coordinate (a and c are minimum positions, b is the saddle point or col).

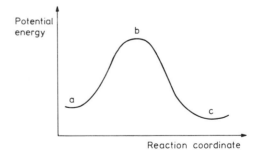

First, it represents only a slice of the potential energy hypersurface. It is the variation along one coordinate, and it is supposed that all the other possible motions of the nuclei, i.e., all the other normal vibrations, are at their optimum value, so their energy is at minimum. Therefore this reaction path can be taken as a *minimum energy path*. If the rearrangements in the reactant molecules or in the supermolecule are really connected only by one particular vibrational mode, then this supposition is reasonable, since the other vibrations do not influence the reaction process. The vibration corresponding to the reaction

Figure 7-3.
Contour diagram of a three-dimensional potential energy surface. The bold line shows the reaction path. Adapted with permission from [7-14]. Copyright (1963) John Wiley & Sons.

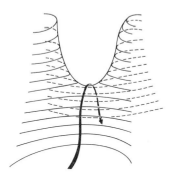

coordinate is orthogonal to all the other vibrations. In other words, if we would try to leave the reaction path sideways, i.e., along some other vibrational mode, the energy would always increase.

Fig. 7-3 illustrates this point after [7-14] with a simple case when only one other vibrational mode exists beside the reaction coordinate. The bold line shows the reaction path. It goes through a maximum point which is the reaction barrier. The surface, however, rises on both sides of the reaction coordinate. Thus, with respect to the energy of the other vibrational coordinates, the reaction follows a minimum energy path indeed.

7.1.2 Symmetry Rules for the Reaction Coordinate

The prediction of molecular structures and possible reaction mechanisms through analysis of the reaction coordinate is mainly due to Pearson [7-6]. The energy variation along the reaction path can be characterized in the following way:

The energy of all vibrational modes except the reaction coordinate is minimal all along the path, i.e.

$$\frac{\partial E}{\partial Q_i} = 0 \text{ and } \frac{\partial^2 E}{\partial Q_i^2} > 0. \tag{7-1}$$

where Q_i is any coordinate except Q_r, the reaction coordinate. With respect to symmetry these vibrations are unrestricted. (Of course, every normal mode must belong to one or another irreducible representation of the molecular point group.)

The energy variation of the reaction coordinate is different. At every point, except at the maximum and minimum values, it is nonzero,

$$\frac{\partial E}{\partial Q_r} \neq 0. \tag{7-2}$$

This is simply the slope of the curve on the potential energy diagram. At the minimum points (a and c in Fig. 7-2):

$$\frac{\partial E}{\partial Q_r} = 0 \quad \text{and} \quad \frac{\partial^2 E}{\partial Q_r^2} > 0. \tag{7-3}$$

At the saddle point (b in Fig. 7-2):

$$\frac{\partial E}{\partial Q_r} = 0 \quad \text{and} \quad \frac{\partial^2 E}{\partial Q_r^2} < 0. \tag{7-4}$$

This result has important consequences regarding the symmetry properties of the reaction coordinate (see [7-6] for details). Calculation of the energy of the reaction involves two different energy integrals:

$$\left\langle \psi_o \left| \frac{\partial E}{\partial Q_r} \right| \psi_o \right\rangle \quad \text{and} \quad \left\langle \psi_o \left| \frac{\partial E}{\partial Q_r} \right| \psi_i \right\rangle \tag{7-5}$$

where ψ_o and ψ_i are the wave function of the ground state and an excited state, respectively. In the actual calculations, these wave functions are approximated by molecular orbitals, but their relationship remains the same.

Examine now the two energy integrals separately, bearing in mind what was said about the conditions necessary for an integral to have nonzero value (Chapter 4). The first integral contains only the ground state wave function. It will have nonzero value only if

$$\Gamma_{\psi_o} \cdot \Gamma_{\psi_o} \subset \Gamma_{Q_r} \tag{7-6}$$

that is, if the direct product of the representation of ψ_o with itself (a function with the *same* symmetry) contains the representation of Q_r. But we know that the direct product of two functions with the same symmetry always contains the totally symmetric representation. Therefore, Q_r must belong to the totally symmetric irreducible representation of the molecular point group so that the integral will have a nonzero value. We can conclude that, except at a maximum or at a minimum, the reaction coordinate belongs to the totally symmetric irreducible representation of the molecular point group.

The reaction coordinate is just one particular normal mode in the simplest case. It must always be, however, a symmetric mode, and this is so even if a more complicated nuclear motion is considered for the reaction coordinate. Such a motion can always be written as a sum of normal modes. Of these modes, however, only those which are totally symmetric will contribute to the reaction coordinate. The nonsymmetric modes may contribute only at the extremes of the potential energy function.

The other integral in Eq. (7-5) expresses the mixing of states during the reaction:

$$\left\langle \psi_o \left| \frac{\partial E}{\partial Q_r} \right| \psi_i \right\rangle. \tag{7-7}$$

This integral will be nonzero if the direct product of the representations of the wave functions ψ_o and ψ_i contains the representation to which the reaction coordinate belongs,

$$\Gamma_{\psi_o} \cdot \Gamma_{\psi_i} \subset \Gamma_{Q_r}. \qquad (7\text{-}8)$$

This expression contains information regarding the symmetry of the excited states as well. Only those excited states can participate in the reaction whose symmetry matches the symmetry of both the ground state and the reaction coordinate. But we already know that Q_r belongs to the totally symmetric irreducible representation except at the extreme points. This implies that only those excited states can participate in the reaction whose symmetry is the same as that of the ground state. This information is instrumental in the construction of correlation diagrams, as will be seen later.

The reaction coordinate can possess any symmetry at the extreme points provided that the condition of Eq. (7-8) is fulfilled. This also means that at the maximum point the symmetry of the excited state may differ from that of the ground state. However, any minute distortion will remove the system from the saddle point. The reaction coordinate must then become again totally symmetric. How can this happen? Obviously, by changing the point group of the system. By reducing the symmetry, nonsymmetric vibrational modes may become symmetric, and the reaction coordinate may become totally symmetric. This reasoning may even help in predicting how the symmetry will be reduced; we just have to find the point group in which the reaction coordinate becomes totally symmetric.

Two examples will illustrate how these rules work. One involves the reduction of symmetry which occurs when a linear molecule becomes bent [7-6]. The other example involves transforming a planar molecule into a pyramidal one.

For a linear AX_2 molecule of $D_{\infty h}$ symmetry the normal mode that reduces it to C_{2v} is the Π_u bending mode (Fig. 7-4a). In the C_{2v} point group this becomes totally symmetric. (The other component of the Π_u mode becomes the rotation of the molecule.)

Figure 7-4.
The effect of symmetry lowering on the reaction coordinate. (a) Bending of a linear AX_2 molecule $(v_2(\pi_u) \rightarrow v_2(A_1))$

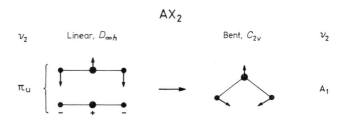

For an AX_3 planar molecule the symmetry is D_{3h}. The puckering mode (Fig. 7-4b) of A_2'' symmetry reduces it to C_{3v}. In the C_{3v} point group, the symmetry of this vibration is A_1.

(b) Puckering of a planar AX_3 molecule $(\nu_2(A_2'') \rightarrow \nu_2(A_1))$.

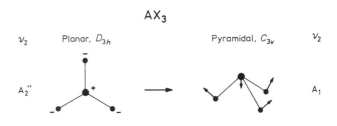

7.2 Electronic Structure

7.2.1 Changes During a Chemical Reaction

A chemical reaction is a consequence of an interaction between molecules. The electronic aspects of these interactions can be discussed in much the same way as the interaction of atomic electron distributions forming molecules. The difference is that while MOs are constructed from the AOs of the constituent atoms, in describing a chemical reaction the MOs of the product(s) are constructed from the MOs of the reactants. Before a reaction takes place (i.e., while the reacting molecules are still far apart), their electron distribution is unperturbed. When they approach each other, their orbitals begin to overlap, and distortion of the original electron distribution takes place. There are two requirements for a constructive interaction between molecules: symmetry matching and energy matching. These two factors can be treated in different ways. The approaches of Fukui [7-1, 7-2], and of Woodward and Hoffmann [7-3, 7-4] differ somewhat. Since those are two most successful methods in this field, we shall concentrate on them. First the basis of each method will be presented briefly, followed by a few classical examples, each of which will be treated in some detail.

7.2.2 Frontier Orbitals: HOMO and LUMO

A successful chemical reaction requires both energy and symmetry matching between the MOs of the reactants. The requirements are essentially the same as in case of constructing MOs from AOs; only orbitals of the same symmetry and comparable energy can overlap successfully. The strongest interactions occur between those orbitals which are close to each other in energy. Nevertheless, the interaction between filled MOs will not contribute to the change in the total energy of the system since one orbital increases in energy about as much as the other decreases (see Fig. 7-5). The most important interactions occur between the filled orbitals of one molecule and the vacant orbitals of the other. Moreover, since the interaction is strongest for energetically similar orbitals, the most significant interactions can be expected between the highest occupied molecular orbital (HOMO) of one molecule and the lowest unoccupied molecular orbital (LUMO) of the other (Fig. 7-6). The labels HOMO and LUMO were incorporated by Fukui into a descriptive collective name: *frontier orbitals*. The first article in this topic appeared in 1952 [7-15], and the idea has been extended to a host of different reactions in the succeeding years (see, e.g., [7-1, 7-2]).

Figure 7-5.
Interaction of two filled orbitals. There is no energy gain, the reaction will not occur.

Figure 7-6.
Interaction of the highest occupied MO (HOMO) of one molecule with the lowest unoccupied MO (LUMO) of another molecule.

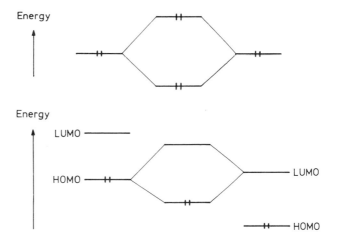

Fukui [7-1] recognized the importance of the symmetry properties of HOMOs and LUMOs perhaps for the first time in connection with the Diels-Alder reaction. According to his Nobel Lecture [7-11], however, it was only after the appearance of the papers by Woodward and Hoffmann in 1965 that he "became fully aware that not only the density distribution but also the nodal property" – that is, symmetry – "of the particular orbitals have significance in ... chemical reactions."

The concept of frontier orbitals simplifies the MO description of chemical reactions enormously, since only these MOs of the reactant molecules need to be considered. Several examples of this approach will be given.

7.2.3 Conservation of Orbital Symmetry

The first papers outlining and utilizing this idea appeared in 1965 [7-16 to 7-18]. Salem [7-7] called the discovery of orbital symmetry conservation a revolution in chemistry. "It was a major breakthrough in the field of chemical reactions in which notions preexisting in other fields (orbital correlations by Mulliken, and nodal properties of orbitals in conjugated systems, by Coulson and Longuet-Higgins) were applied with great conceptual brilliance to a far-reaching problem. Chemical reactions were suddenly adorned with novel significance." [7-7]

The idea and the principles of drawing correlation diagrams follows directly from the atomic correlation diagrams of Hund and Mulliken [7-19]. They are very useful for predicting the "allowedness" of a given concerted reaction. In constructing correlation diagrams, both the energy and the symmetry aspects of the problem must be considered. On one side of the diagram the approximate energy levels of the reactants are drawn, while on the other side those of the product(s) are indicated. A particular geometry of approach must be assumed. Furthermore, the symmetry properties of the molecular orbitals must be considered in the framework of the point group of the supermolecule. In contrast to the frontier orbital method, it is not necessarily the HOMOs and LUMOs which are considered. Instead, attention is focused upon those molecular orbitals which are associated with bonds that are broken or formed during the chemical reaction. We know that each acceptable molecular orbital must belong to one irreducible representation of the point group of the system. At least for nondegenerate point groups, this MO must be either symmetric or antisymmetric with respect to any symmetry element that may be present. (The character under any operation is either 1 or –1.)

Among all possible symmetry elements, those must be considered which are maintained throughout the approach and which bisect bonds that are either formed or broken during the reaction. There must always be at least one such symmetry element. The next step is to connect levels of like symmetry without violating the so-called *noncrossing rule*. According to this rule, two orbitals of the same symmetry cannot intersect [7-20]. Thus the correlation diagram is completed. These diagrams yield valuable information about the transition state of the chemical reaction. The method will be illustrated with examples.

7.2.4 Analysis in Maximum Symmetry

In this approach two points are considered when predicting whether or not a chemical reaction can occur. One such point involves the allowedness of an electron transfer from one orbital to another. The other involves consideration of the reaction-decisive normal vibration. In both cases symmetry arguments are used. This approach is thorough and is similar in part to the Pearson method and in part to the Woodward-Hoffmann method. It incorporates several features of each method in a rigorous group-theoretical way. First, the transformation of *both* the molecular orbitals (electronic structure) and the displacement coordinates (vibration) are examined in the context of the full symmetry group of the reacting system. All ways of breaking the symmetry of the system are explored, and no symmetry elements which are retained along the pathway are ignored. The correlation diagrams are called "correspondence diagrams" in this approach to distinguish them from the Woodward-Hoffmann diagrams.

It was pointed out [7-21] that this method gives the same result as the Woodward-Hoffmann procedure if a given reaction is predicted to be forbidden. However, it shows some reactions to be forbidden which would be allowed by the Woodward-Hoffmann rules.

As was mentioned before, analysis of electronic transitions and of the reaction coordinate has been developed mainly by Pearson [7-6]. The orbital correspondence analysis in maximum symmetry method was devised by Halevi [7-21, 7-22]. These are probably the most rigorous ways to determine whether a chemical reaction is allowed or forbidden since they consider both the electronic and vibrational changes in the molecule. Of course, their degree of rigor may render their application more complicated as compared with the methods which focus only upon changes in the electronic structure. The approaches introduced by Fukui and Woodward and Hoffmann seem to have received more widespread acceptance and utilization.

7.3 Examples

7.3.1 Cycloaddition

7.3.1.1 Ethylene Dimerization

The interaction of two ethylene molecules will be considered in two geometrical arrangements. The two molecules adopt a mutually parallel approach in one arrangement and a mutually perpendicular approach in the other. Applications of various methods will be considered briefly.

Parallel approach, HOMO-LUMO

According to the frontier orbital method only the HOMOs and the LUMOs of the two ethylene molecules need to be considered. A further simplification is introduced in the pictorial description of the interactions. Although the molecular orbitals of the reactants are used to construct the MOs of the products, the former are usually drawn schematically as atomic orbitals from which they are built. The reason is that the form of the atomic orbitals is better defined and better understood than is the form of the molecular orbitals without resorting to actual calculations.

The MOs of ethylene can be constructed according to the principles given in the preceding chapter. The HOMO of ethylene is the bonding MO, and the LUMO is the antibonding MO composed of the two p_z orbitals of carbon. These MOs are of b_{1u} and b_{2g} symmetry, respectively, in the D_{2h} point group. Fig. 7-7 shows them both in a simplified way along with the corresponding contour diagrams.

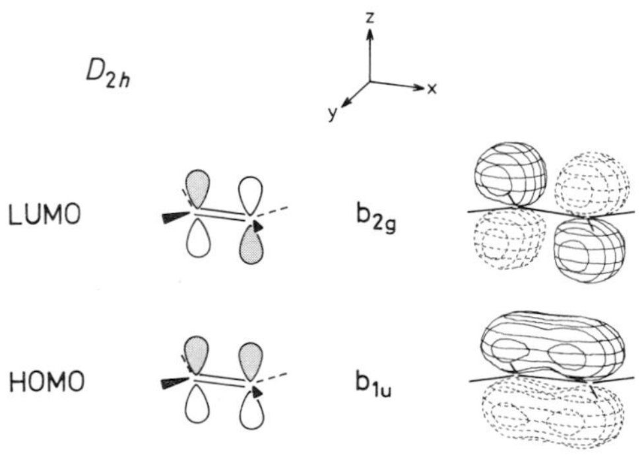

Figure 7-7.
The HOMO and LUMO of ethylene. The contour diagrams are reproduced by permission from [7-23]. Copyright (1973) Academic Press.

294

Consider first the frontier orbital interactions between two ethylene molecules that approach one another in parallel planes ("face to face"). Their HOMOs and LUMOs are indicated in Fig. 7-8 on the left and right, respectively. Also shown is the behavior of these orbitals with respect to the symmetry plane bisecting the two breaking π bonds. Since the HOMOs are symmetric and the LUMOs are antisymmetric with respect to this operation, there is a symmetry mismatch between the HOMO of one molecule and the LUMO of the other. The symmetry allowed combination is between the two filled HOMOs. Since the interaction of two filled molecular orbitals of the same energy is destabilizing, the reaction will not occur thermally.

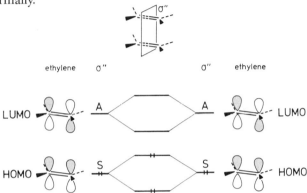

Figure 7-8.
Frontier orbital interactions in the face to face approach of two ethylene molecules. S indicates symmetric and A indicates antisymmetric behavior with respect to the σ'' symmetry plane.

Parallel approach, correlation diagram

Now consider the ethylene dimerization using the Woodward-Hoffmann approach. There is again the important condition mentioned before which must be fulfilled: for the whole reacting system at least one symmetry element must persist throughout the entire process. Let us consider the reaction in this respect. Each separated ethylene molecule has D_{2h} symmetry. When two of these molecules approach one another with their molecular planes parallel as shown in Fig. 7-9, the whole system retains this symmetry. Finally, the product cyclobutane is of D_{4h} symmetry. Since D_{2h} is a subgroup of D_{4h}, the symmetry elements of D_{2h} persist.

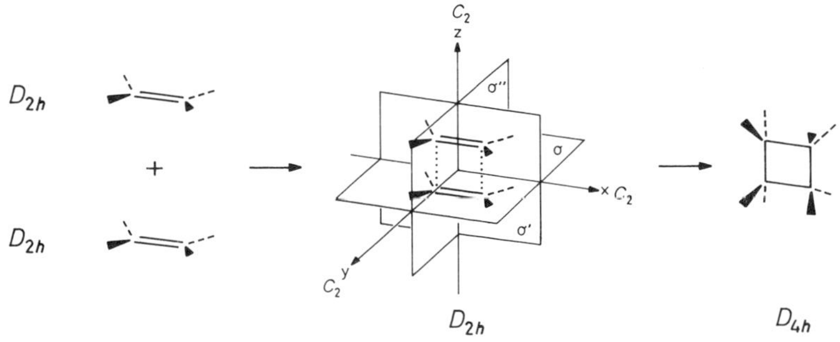

Figure 7-9.
The symmetry of reactants, transition structure, and product in the face to face dimerization of ethylene.

295

One of the symmetry elements in the D_{2h} point group is the symmetry plane σ' (Fig. 7-9). All of the MOs considered in this reaction, that is, those associated with the broken π bonds of the two ethylene molecules and the two new σ bonds of cyclobutane lie in the plane of this symmetry element. All of them will be symmetric to reflection in this plane. There will be no change in their behavior with respect to this symmetry operation during the reaction. This brings us back to a very important point in the construction of correlation diagrams: the symmetry element chosen to follow the reaction must bisect bonds broken or made during the process. Adding extra symmetry elements, like σ' above, will not change the result. It is not wrong to include them; it is just not necessary. Considering, however, only such symmetry elements could lead to the erroneous conclusion that every reaction is symmetry allowed.

As was found to be the case when constructing MOs from AOs, the symmetry of the reacting system as a whole must be considered rather than just the symmetry of the individual molecules alone. Fig. 7-10 illustrates this point with respect to one of the reflection planes. The σ plane transforms the MO drawn as the two p_z orbitals of the two carbon atoms of one ethylene molecule into the molecular orbital of the other ethylene molecule. Thus, each MO of the reacting system has a contribution from each p_z orbital.

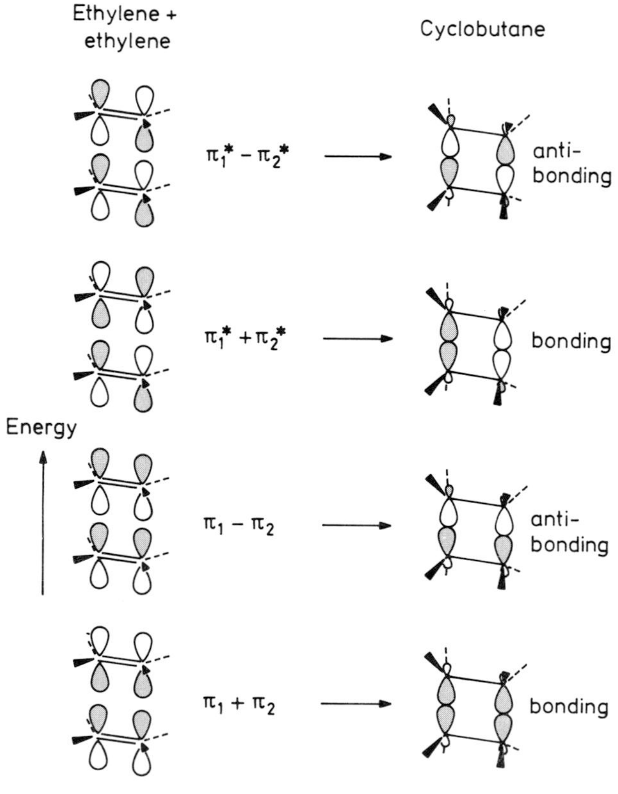

Figure 7-10.
The π MO of one ethylene molecule alone does not belong to any irreducible representation of the point group of the system of two ethylene molecules.

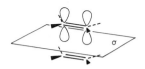

Figure 7-11.
Molecular orbitals of the ethylene-ethylene system and the construction of molecular orbitals of cyclobutane. (The energy scale refers to the reactant orbitals only.)

Ethylene + ethylene   Cyclobutane

$\pi_1^* - \pi_2^*$ ⟶ anti-bonding

$\pi_1^* + \pi_2^*$ ⟶ bonding

Energy

$\pi_1 - \pi_2$ ⟶ anti-bonding

$\pi_1 + \pi_2$ ⟶ bonding

296

The possible combinations of the π and π^* orbitals of the two ethylene molecules are presented on the left side of Fig. 7-11 in order of increasing energy. Consideration of these MOs shows that $\pi_1 + \pi_2$ and $\pi_1^* + \pi_2^*$ are in proper phase to form a bonding MO (that is, closing the ring). The right side of Fig. 7-11 illustrates this, together with the formation of the antibonding orbitals of cyclobutane.

The construction of the correlation diagram is shown in Fig. 7-12. The two crucial symmetry elements are indicated in the upper part of the figure. The molecular orbitals of the

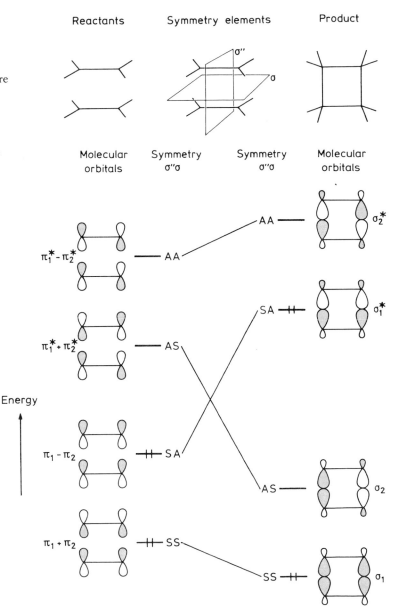

Figure 7-12.
Construction of the correlation diagram for ethylene dimerization with parallel approach. Adaptation of Figure 10.21 from [7-9]. Copyright 1981 by Thomas H. Lowry and Kathleen Schueller Richardson. Reprinted by permission of Harper & Row, Publishers, Inc.

reactants and their behavior with respect to these symmetry elements are indicated in order of increasing energy at the left side of the diagram; the corresponding product MOs are shown at the right in this same figure.

Since σ and σ'' are maintained throughout the reaction, there must be a continuous correlation of orbitals of the same symmetry type. Therefore, orbitals of like symmetry correlate with one another, and they can be connected. This, the crucial idea of the Woodward-Hoffmann method, is shown in the central part of the diagram.

Inspection of this correlation diagram immediately reveals that there is a problem. One of the bonding orbitals at the left correlates with an antibonding orbital on the product side. Consequently, if orbital symmetry is to be conserved, two ground state ethylene molecules cannot combine via face to face approach to give a ground-state cyclobutane, or *vice versa*. This concerted reaction is *symmetry forbidden*.

State correlation

The correlation diagram in Fig. 7-12 refers to molecular orbitals. The molecular orbitals and the corresponding electronic configurations are, however, only substitutes for the real wave functions which describe the actual electronic states. It is the electronic states that have definite energy and not the electronic configurations (cf. preceding chapter). Since electronic transitions occur physically between electronic states, the correlation of these states is of interest.

The rules for the state correlation diagrams are the same as for the orbital correlation diagrams; only states that possess the same symmetry can be connected. In order to determine the symmetries of the states, first the symmetries of the MOs must be determined. These are given for the face to face dimerization of ethylene in Table 7-1. The D_{2h} character table (Table 7-2) shows that it is sufficient to examine the behavior of the molecular orbitals with respect to the two crucial symmetry planes, $\sigma(xy)$ and $\sigma''(yz)$. The MOs are all symmetric with respect to the third plane, $\sigma'(xz)$ (*vide supra*). These three symmetry operations will unambiguously determine their symmetry. Another possibility is to take the simplest subgroup of D_{2h}, which contains the two crucial symmetry operations, that is, the C_{2v} point group (cf., [7-24]). In these two approaches only the designation of the orbitals and states is different: the outcome, i.e., the state correlation diagram, is the same.

Table 7–1. The Symmetry of Molecular Orbitals in the Face to Face Dimerization of Ethylene[a]

Ethylene + Ethylene						Cyclobutane		
Character under D_{2h} σ' σ σ'' (xz) (xy) (yz)			Orbital Occupation			Character under D_{2h} σ' σ σ'' (xz) (xy) (yz)		
1	−1	−1	b_{2g} ——		b_{2g} ——	1	−1	−1
1	1	−1	b_{3u} ——		b_{1u} ——	1	−1	1
1	−1	1	b_{1u} ⥮	⥮	b_{3u}	1	1	−1
1	1	1	a_g ⥮	⥮	a_g	1	1	1

[a] The orientation of the coordinate axes is given in Fig. 7–9.

Table 7–2. The D_{2h} Character Table

D_{2h}	E	$C_2(z)$	$C_2(y)$	$C_2(x)$	i	$\sigma(xy)$	$\sigma(xz)$	$\sigma(yz)$		
A_g	1	1	1	1	1	1	1	1		x^2, y^2, z^2
B_{1g}	1	1	−1	−1	1	1	−1	−1	R_z	xy
B_{2g}	1	−1	1	−1	1	−1	1	−1	R_y	xz
B_{3g}	1	−1	−1	1	1	−1	−1	1	R_x	yz
A_u	1	1	1	1	−1	−1	−1	−1		
B_{1u}	1	1	−1	−1	−1	−1	1	1	z	
B_{2u}	1	−1	1	−1	−1	1	−1	1	y	
B_{3u}	1	−1	−1	1	−1	1	1	−1	x	

In determining the symmetries of the states (see preceding chapter), we must remember that states with completely filled orbitals are always totally symmetric. In other cases, the symmetry of the state is determined by the direct product of the incompletely filled orbitals.

The ground state configuration of the two-ethylene system is $a_g^2 b_{1u}^2$ (see Table 7-1). This state is totally symmetric, A_g. The excitation of an electron from the HOMO to the LUMO will give an electron configuration: $a_g^2 b_{1u} b_{3u}$. The direct product is:

$$b_{1u} \cdot b_{3u} = b_{2g}$$

This yields a state of B_{2g} symmetry. The electronic configuration of the product is $a_g^2 b_{3u}^2$, again with A_g symmetry. This electron configuration corresponds to a doubly excited state of the reactants. Finally, the state correlation diagram can be drawn (Fig. 7-13).

Figure 7-13.
State correlation diagram for
the ethylene dimerization.

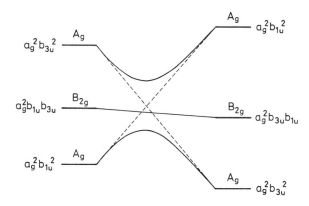

An obvious connection between states that possess the same electronic configuration would be the one indicated by dashed lines in Fig. 7-13. This does not occur, however, because states of the same symmetry cannot cross. This is again a realization of the noncrossing rule which applies to electronic states as well as to orbitals. Instead of crossing, when two states are coming too close to each other, they will turn away, and so the two ground states, both of A_g symmetry and also two A_g symmetry excited states will each mutually correlate.

The full line connecting the two ground states in Fig. 7-13 indicates that there is a substantial energy barrier to the ground-state-to-ground-state process; this reaction is said to be "thermally forbidden".

Consider now one electron in the reactant system excited photochemically to the B_{2g} state. Since this state correlates directly with the B_{2g} state of the product, this reaction does not have any energy barrier and may occur directly. It is said that the reaction is "photochemically allowed". Indeed, it is an experimental fact that olefin dimerization occurs smoothly under irradiation.

This observation can be generalized. *If a concerted reaction is thermally forbidden, it is photochemically allowed and vice versa; if it is thermally allowed then it is photochemically forbidden.*

Although the state correlation diagram is physically more meaningful than the orbital correlation diagram, usually the latter is used because of its simplicity. It is similar to the kind of approximation made when the electronic wave function is replaced by the products of one-electron wave functions in MO theory. The physical basis for the rule that only orbitals of the same symmetry can correlate is that only in this case can constructive overlap occur. This again has its analogy in the construction of molecular orbitals. The physical basis for the noncrossing rule is electron repulsion. It is important that this applies to orbitals – or states – of the same symmetry only. Orbitals of different symmetry cannot interact anyway, so their correlation lines are allowed to cross.

Parallel approach, orbital correspondence analysis

It is worthwhile to see what additional information can be learned from orbital correspondence analysis [7-21, 7-7]. The correspondence diagram of the ethylene dimerization reaction is drawn in Fig. 7-14. It is essentially the same as the correlation diagram in Fig. 7-12 with the following difference: Here the maximum symmetry of the system, D_{2h}, is taken into consider-

Figure 7-14.
Correspondence diagram of the face to face dimerization of ethylene. After [7-21], reproduced with permission.

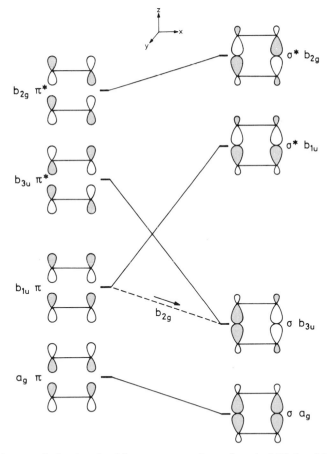

ation, and the irreducible representation of each MO in this point group is shown. The full lines of the diagram connect molecular orbitals of the same symmetry. This is the same as the correlation diagram derived from consideration of the crucial symmetries. In addition we can see that the required transition toward producing a stable ground state cyclobutane would be

301

from an MO of b_{1u} symmetry to another MO of b_{3u} symmetry. The symmetry of the necessary vibration is given by the direct product of these MOs:

$$b_{1u} \cdot b_{3u} = b_{2g}$$

The B_{2g} symmetry motion of a rectangle of D_{2h} symmetry would be an in-plane vibration that shortens one of the diagonals and lengthens the other:

This result suggests a stepwise mechanism. The first step is the formation of a transoid tetramethylene biradical. Then, this intermediate rotates, thereby permitting closure of the cyclobutane ring in a second step. Some quantum chemical calculations [7-25] seem to support this mechanism. The reverse of ethylene dimerization, the pyrolysis of cyclobutane, is experimentally observed [7-26]. Both quantum chemical calculations [7-27] and thermochemical considerations [7-28] suggest that the pyrolysis proceeds through a 1,4-biradical intermediate. This shows the value of the additional information yielded by the orbital correspondence approach.

Ethylene dimerization and its reverse have been studied extensively. For some pertinent results and references, see [7-6].

Orthogonal approach

Let us consider ethylene dimerization in yet another approach. Assume that the orientation of the two molecules is orthogonal:

There is one symmetry element that is maintained in this arrangement, i.e., the C_2 rotation. Considering the behavior of the reactant π MOs and the product σ MOs under the C_2 operation, the correlation diagram shown in Fig. 7-15 can be drawn. It shows that both bonding MOs of the reactant side correlate with a bonding MO on the product side. There is a net energy gain in the reaction, and the process is "thermally allowed".

Figure 7-15.
Correlation diagram for the orthogonal orientation of two ethylene molecules at the dimerization reaction. Adaptation of Figure 10.24 from [7-9]. Copyright 1981 by Thomas H. Lowry and Kathleen Schueller Richardson. Reprinted by permission of Harper & Row, Publishers, Inc.

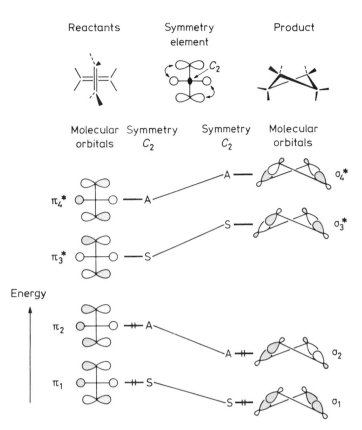

One of the ethylene molecules enters the above reaction *antarafacially;* this means that the two new bonds are formed on opposite sides of this molecule:

The other ethylene molecule enters the reaction *suprafacially;* this means that the two new bonds are formed on the same side of this second molecule:

Thus in the orthogonal approach the two molecules enter the reaction differently: one of them antarafacially and the other suprafacially. On the other hand, in the parallel approach of two ethylenes, both molecules enter the reaction suprafacially:

The following abbreviation is often used in the literature: $_\pi 2_s + _\pi 2_s$ means that both ethylene molecules are approaching in a suprafacial manner, while $_\pi 2_s + _\pi 2_a$ indicates that the same molecules are reacting in a process which is suprafacial for one component and antarafacial for the other. The number $_\pi 2$ indicates that two π electrons are contributed by each ethylene molecule.

Just for the sake of completeness it is worthwhile mentioning that, according to the orbital correspondence analysis, this $_\pi 2_s + _\pi 2_a$ cycloaddition of ethylene is also thermally forbidden [7-21]. Quantum chemical calculations [7-25] show this process to be symmetry allowed, although considerable steric repulsion among some of the hydrogen atoms was indicated. This may account for the fact that this symmetry allowed process is not observed experimentally. On the other hand, there is experimental evidence that the $_\pi 2_s + _\pi 2_a$ type of cycloaddition is possible with twisted ethylenes. Bicyclo-[4.2.2]deca-trans-3-cis-7,9-triene dimerizes spontaneously [7-29].

Quantum chemical calculations [7-25] suggest that steric hindrance is reduced here as compared with the case of cyclo-dimerization of two ethylenes, and the symmetry allowed process actually occurs.

7.3.1.2 Diels-Alder Reaction

HOMO-LUMO interaction

Another famous example that demonstrates the applicability of symmetry rules in determining the course of chemical reactions is the Diels-Alder reaction. It was discussed in Fukui's seminal paper [7-1] on the frontier orbital method. Fig. 7-16 illustrates the HOMOs and LUMOs of ethylene (dienophile) and butadiene (diene). Part (a) of this figure indicates the orbital interactions in the conventional atomic orbital representation, while Part (b) reproduces the calculated contour diagrams [7-11]. The only symmetry element common to both the diene and dienophile is the reflection plane that passes through the central 2,3-bond of the diene and the double bond of the dienophile. The symmetry behavior of the MOs with respect to this symmetry element is also shown.

Figure 7-16.
HOMO-LUMO interaction in
the Diels-Alder reaction.

Symmetry

(a) Conventional representation.

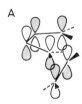

(b) Calculated contour diagrams. Adapted after [7-11] with permission. Copyright 1982 by the American Association for the Advancement of Science.

There are two favorable interactions here. One is between the HOMO of ethylene and the LUMO of butadiene and the other is between the HOMO of butadiene and the LUMO of ethylene. These two interactions occur simultaneously. There is, however, a difference in the role of these two interactions because of their different symmetry behavior. The HOMO of ethylene and the LUMO of butadiene are symmetric with respect to the symmetry element that is maintained throughout the reaction. There is no nodal plane at this symmetry element, so the electrons can be delocalized over the whole new bond. Thus both carbon atoms of ethylene are bound synchronously to both terminal atoms of butadiene. This indeed can be seen on the left-hand side of the contour diagram of Fig. 7-16b.

The situation is different with the other HOMO-LUMO interaction. These orbitals are antisymmetric with respect to the symmetry element, and the two ends of the new linkage are separated by a nodal plane. Therefore, two separated chemical bonds will form, each connecting an ethylene carbon atom with a terminal butadiene carbon atom. From this consideration, it follows that the first symmetric interaction is the dominant one.

Orbital correlation diagram

The ethylene and the butadiene molecules must approach each other in the manner indicated at the top of Fig. 7-17 in order to participate in a concerted reaction. There is only one

305

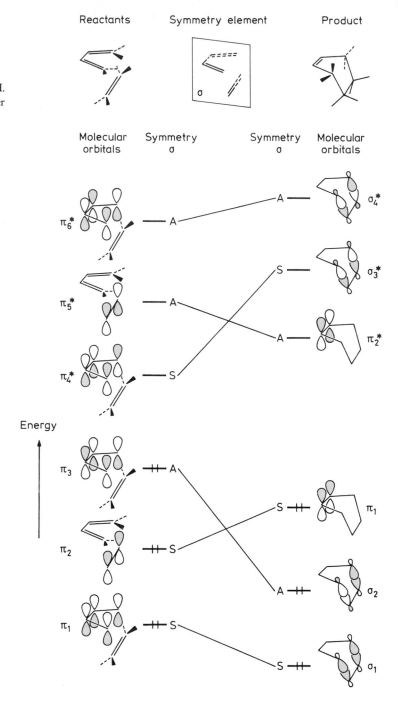

Figure 7-17.
Orbital correlation diagram for the ethylene-butadiene cycloaddition. Adaptation of Figure 10.22 from [7-9]. Copyright 1981 by Thomas H. Lowry and Kathleen Schueller Richardson. Reprinted by permission of Harper & Row, Publishers, Inc.

persisting symmetry element in this arrangement, viz. the σ plane which bisects the 2,3-bond of the diene and the double bond of the dienophile. The orbitals affected by the reaction are the π orbitals of the reactants which will be broken; two new σ bonds and one new π bond are formed in the product. The π orbitals and their antibonding pairs for the reactants are shown on the left side of Fig. 7-17. The new σ and π orbitals, both bonding and antibonding, of the product cyclohexene are on the right side of this figure. These are the orbitals which are affected by the reaction. The behavior of these orbitals with respect to the vertical symmetry plane is also indicated. The correlation diagram shows that all the filled bonding orbitals of the reactants correlate with filled ground state bonding orbitals of the product. The reaction, therefore, is symmetry allowed. The predictions that arise by application of the correlation method and by application of the HOMO-LUMO treatment are identical.

The ethylene-butadiene cycloaddition is a good example which illustrates that symmetry allowedness does not necessarily mean that the reaction occurs easily. This reaction has a comparatively high activation energy, 144 kJ/mol [7-7]. Salem and co-workers [7-30] examined this reaction by *ab initio* calculations. The results suggest concertedness, but the three double bonds of the reactants must stretch in the initial phase of the reaction and this may account for the high activation energy. According to another theoretical study [7-31], the barrier to thermally allowed reactions is often due to the substantial distortion that must occur in the reactants before frontier orbital interactions can stabilize the product.

7.3.2 Intramolecular Cyclization

Orbital correlation for the butadiene/cyclobutene interconversion

The electrocyclic interconversion between an open chain conjugated polyene and a cyclic olefin is another example for the application of the symmetry rules. The simplest case is the interconversion of butadiene and cyclobutene:

This process can occur in principle in two ways. In one the two ends of the open chain turn in the opposite direction into the transition state. This is called a *disrotatory* reaction.

The other possibility is a *conrotatory* process in which the two ends of the open chain turn in the same direction.

The ring opening of substituted cyclobutenes proceeds at relatively low temperatures and always in conrotatory fashion [7-3, 7-32], as is illustrated by the isomerization of *cis-* and *trans-*3,4-dimethylcyclobutene [7-32]:

This stereospecificity is well accounted for by the correlation diagrams constructed for the unsubstituted butadiene/cyclobutene isomerization in Figs. 7-18 and 7-19. Since two double bonds in butadiene are broken and a new double bond and a single bond are formed during the cyclization, two bonding and two antibonding orbitals must be considered on both sides. The persisting symmetry element is a plane of symmetry in the disrotatory process. The correlation diagram (Fig. 7-18) shows a bonding electron pair moving to an antibonding level in the product, and, thus, the right-hand side corresponds to an excited state configuration. Although this corresponds to the conservation of orbital symmetry principle, it requires so much energy that the reaction cannot occur in a thermal process.

Fig. 7-19 shows the same reaction with conrotatory ring closure. Here, the symmetry element maintained throughout the reaction is the C_2 rotation axis. After connecting orbitals of like symmetry, it is seen that all ground state reactant orbitals correlate with ground state product orbitals, so the process is thermally allowed.

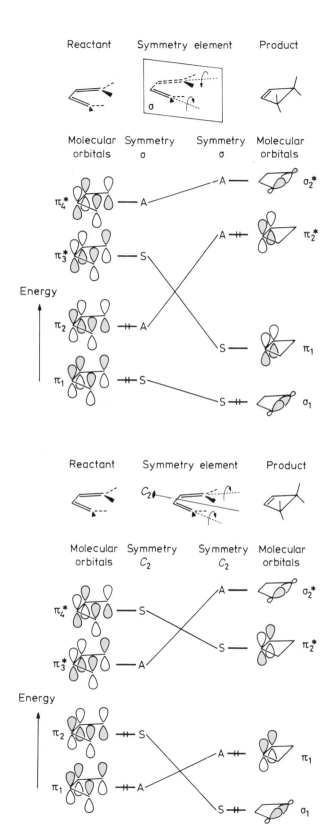

Figure 7-18.
Correlation diagram for the disrotatory closure of butadiene. Adaptation of Figure 10.16 from [7-9]. Copyright 1981 by Thomas H. Lowry and Kathleen Schueller Richardson. Reprinted by permission of Harper & Row, Publishers, Inc.

Figure 7-19.
Correlation diagram for the conrotatory ring closure in the butadiene-cyclobutene isomerization. Adaptation of Figure 10.14 from [7-9]. Copyright 1981 by Thomas H. Lowry and Kathleen Schueller Richardson. Reprinted by permission of Harper & Row, Publishers, Inc.

309

It is interesting to consider the butadiene-cyclobutene reaction from a somewhat different viewpoint, to determine whether the symmetry of the reaction coordinate does indeed predict the proper reaction. Let us look at the reaction in the opposite direction, i.e., the cyclobutene ring opening process. From the symmetry point of view, this change of direction is irrelevant.

The symmetry group of both cyclobutene and butadiene is C_{2v}, but the transition state is of C_2 symmetry in the conrotatory and C_s in the disrotatory mode. Pearson [7-6] suggested that this reaction might be visualized in the following way: In the cyclobutene-butadiene transition, two bonds of cyclobutene are destroyed, to wit the ring-closing σ and the opposite π bonds. Hence, four orbitals are involved in the change, the filled and empty σ and σ^* orbitals and the filled and empty π and π^* orbitals. These orbitals are indicated in Fig. 7-20. Their symmetry is also given for the three point groups involved.

Figure 7-20.
The molecular orbitals participating in the cyclobutene ring opening.

		Symmetry		
		C_{2v}	C_2	C_s
σ^*		b_2	b	a″
π^*		a_2	a	a″
π		b_1	b	a′
σ		a_1	a	a′

Fig. 7-21 demonstrates the nuclear movements involved in the conrotatory and disrotatory ring opening. These movements define the reaction coordinate, and they belong to the A_2 and B_1 representation of the C_{2v} point group, respectively.

Figure 7-21.
The symmetry of the reaction coordinate in the conrotatory and disrotatory ring opening of cyclobutene.

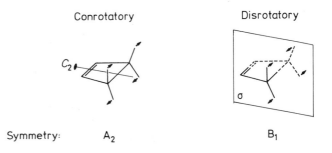

Conrotatory   Disrotatory

Symmetry: A_2 B_1

The two bonds of cyclobutene can be broken either by removing electrons from a bonding orbital or by putting electrons into an antibonding orbital. Consider the $\sigma \rightarrow \pi^*$ and the $\pi \rightarrow \sigma^*$ transitions. According to Pearson [7-6], the direct product of the two representations must contain the reaction coordinate:

$$\sigma \rightarrow \pi^*: \quad a_1 \cdot a_2 = a_2$$
$$\pi \rightarrow \sigma^*: \quad b_1 \cdot b_2 = a_2$$

A_2 is the irreducible representation of the conrotatory ring opening motion, so this type of ring opening seems to be possible. We can test the rules further: during a conrotatory process, the symmetry of the system decreases to C_2. The symmetry of the relevant orbitals also changes in this point group (see Fig. 7-20). Both a_1 and a_2 become a, and both b_1 and b_2 become b. Therefore these orbitals are able to mix. Also, the symmetry of the reaction coordinate becomes A. This is consistent with the rule which says that the reaction coordinate, except at maxima and minima, must belong to the totally symmetric representation of the point group.

The next step is to test the possibility of the disrotatory ring opening. The $\sigma \rightarrow \pi^*$ and $\pi \rightarrow \sigma^*$ transitions obviously cannot be used here, since they correspond to the conrotatory ring opening of A_2 symmetry. Let us consider the $\sigma \rightarrow \sigma^*$ and the $\pi \rightarrow \pi^*$ transitions:

$$\sigma \rightarrow \sigma^*: \quad a_1 \cdot b_2 = b_2$$
$$\pi \rightarrow \pi^*: \quad b_1 \cdot a_2 = b_2$$

Both direct products contain the B_2 irreducible representation. It corresponds to an in-phase asymmetric distortion of the molecule which cannot lead to ring opening. The symmetry of the disrotatory reaction coordinate is B_1 (Fig. 7-21). Moreover, if we consider the symmetry of the orbitals in the C_s symmetry point group of the disrotatory transition, it appears that σ and σ^* as well as π and π^* belong to different irreducible representations. Hence, their mixing would not be possible anyway. The prediction from this method is the same as the prediction from the orbital correlation diagrams. While examination of the reaction coordinate gives more insight into what is actually happening during a chemical reaction, it is somewhat more complicated than using orbital correlation diagrams.

An *ab initio* calculation of the cyclobutene to butadiene ring opening [7-33] led to the following observation: In the conrotatory process, first the C-C carbon single bond lengthens followed by twisting of the methylene groups. The C-C bond lengthening is a symmetric stretching mode with A_1 symmetry, and the methylene twist is an A_2 symmetry process which was earlier supposed to be the reaction coordinate. This apparent controversy was resolved by Pearson [7-6], who emphasized the special role of the totally symmetric reaction coordinate. The effect of the C-C stretch is shown in Fig. 7-22. The energies

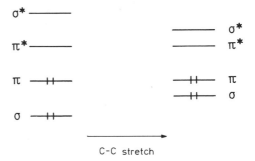

Figure 7-22.
The effect of C-C stretching on the energies of the critical orbitals in the cyclobutene ring opening.

of the σ and σ^* orbitals increase and decrease, respectively, as a consequence of the bond lengthening. The A_1 symmetry vibration mode does not change the molecular symmetry. The crucial $\sigma \rightarrow \pi^*$ and $\pi \rightarrow \sigma^*$ transitions, which are symmetry related to the A_2 twisting mode, occur more easily. Apparently, the large energy difference between these orbitals is the determining factor in the actual process.

7.3.3 Generalized Woodward-Hoffmann Rules

The selection rules for chemical reactions derived by using symmetry arguments show a definite pattern. Woodward and Hoffmann generalized the selection rules [7-3] on the basis of orbital symmetry considerations applied to a large number of systems. Two important observations are summarized here; we refer to the literature [7-3, 7-9] for further details.

Cycloaddition. The reaction between two molecules is thermally allowed if the total number of electrons in the system is $4n + 2$ (n is an integer), and both components are either suprafacial or antarafacial. If one component is suprafacial and the other is antarafacial, the reaction will be thermally allowed if the total number of electrons is $4n$.

Electrocyclic reactions. The rules are similar to those given above. A disrotatory process is thermally allowed if the total number of electrons is $4n + 2$, and a conrotatory process is allowed thermally if the number of delocalized electrons is $4n$. For a photochemical reaction, both sets of rules are reversed.

7.4 Hückel-Möbius Concept

There are a number of other methods used to predict and interpret chemical reactions without relying upon symmetry arguments. It is worthwhile to compare at least some of them with symmetry based approaches.

The so-called "aromaticity rules" are chosen for comparison, as they provide a beautiful correspondence with the symmetry based Woodward-Hoffmann rules. A detailed analysis showed [7-34] the equivalence of the generalized Woodward-Hoffmann selection rules and the aromaticity-based selection rules for pericyclic reactions. Zimmermann [7-35] and Dewar [7-36] have made especially important contributions in this field.

The word "aromaticity" usually implies that a given molecule is stable, compared to the corresponding open chain hydrocarbon. The aromaticity rules are based on the Hückel-Möbius concept. A cyclic polyene is called a Hückel system if its constituent p orbitals overlap everywhere in phase, i.e., the p orbitals all have the same sign above and below the nodal plane (Fig. 7-23). According to Hückel's rule [7-37], if such a system has $4n + 2$ electrons, the molecule will be aromatic and stable. On the other hand, a Hückel ring with $4n$ electrons will be antiaromatic.

Figure 7-23.
Illustration of a Hückel ring. Reproduced by permission [7-39].

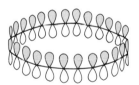

As discovered by Heilbronner [7-38], if the Hückel ring is twisted once, as shown in Fig. 7-24a, the situation is reversed. Therefore, Dewar [7-36] referred to this twisted ring as an "anti-Hückel system". It is also called a "Möbius system" [7-35, 7-39],

Figure 7-24.
(a) Illustration of a Möbius ring. Reproduced by permission [7-39].

an appropriate name indeed. A Möbius strip is a continuous, one-sided surface which is formed by twisting the strip by 180° around its own axis and then attaching its two ends together. The Möbius strip of Fig. 7-24b shows the one phase inversion in the front of the ring just as Fig. 7-24a does with *p* orbitals. Fig. 7-24c depicts yet another Möbius strip.

(b) Möbius strip. Drawing and courtesy by György Doczi, Seattle, WA.

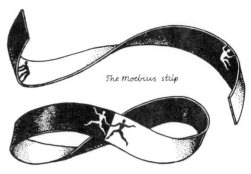

(c) Möbius strip on the facade of a Moscow scientific institute. Photograph by the authors.

According to Zimmermann [7-35] and Dewar [7-36], the allowedness of a concerted pericyclic reaction can be predicted in the following way: A cyclic array of orbitals belongs to the Hückel system if it has zero or even number phase inversions. For such a system, a transition state with $4n + 2$ electrons will be thermally allowed due to aromaticity, while the transition state with $4n$ electrons will be thermally forbidden due to anti-aromaticity.

A cyclic array of orbitals is a Möbius system if it has an odd number of phase inversions. For a Möbius system, a transition state with $4n$ electrons will be aromatic and thermally allowed, while that with $4n + 2$ electrons will be antiaromatic and thermally forbidden. For a concerted photochemical reaction, the rules are exactly the opposite to those for the corresponding thermal process.

Each of these rules has its counterpart among the Woodward-Hoffmann selection rules. There was a marked difference between the suprafacial and antarafacial arrangements in the application of the Woodward-Hoffmann treatment of cycloadditions. The disrotatory and conrotatory processes in electrocyclic reactions presented similar differences. The suprafacial arrangement in both components as well as the disrotatory ring closure in Fig. 7-25 correspond to the Hückel system. On the other hand, the suprafacial-antarafacial arrangement as

Figure 7-25.
Comparison of the disrotatory ring closure and the $_\pi 2_s + {}_\pi 2_s$ reaction with the Hückel ring.

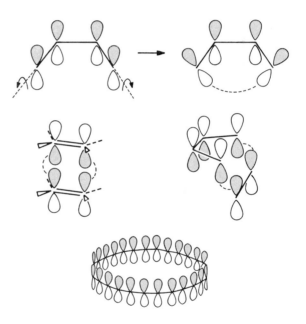

well as the conrotatory cyclization have a phase inversion (Fig. 7-26), and they can be regarded as Möbius systems. All the selection rules mentioned above are summarized in Table 7-3; their mutual correspondence is evident.

Both the Woodward-Hoffmann approach and the Hückel-Möbius concept are useful for predicting the course of concerted reactions. They both have their limitations as well. The application of the Hückel-Möbius concept is probably preferable for systems with low symmetry. On the other hand, this concept can only be applied when there is a cyclic array of orbitals. The conservation of orbital symmetry approach does not have this limitation.

Figure 7-26.
Comparison of the conrotatory ring closure and the $_\pi 2_s$ + $_\pi 2_a$ reaction with the Möbius ring.

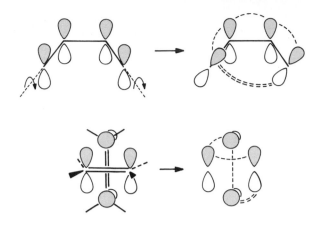

Table 7–3. Selection Rules for Chemical Reactions from Different Approaches

	Reaction	Thermally allowed	Thermally forbidden
1	s + s a + a	$4n + 2$	$4n$
	s + a	$4n$	$4n + 2$
2	disrotatory	$4n + 2$	$4n$
	conrotatory	$4n$	$4n + 2$
3	Hückel system: sign inversion even or 0	$4n + 2$	$4n$
	Möbius system: sign inversion odd	$4n$	$4n + 2$

1 Woodward-Hoffmann cycloaddition
2 Woodward-Hoffmann electrocyclic reaction
3 Hückel-Möbius concept

316

7.5 Isolobal Analogy [7-10]

Thus far, our discussion of reactions has been restricted to organic molecules. However, all of the main ideas are applicable to inorganic systems as well. Thus, for example, the formation of inorganic donor-acceptor complexes may be conveniently described by the HOMO-LUMO concept. A case in point is the formation of the aluminum trichloride-ammonia complex (cf. Fig. 3-26). This complex can be considered to result from interaction between the LUMO of the acceptor ($AlCl_3$) and the HOMO of the donor (NH_3).

The potential of a unified treatment of organic and inorganic systems has been expressed eloquently in Roald Hoffmann's Nobel lecture [7-10] entitled "Building Bridges between Inorganic and Organic Chemistry".

The main idea is to examine the similarities between the structures of relatively complicated inorganic complexes and relatively simple and well understood organic molecules. Then the structure and possible reactions of the former can be understood and even predicted by the considerations working so well for the latter. Two important points were stressed [7-10]:

1. "It is the resemblance of the frontier orbitals of inorganic and organic moieties that will provide the bridge that we seek between the subfields of our science." [7-10]

2. Many aspects of the electronic structure of the molecules discussed and compared are heavily simplified, but "the time now, here, is for building conceptual frameworks, and so similarity and unity take temporary precedence over difference and diversity." [7-10]

One of the fastest growing areas of inorganic chemistry is transition metal organometallic chemistry. In a general way, the structure can be thought of as containing a transition metal-ligand fragment, such as $M(CO)_5$, $M(PF_3)_5$, $M(allyl)$ and MCp, or in general, ML_n. All these fragments may be derived from an octahedral arrangement

In describing the bonding in these fragments, first the six octahedral hybrid orbitals on the metal atom are constructed. Hybridization is not discussed here, but symmetry considerations are used in constructing hybrid orbitals just as in constructing molecular orbitals [7-24]. In an octahedral complex, the six hybrid orbitals point towards the ligands, and together they can be used as a basis for a representation of the point group. The O_h character table and the representation of the six hybrid orbitals are given in Table 7-4. The representation reduces to

$$\Gamma_h = A_{1g} + E_g + T_{1u}$$

Inspection of the O_h character table shows that the only possible combination from the available nd, $(n+1)s$ and $(n+1)p$ orbitals of the metal is:

$$s, \qquad p_x, p_y, p_z, \qquad d_{x^2-y^2}, d_{z^2}$$
$$a_{1g} \qquad t_{1u} \qquad e_g$$

These six orbitals will participate in the hybrid, and the remaining t_{2g} symmetry orbitals (d_{xz}, d_{yz}, and d_{xy}) of the metal will be nonbonding.

Table 7–4. The O_h Character Table and the Representation of the Hybrid Orbitals of the Transition Metal in an ML_6 Complex

O_h	E	$8C_3$	$6C_2$	$6C_4$	$3C_2$ $(=C_4{}^2)$	i	$6S_4$	$8S_6$	$3\sigma_h$	$6\sigma_d$		
A_{1g}	1	1	1	1	1	1	1	1	1	1		$x^2+y^2+z^2$
A_{2g}	1	1	−1	−1	1	1	−1	1	1	−1		
E_g	2	−1	0	0	2	2	0	−1	2	0		$(2z^2-x^2-y^2,$ $x^2-y^2)$
T_{1g}	3	0	−1	1	−1	3	1	0	−1	−1	(R_x, R_y, R_z)	
T_{2g}	3	0	1	−1	−1	3	−1	0	−1	1		(xz, yz, xy)
A_{1u}	1	1	1	1	1	−1	−1	−1	−1	−1		
A_{2u}	1	1	−1	−1	1	−1	1	−1	−1	1		
E_u	2	−1	0	0	2	−2	0	1	−2	0		
T_{1u}	3	0	−1	1	−1	−3	−1	0	1	1	(x, y, z)	
T_{2u}	3	0	1	−1	−1	−3	1	0	1	−1		
Γ_h	6	0	0	2	2	0	0	0	4	2		

Six ligands approach the six hybrid orbitals of the metal in forming an octahedral complex. These ligands are supposed to be donors, or, in other words, Lewis bases with even numbers of electrons. Six bonding σ orbitals and six antibonding σ^* orbitals are formed, with the ligand electron pairs occupying the bonding orbitals as seen in Fig. 7-27. As a consequence of the strong interaction, all six hybrid orbitals of the metal are removed from the frontier orbital region, and only the unchanged metal t_{2g} orbitals remain there.

Figure 7-27.
Molecular orbital construction in an ideal octahedral complex formation. Reproduced by permission from Hoffmann [7-10]. © The Nobel Foundation 1982.

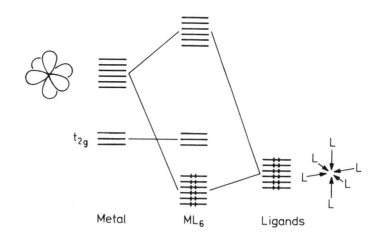

We can also deduce the changes which will occur in the five, four and three ligand fragments as compared to the ideal six-ligand case with the help of Fig. 7-27. The situation is illustrated in Fig. 7-28. With five ligands, only five of the six metal hybrid

Figure 7-28.
Molecular orbitals in different ML_n transition metal-ligand fragments. Adapted with permission [7-10]. © The Nobel Foundation 1982.

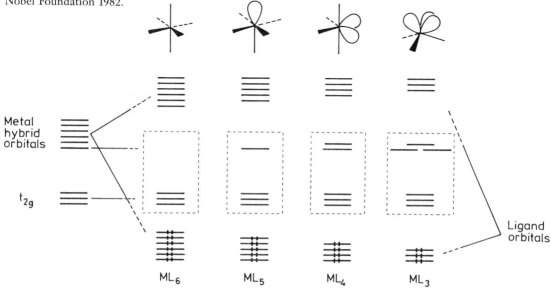

319

orbitals will interact; the sixth orbital, the one pointing to where no ligand is, will be unchanged. Consequently, this orbital will remain in the frontier orbital region, together with the t_{2g} orbitals. With four ligands, two of the six hybrid orbitals remain unchanged and with three ligands, three. Always, those metal hybrid orbitals, which point towards the missing ligands in the octahedral site, remain unchanged.

Now we shall seek analogies between transition metal complexes and simple, well-studied organic molecules or fragments. In principle, any hydrocarbon can be constructed from methyl groups, (CH_3), methylenes (CH_2), methynes (CH), and quaternary carbon atoms. They can be imagined as being derived from the methane molecule itself, which has a tetrahedral structure:

The essence of the "isolobal analogy" concept is to establish similarities between these simple organic fragments and the transition metal ligand fragments, and then to build up the organometallic compounds. "Two fragments are called *isolobal*, if the number, symmetry properties, approximate energy and shape of the frontier orbitals and the number of electrons in them are similar – not identical, but similar" [7-10]. However, the molecules involved are not and need not be either isoelectronic or isostructural.

The first analogy considered here is a d^7-metal-ligand fragment, for example, $Mn(CO)_5$ and the methyl radical, CH_3

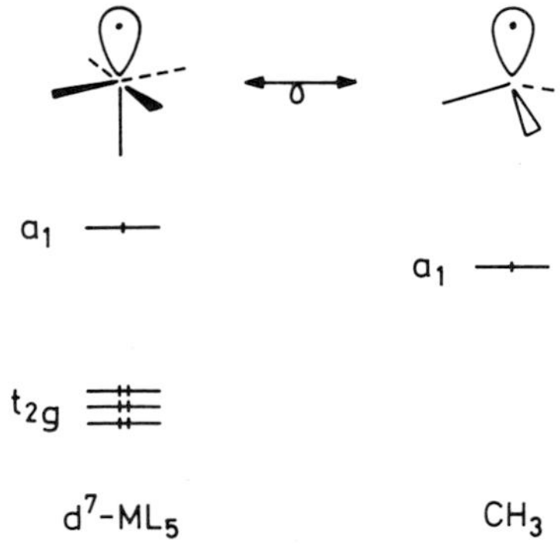

a_1 —•—

a_1 —•—

t_{2g} (filled orbitals)

d^7-ML_5 CH_3

Though the two fragments belong to different point groups, C_{4v} and C_{3v}, respectively, the orbitals that contain unpaired electron belong to the totally symmetric representation in both cases. Since the three occupied t_{2g} orbitals of the ML_5 fragment are comparatively low lying, the frontier orbital pictures of the two fragments should be similar. If this is so, then they are expected to show some similarity in their chemical behavior, notably in reactions. Indeed, both of them dimerize [7-10], and even the organic and inorganic fragments can codimerize, giving $(CO)_5MnCH_3$

Following this analogy, the four-ligand d^8-ML_4 fragments (e.g., $Fe(CO)_4$) are expected to be comparable with the methylene radical, CH_2

$$a_1 \quad \text{—} \qquad b_2 \quad \text{—}$$
$$b_2 \quad \text{—} \qquad a_1 \quad \text{—}$$

$$t_{2g}$$

$$d^8\text{-}ML_4 \qquad\qquad CH_2$$

Both fragments belong to the C_{2v} point group, and the representation of the two hybrid orbitals with the unpaired electrons is:

C_{2v}	E	C_2	σ	σ'
Γ	2	0	0	2

This reduces to $a_1 + b_2$. Although the energy ordering differs in the two fragments, this fact is not important, since both will participate in bonding when they interact with another ligand, and their original ordering will thereby change anyway.

Consider the possible dimerization process: two methylene radicals give ethylene, which is a known reaction. Similarly, the mixed product, $(CO)_4FeCH_2$, or at least its derivatives, can be prepared. The $Fe_2(CO)_8$ dimer, however, is unstable and has only been observed in matrix [7-40]. This illustrates that the isolobal analogy suggests only the possible consequences of the similarity in the electronic structure of two fragments. It says nothing, however, about the thermodynamic and kinetic stability of any of the possible reaction products.

Although $Fe_2(CO)_8$ is unstable, it can be stabilized by complexation. The molecule in Fig. 7-29a consists of two $Fe_2(CO)_8$ units connected through a tin atom [7-41]. Using the inorganic/organic analogy, this molecule can be compared to spiropentane (see Fig. 7-29b).

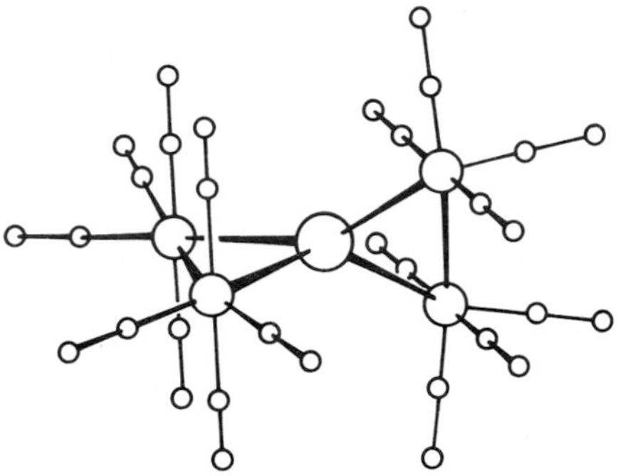

Figure 7-29.
(a) The molecular geometry of $Sn[Fe_2(CO)_8]_2$. Reproduced by permission from Hoffmann [7-10]. © The Nobel Foundation 1982.

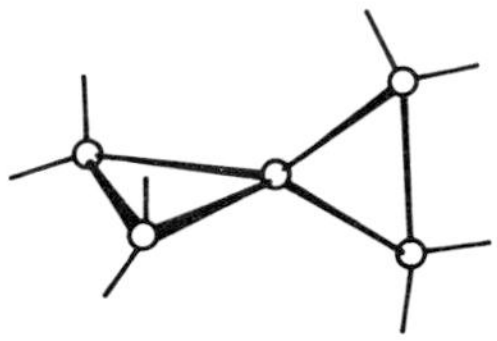

(b) The organic analog: spiropentane.

An example of a d^9-ML_3 fragment is $Co(CO)_3$. This is isolobal to a methyne radical, CH

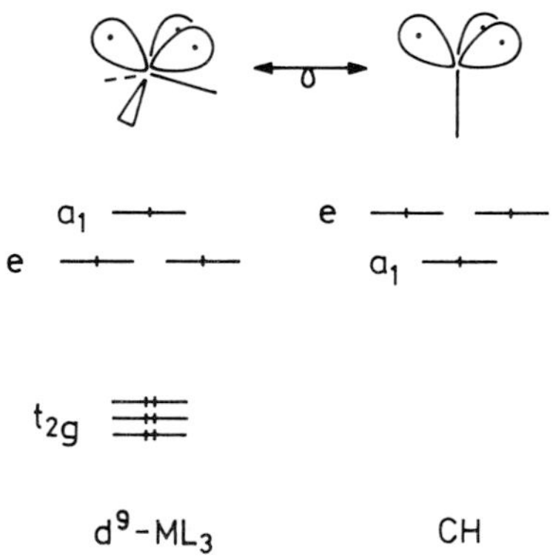

$$d^9\text{-}ML_3 \qquad\qquad CH$$

Both of them have C_{3v} symmetry. The representation of the three hybrid orbitals with an unpaired electron is:

C_{3v}	E	$2C_2$	$3\sigma_v$
Γ	3	0	1

It reduces to $a_1 + e$. Again, the ordering is different, but the similarity between their electronic structure is obvious. A series of molecules and their similarities is revealed in Fig. 7-30. The first molecule is tetrahedrane and the last one is a cluster with metal-metal bonds which can be considered as being the inorganic analog of tetrahedrane [7-10].

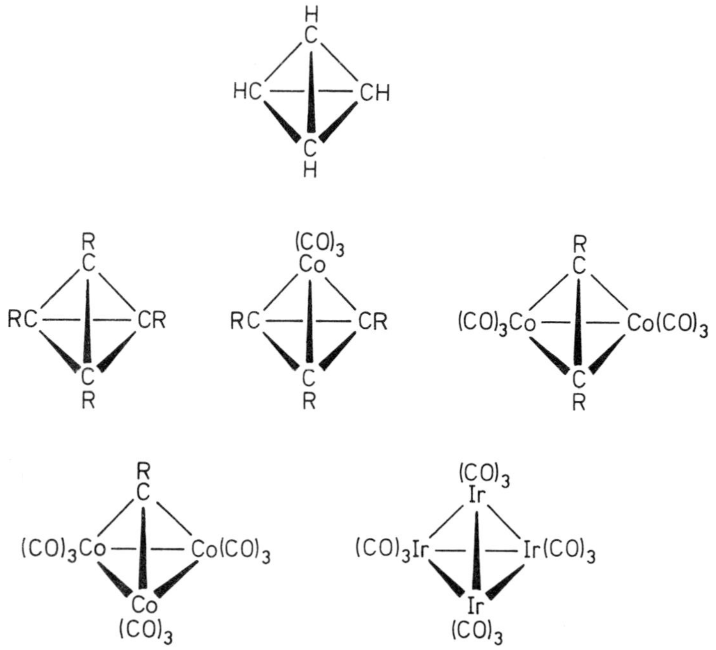

Figure 7-30.
Molecular geometries from tetrahedrane to its inorganic analog. Adapted with permission [7-10]. © The Nobel Foundation 1982.

Only a few examples have been given to illustrate the isolobal analogy. Hoffmann and his co-workers have extended this concept to other metal-ligand fragment compositions with various d orbital participations. Some of these analogies are summarized in Table 7-5. Hoffmann's Nobel lecture [7-10] contained several of them and many more can be found in the references given therein.

Table 7-5. Isolobal Analogies

Organic Fragment	Transition Metal Coordination Number				
	9	8	7	6	5
CH_3	d^1–ML_8	d^3–ML_7	d^5–ML_6	d^7–ML_5	d^9–ML_4
CH_2	d^2–ML_7	d^4–ML_6	d^6–ML_5	d^8–ML_4	d^{10}–ML_3
CH	d^3–ML_6	d^5–ML_5	d^7–ML_4	d^9–ML_3	

References

[7-1]   K. Fukui, in *Molecular Orbitals in Chemistry, Physics and Biology,* P. O. Löwdin and B. Pullmann, eds., Academic Press, New York, 1964.

[7-2]   K. Fukui, *Theory of Orientation and Stereoselection,* Springer-Verlag, Berlin, 1975; Top. Curr. Chem. **15**, 1 (1970).

[7-3]   R. B. Woodward and R. Hoffmann, *The Conservation of Orbital Symmetry,* Verlag Chemie, Weinheim, 1970.

[7-4]   R. B. Woodward and R. Hoffmann, Angew. Chem. Int. Ed. Engl. **8**, 781 (1969).

[7-5]   *Orbital Symmetry Papers,* H. E. Simmons and J. F. Bunnett, eds., American Chemical Society, Washington, D.C., 1974.

[7-6]   R. G. Pearson, *Symmetry Rules for Chemical Reactions, Orbital Topology and Elementary Processes,* Wiley-Interscience, New York, 1976; R. G. Pearson, in *Special Issues on Symmetry,* Computers & Mathematics with Applications **12 B**, 229 (1986).

[7-7]   L. Salem, *Electrons in Chemical Reactions: First Principles,* Wiley-Interscience, New York, 1982.

[7-8]   A. P. Marchand and R. E. Lehr, eds., *Pericyclic Reactions,* Vols. 1 and 2, Academic Press, New York, 1977.

[7-9]   T. H. Lowry and K. S. Richardson, *Mechanism and Theory in Organic Chemistry,* Second Edition, Harper & Row, Publishers, New York, 1981.

[7-10]   R. Hoffmann, Angew. Chem. Int. Ed. Engl. **21**, 711 (1982).

[7-11]   K. Fukui, Science **218**, 747 (1982).

[7-12]   E. Wigner and E. E. Witmer, Z. Phys. **51**, 859 (1928).

[7-13]   (a) B. Solouki and H. Bock, Inorg. Chem. **16**, 665 (1977); (b) F. Bernardi, I. G. Csizmadia, A. Mangini, H. B. Schlegel, M.-H. Whangbo, and S. Wolfe, J. Am. Chem. Soc. **97**, 2209 (1975).

[7-14]   J. E. Leffler and E. Grunwald, *Rates and Equilibria of Organic Reactions,* John Wiley & Sons, New York, 1963.

[7-15]   K. Fukui, T. Yonezawa, and H. Shingu, J. Chem. Phys. **20**, 722 (1952).

[7-16]   R. B. Woodward and R. Hoffmann, J. Am. Chem. Soc. **87**, 395 (1965).

[7-17]   R. Hoffmann and R. B. Woodward, J. Am. Chem. Soc. **87**, 2046 (1965).

[7-18]   R. B. Woodward and R. Hoffmann, J. Am. Chem. Soc. **87**, 2511 (1965).

[7-19]   F. Hund, Z. Phys. **40**, 742 (1927); **42**, 93 (1927); **51**, 759 (1928); R. S. Mulliken, Phys. Rev. **32**, 186 (1928).

[7-20]   J. von Neumann and E. Wigner, Phys. Z. **30**, 467 (1929); E. Teller, J. Phys. Chem. **41**, 109 (1937).

[7-21]   E. A. Halevi, Helv. Chim. Acta **58**, 2136 (1975).

[7-22]   J. Katriel and E. A. Halevi, Theor. Chim. Acta **40**, 1 (1975).

[7-23]   W. L. Jorgensen and L. Salem, *The Organic Chemist's Book of Orbitals,* Academic Press, New York, 1973.

[7-24]   F. A. Cotton, *Chemical Applications of Group Theory,* Second Edition, Wiley-Interscience, New York, 1971.

[7-25]   T. Okada, K. Yamaguchi, and T. Fueno, Tetrahedron **30**, 2293 (1974).

[7-26]   R. W. Carr, Jr. and W. D. Walters, J. Phys. Chem. **67**, 1370 (1963).

[7-27]   R. Hoffmann, S. Swaminathan, B. G. Odell, and R. Gleiter, J. Am. Chem. Soc. **92**, 7091 (1970).

[7-28]   H. E. O'Neal and S. W. Benson, J. Phys. Chem. **72**, 1866 (1968).

[7-29]   K. Kraft and G. Koltzenburg, Tetrahedron Lett. 4357 (1967).

[7-30]   R. E. Townshend, G. Ramunni, G. Segal, W. J. Hehre, and L. Salem, J. Am. Chem. Soc. **98**, 2190 (1976).

[7-31]   K. N. Houk, R. W. Gandour, R. W. Strozier, N. G. Rondan, and L. A. Paquette, J. Am. Chem. Soc. **101**, 6797 (1979).

[7-32]   R. E. K. Winter, Tetrahedron Lett. 1207 (1965).

[7-33]   K. Hsu, R. J. Buenker, and S. D. Peyerimhoff, J. Am. Chem. Soc. **93**, 2117 (1971).

[7-34]   A. C. Day, J. Am. Chem. Soc. **97**, 2431 (1975).

[7-35]   H. E. Zimmermann, J. Am. Chem. Soc. **88**, 1564, 1566 (1966); Acc. Chem. Res. **4**, 272 (1971).

[7-36]   M. J. S. Dewar, Tetrahedron Suppl. **8**, 75 (1966); Angew. Chem. Int. Ed. Engl. **10**, 761 (1971); *The Molecular Orbital Theory of Organic Chemistry,* McGraw-Hill, New York, 1969.

[7-37]   E. Hückel, Z. Phys. **70**, 204 (1931); **76**, 628 (1932); **83**, 632 (1933).

[7-38]   E. Heilbronner, Tetrahedron Lett. 1923 (1964).

[7-39]   K.-w. Shen, J. Chem. Educ. **50**, 238 (1973).

[7-40]   M. Poliakoff and J. J. Turner, J. Chem. Soc. A **1971**, 2403.

[7-41]   J. D. Cotton, S. A. R. Knox, I. Paul, and F. G. A. Stone, J. Chem. Soc. A **1967, 264**.

8 Space-Group Symmetries

8.1 Expanding to Infinity

Up to this point structures of mostly finite objects have been discussed. Thus point groups were applicable. A simplified compilation of various symmetries was presented in Fig. 2-52 and Table 2-2. The point-group symmetries are characterized by the lack of periodicity in any direction. Periodicity may be introduced by translational symmetry. If periodicity is present, space groups are applicable for the symmetry description. There is a slight inconsistency here in the terminology. Even a three-dimensional object may have point-group symmetry. On the other hand, the so-called dimensionality of the space group is not determined by the dimensionality of the object. Rather, it is determined by its periodicity. The following groups are space-group symmetries where the superscript refers to the dimensionality of the object, and the subscript to the periodicity.

$$G_1^1$$
$$G_1^2 \qquad G_2^2$$
$$G_1^3 \qquad G_2^3 \qquad G_3^3$$

Objects or patterns which are periodic in one, two, and three directions will have one-, two-, and three-dimensional space groups, respectively. The dimensionality of the object/pattern is merely a necessary but not a satisfactory condition for the "dimensionality" of their space groups. We shall first describe a planar pattern after Budden [8-1] in order to get the flavor of space-group symmetry. Also some new symmetry elements will be introduced. Later in this chapter the simplest one-dimensional and two-dimensional space groups will be presented. A separate chapter, the next one, will be devoted to the obviously most important three-dimensional space groups which characterize the crystal structures.

A pattern expanding to infinity always contains a basic unit, a motif which is then repeated infinitely throughout the pattern. Fig. 8-1a presents a planar decoration. The pattern shown in this figure is only part of the whole as the latter expands to infinity! The pattern is obviously highly symmetrical. Fig. 8-1b shows the system of mutually perpendicular symmetry planes by solid lines. Some of the four-fold and two-fold rotation axes are also indicated in this figure. A new symmetry element in our discussion is the glide reflection which is shown by a dashed line. Some of these glide reflections are separately indicated in Fig. 8-1c. A glide-mirror plane is a combination of translation and

Figure 8-1.
(a) Part of a planar decoration with two-dimensional space group (cf. [8-1]).

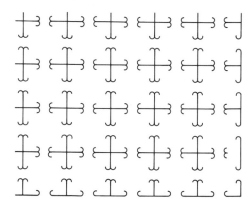

(b) Symmetry elements of the pattern shown in Part (a) of this figure.

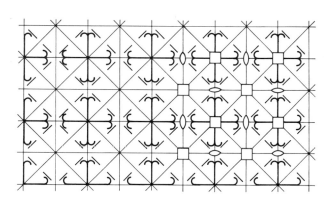

(c) Some of the glide reflection axes and their effects in the pattern shown in Part (a) of this figure.

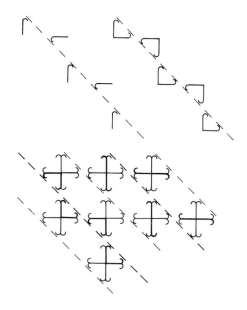

328

reflection. It is a symmetry element which can be present in space groups only. The glide-mirror plane involves an infinite sequence of consequent translations and reflections. Glide mirrors are illustrated in Fig. 8-2. However, they can be considered to be a symmetry element only if extended to infinite

Figure 8-2.
Illustrations for glide mirrors.
(a) Pillow-edge from Buzsák.

(b) Bach's music.

(c) Wood-carving from a
Budapest street. Photograph
by the authors.

repetition, at least in our imagination. The propagation axis of the sine curve is obviously a glide-reflection axis as shown in Fig. 8-3.

Figure 8-3.
Function describing simple harmonic motion.

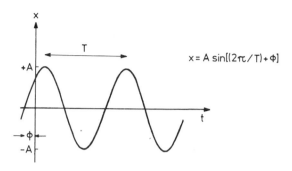

$$x = A \sin[(2\pi/T) + \phi]$$

Simple *translation* is the most obvious and most important symmetry element of the space groups. The translation brings the pattern into congruence with itself over and over again. The shortest displacement through which this translation brings the pattern into coincidence with itself is the elementary translation or elementary period. Sometimes it is also called the identity period. The presence of translation is seen well in the pattern in Fig. 8-1. The symmetry analysis of the whole pattern is called by Budden [8-1] the analytical approach. The reverse procedure is the synthetic approach [8-1] in which the infinite and often complicated pattern is built up from the basic motif. Thus the pattern of Fig. 8-1a may be built up from a single crochet. There are several ways to proceed. Thus for example the crochet may be subjected to simple translation, then reflection, and then transverse reflection as these steps are illustrated in Fig. 8-4a. The horizontal array obtained this way is a one-dimensional pattern. It can be extended to a two-dimensional pattern by simple translation as is done in Fig. 8-4a, or by glide reflection as is shown in Fig. 8-4b. Eventually the complete two-dimensional pattern of Fig. 8-1 can be obtained containing, of course, all its symmetry elements. In this synthetic approach instead of the single crochet, any other motif combined from it could be selected for the start. If the cross-like motif were chosen, which contains eight of these crochets, then only translations in two directions would be needed to build up the final pattern. To learn most about the structure of a pattern, it is advantageous to select the smallest possible motif for the start.

The one-dimensional space groups are the simplest of the space groups. They have periodicity only in one direction. They may refer to one-dimensional, two-dimensional, or three-dimensional objects, cf. G_1^1, G_1^2, and G_1^3, of Table 2-2, respectively. The "infinite" carbon chains of the carbide molecules

Figure 8-4.
Pattern generation.
(a) Starting with a single crochet, then applying horizontal translation/reflection/transverse reflection/vertical translation.

(b) Application of glide reflection.

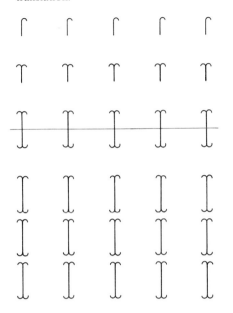

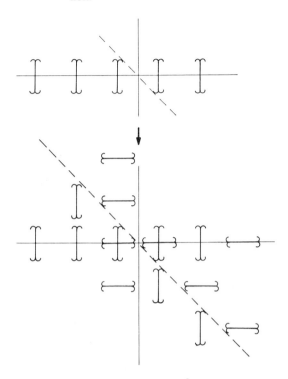

present one-dimensional patterns as are drawn in Fig. 8-5. The elementary translation or identity period is the length of the carbon-carbon double bond (r) in the uniformly bonded chain while it is the sum of the lengths of the two-different bonds $(r_1 + r_2)$ in the chain consisting of alternating bonds. As the chain molecule extends along the axis of the carbon-carbon bonds, this axis can be called the translation axis. However, it is the spatial direction of the axis rather than its location that is important. So the translation axis may be any line parallel to the carbon chain. The carbon-carbon axis is a singular axis and it is not polar as the two directions along the chain are equivalent. Earlier we have seen the binary array ... ABAB ... in a crystal. The unequal spacings between the atom A and the two adjacent atoms B produced a polar axis (cf. Sect. 2.6 on polarity).

Figure 8-5.
"Infinite" chains of carbide molecules.
(a) Uniformly bonded system; the elementary translation is one bond length (r).
(b) Alternating bonds; the elementary translation is the sum of the two different bond lengths $(r_1 + r_2)$.

$$\cdots=C\overset{r}{\underset{}{=}}C=C=C=C=\cdots$$

$$\cdots-C\overset{r_1}{\underset{}{=}}\overset{r_2}{C}C=C-C=C-\cdots$$

8.2 One-Sided Bands

Kindergarten-age children and first-graders often draw line-patterns similar to those in Fig. 8-6. These patterns extend in two dimensions with periodicity in one direction (G_1^2). These

Figure 8-6.
Line-patterns drawn by Eszter Hargittai, 1980.

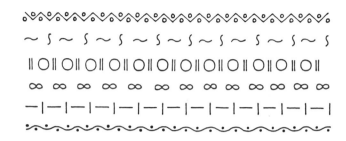

Figure 8-7.
Greek patterns with one-dimensional space-group symmetry with (a) and without (b) polar axes.

patterns have a singular axis and for those in this figure this axis is nonpolar. They could also have a polar rather than a nonpolar axis. Fig. 8-7 presents two Greek band decorations; one of them has a polar axis while the other has a nonpolar axis. An important feature of these patterns in Figs. 8-6 and 8-7 is that they all have a polar singular plane which is the plane of the drawing. This plane is left unchanged during the translation. Such two-dimensional patterns with periodicity in one direction are called one-sided bands [8-2]. Fig. 8-8 presents three more one-sided band patterns with markedly polar axes.

Figure 8-8.
One-sided band patterns with polar axes.

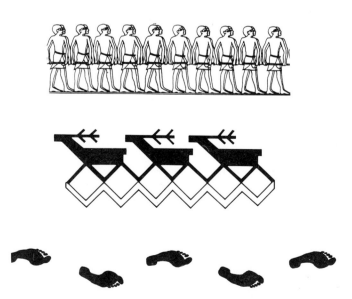

There are altogether seven symmetry classes of one-sided bands (cf. [8-2]). They are illustrated in Fig. 8-9 for a suitable

Figure 8-9.
The seven symmetry classes of one-sided bands.

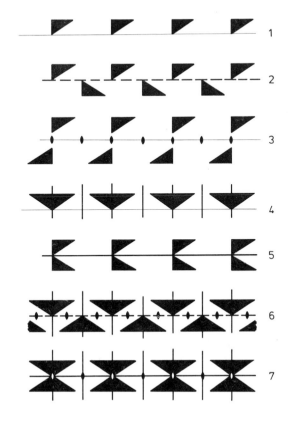

motif, a black triangle. A brief characterization of the seven classes is given here:

1. Notation (*a*). The only symmetry element is the translation axis. The translation period is the distance between two identical points of the consecutive black triangles.

2. Notation (*a*) · *ā*. This pattern is characterized by a glide-reflection plane (*ā*). The black triangle comes into coincidence with itself after translation through half of the translation period (*a*/2) and reflection in the plane perpendicular to the plane of the drawing.

3. Notation (*a*):2. There is translation and rotation by 180° in this pattern. The two-fold rotation axis is perpendicular to the plane of the one-sided band.

4. Notation (*a*):*m*. In this pattern the translation is achieved by transverse symmetry planes.

333

5. Notation $(a) \cdot m$. Here the translation axis is combined with a longitudinal symmetry plane.
6. Notation $(a) \cdot \tilde{a} : m$. The symmetry of this pattern may be characterized by the combination of a glide-reflection plane with transverse symmetry planes. There are also a translation axis and two-fold rotation axes perpendicular to the plane of the drawing. All these later elements are generated by the earlier ones. An alternative description of this symmetry class could be by combining the glide-reflection plane with the two-fold axes, and the corresponding notation would be $(a) : 2 \cdot \tilde{a}$.
7. Notation $(a) \cdot m : m$. This pattern has the highest symmetry achieved by a combination of translation axis with transverse and longitudinal symmetry planes. In this description the two-fold axes perpendicular to the plane of the drawing are generated by the other symmetry elements. An alternative description is $(a) : 2 \cdot m$.

The seven one-dimensional symmetry classes for the one-sided bands are illustrated by patterns of Hungarian needlework in Fig. 8-10. This kind of needlework is a real "one-sided band" and is ideally suited for this purpose.

Figure 8-10.
Illustration of the seven symmetry classes of one-sided bands by Hungarian needlework [8-3]. The numbering corresponds to that of Fig. 8-9. A brief description of the origin of the needlework is given here:
1. Edge decoration of table cover from Kalocsa, Southern Hungary.

2. Pillow-end decoration from Tolna county, Southwest Hungary.

3. Decoration patched onto a long embroidered felt coat of Hungarian shepherds in Bihar county, Eastern Hungary.

1

2
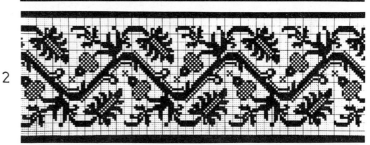

3

334

4. Embroidered edge-decoration of bed-sheet from the 18th century. Note the deviations from the described symmetry in the lower stripes of the pattern.

4

5. Decoration of shirt-front from Karád, Southwest Hungary.

5

6. Pillow decoration pattern from Torocko (Rimetea), Transsylvania, Romania.

6

7. Grape-leaf pattern from the territory east of the river Tisza.

7

8.3 Two-Sided Bands

If the singular plane of a band is not polar, the band is two-sided. Bands have altogether 31 symmetry classes [8-2] of which 7 characterize the one-sided bands only. Fig. 8-11a shows a one-sided band generated by translation of a leaf motif. The (b) part of the figure is a two-sided band characterized by a glide-reflection plane. There is a translation by half of the translation period and then a reflection in the plane of the drawing. The leaf patterns are paralleled by patterns of the black triangle in Fig. 8-11. A new symmetry element is illustrated in Fig. 8-11c,

Figure 8-11.
(a) One-sided bands generated by simple translation of the leaf motif and the black triangle motif. The plane of the drawing is a polar singular plane.
(b) Two-sided bands generated from the one-sided bands by introducing a glide-symmetry plane. The singular plane in the plane of the drawing is no longer polar. The glide-symmetry plane coinciding with the plane of the drawing is labeled $\tilde{a}_{11}$ [8-2]. Note that the two sides of the leaves are of different color (black and white).
(c) Two-sided bands generated from the one-sided bands by introducing a screw axis of the second order.

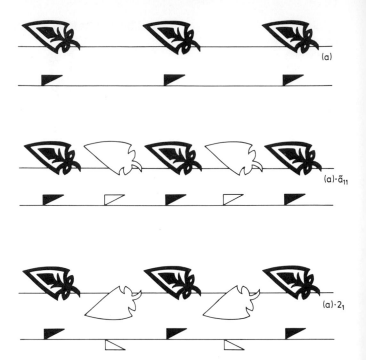

the second order screw axis, 2_1. The corresponding transformation is a translation by half the translation period and a 180° rotation. All the 31 symmetry classes of the bands constituting the one-dimensional space groups are shown in Fig. 8-12 (after Shubnikov and Koptsik [8-2]). In addition to the two classes already illustrated in Fig. 8-11b and c, the notation is given only for the 7 special classes which correspond to the one-sided bands:

Numbering (cf. Fig. 8-12)	Noncoordinate notation	Coordinate (international) notation
1	(a)	$p1$
4	$(a)\cdot m$	$p1m1$
5	$(a)\cdot \tilde{a}$	$p1a1$
12	$(a):2$	$p112$
16	$(a):m$	$pm11$
18	$(a)\cdot \tilde{a}:m$	$pma2$
29	$(a)\cdot m:m$	$pmm2$

Figure 8-12.
The 31 symmetry classes of bands. The two sides of the black and white triangles are of different color. The two sides of the dotted triangles are of the same color [8-2]. The bold numbers indicate the special cases of one-sided bands.

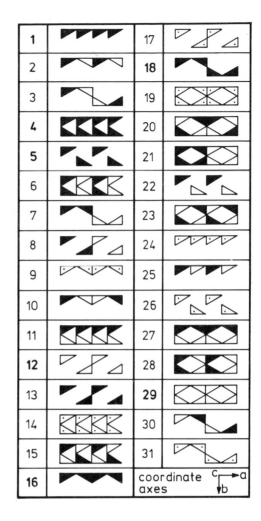

The so-called coordinate or agreed international notation refers to the mutual orientation of the coordinate axes and symmetry elements [8-2]. The notation always starts with the letter p referring to the translation group. Axis a is directed along the band, axis b lies in the plane of the drawing and axis c is perpendicular to this plane. The first, second, and third positions of the symbol after the letter p indicate the mutual orientation of the

Numbering (cf. Fig. 8-12)	Noncoordinate notation	Coordinate (international) notation
22	$(a) \cdot 2_1$	$p2_111$
25	$(a) \cdot \tilde{a}_{11}$	$p11a$

337

symmetry elements with respect to the coordinate axes. If no rotation axis or normal of a symmetry plane coincides with a coordinate axis, the number 1 is placed in the corresponding position in the symbol. The coincidence of a rotation axis, 2 or 2_1, or the normal of a symmetry plane, m or $\tilde{a}$, with one of the coordinate axes is indicated by placing the symbol of this element in the corresponding position in the notation. In addition to the above notations, two more are given here referring to the two-sided bands shown in Fig. 8-11.

8.4 Rods

The "infinite" carbide molecule, of course, is of finite width. It is indeed a three-dimensional construction with periodicity in only one direction. Thus it has one-dimensional space-group symmetry (G_1^3). It is like an infinitely long rod. For a rod the axis is a singular axis and it has no singular plane. All kinds of symmetry axes may coincide with the axis of the rod, viz., translation axis, simple rotation axis, glide-reflection axis, and screw-rotation axis. The screw-rotation axis may be not only of second order as was the case for the bands but of any order. Of course, these symmetry elements, except the simple rotation axis, may characterize the rod only if it expands to infinity indeed. As regards symmetry, a tube, a screw or various rays are as much rods as are the stems of plants, vectors, or spiral stairways. Considering them to be infinite is a necessary assumption in describing their symmetries by space groups. Real objects are not infinite. For symmetry considerations it may be convenient to look only at some portions of the whole, where the ends are not yet in sight, and extend them in thought to infinity. A portion of a spiral stairway displaying screw-axis symmetry is shown in Fig. 8-13. The imaginary-impossible stairway of Fig. 8-14 indeed seems to go on forever. Thus space-group symmetry can be properly assigned to it.

Figure 8-13.
A portion of a spiral stairway displaying screw-axis symmetry. Photograph by the authors.

Figure 8-14.
An impossible stairway with proper space-group symmetry (as we can walk around this stairway ad infinitum). The idea for this drawing originated from a movie poster advertising "Glück im Hinterhaus".

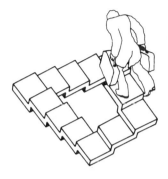

A screw axis brings the infinite rod into coincidence with itself after a translation through a distance t accompanied by a rotation through an angle α. This rod thus has a screw symmetry axis α_t. The screw axis is of the order $n = 360°/\alpha$. It is a special case when n is an integer. For the screw axis of the second order the direction of the rotation is immaterial. For all other screw axes they may be either left-handed or right-handed.

The scattered leaf arrangement around the stems of many plants is a beautiful occurrence of screw-axis symmetry in nature. The stem of Plantago media shown in Fig. 8-15 certainly does not extend to infinity. It has been suggested, however, that for plants the plant/seed/plant/seed ... infinite sequence (at least in time) provides enough justification to apply space groups in their symmetry description. Let us consider now the relative positions of the leaves around the stem of Plantago media. Starting from leaf "0", leaf "8" will be in eclipsed orientation to it. In order to reach leaf "8" from leaf "0" the stem has to be circled three times. The ratio of the two numbers, viz. 3/8 tells us that a new leaf occurs at each 3/8 part of the circumference of the stem. The ratio 3/8 is characteristic in phyllotaxy, as are 1/2, 1/3, 2/5, and even 5/13. Very little is known about the origin of phyllotaxy. What has been noted a long time ago that the numbers occurring in these characteristic ratios, viz.

$$1, 1, 2, 3, 5, 8, 13, ...$$

Figure 8-15.
The scattered leaf arrangement (phyllotaxy) of Plantago media.

are members of the so-called Fibonacci series in which each consecutive number is the sum of the previous two. Fibonacci numbers can be observed also in the numbers of the spirals of the scales as looking at pine cones from below. Fig. 8-16 shows a pine cone in two views. The bottom view clearly displays 13 left-

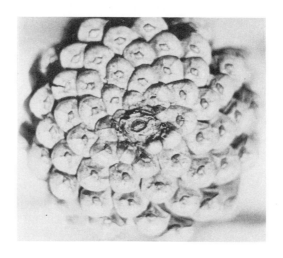

Figure 8-16.
A pine cone in two views. The view from below displays 13 left-bound and 8 right-bound spirals of scales. Photographs by the authors.

bound and 8 right-bound spirals of scales. Left-bound and right-bound spirals in strictly Fibonacci numbers are found in other plants as well. The plate of seeds of the sunflower can be considered as if it were in a compressed scattered arrangement around the stem. Fig. 8-17 shows several examples. What is possibly the most striking is that the continuation of the ratios of the characteristic leaf arrangements eventually leads to an extremely important irrational number 0.381966...., expressing the *golden mean*!

Returning to screw axes, Fig. 8-18 shows an infinite anion with a 10_5 screw axis [8-4]. The most important application of one-dimensional space groups is for polymeric molecules in chemistry [8-5]. Fig. 8-19 illustrates the structure and symmetry elements of an extended polyethylene molecule. The translation or identity period is shown in Fig. 8-19a. It is the distance between two carbon atoms separated by a third one. However, any portion with this length may be selected as the identity period along the polymeric chain. The translational symmetry of polyethylene is characterized by this identity period. In addition, there is a host of other symmetries as shown in Fig. 8-19b.

Figure 8-17.
Spirals in plants: A Singapore
stamp and photographs by
the authors.

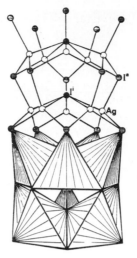

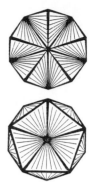

Figure 8-18.
The polymeric anion (Ag_5I_6) with a 10_5 screw axis [8-4]. Reproduced with permission.
(a) View along the screw axis, the sequence of the atoms from inside to outside is I-Ag-I.

(b) View perpendicular to the molecular axis.

(c) Views of the five condensed AgI_4 tetrahedra from the bottom and from the top.

Figure 8-19.
(a) The structure and translation period of the polyethylene chain molecule.
(b) Symmetry elements of the polyethylene chain molecule.

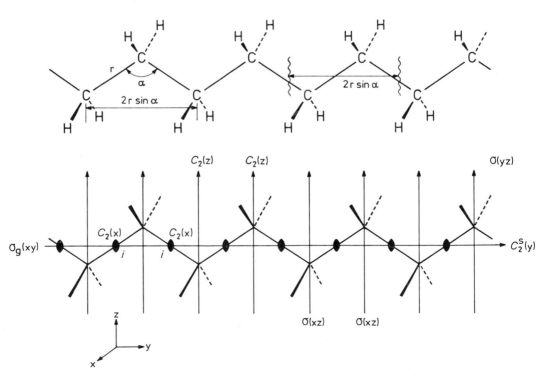

There are two kinds of two-fold axes. One is $C_2(z)$ passing through the carbon atoms in the z direction, and the other is $C_2(x)$ passing through the centers of the C-C bonds in the x direction. These centers of the C-C bonds are also inversion centers, i. There are also two kinds of mirror planes. One is a unique element coinciding with the plane of the carbon chain, $\sigma(yz)$. The other kind is a whole set of planes, $\sigma(xz)$, which are perpendicular to the chain axis and include the $C_2(z)$ two-fold axes. Then there is a glide-symmetry plane, $\sigma_g(xy)$, which is a combination of a symmetry plane perpendicular to the plane of the carbon chain and a translation by half of the identity period ($r\sin\alpha$). Finally, there is a two-fold screw axis, $C_2^s(y)$, along the molecular axis implying a 180° rotation followed by a translation of half of the identity period.

Biological macromolecules are often distinguished by their helical structures to which one-dimensional space-group symmetries are applicable. Fig. 8-20a shows the polypeptide chain of alpha helix, while Fig. 8-20b depicts a polypeptide molecule in solution. The repeating units are the same in the two systems, viz. planar CCONHC skeletons. The linear, rod-like structure of alpha helix is accomplished by the hydrogen bonds, whereas these hydrogen bonds are disrupted in solution [8-6].

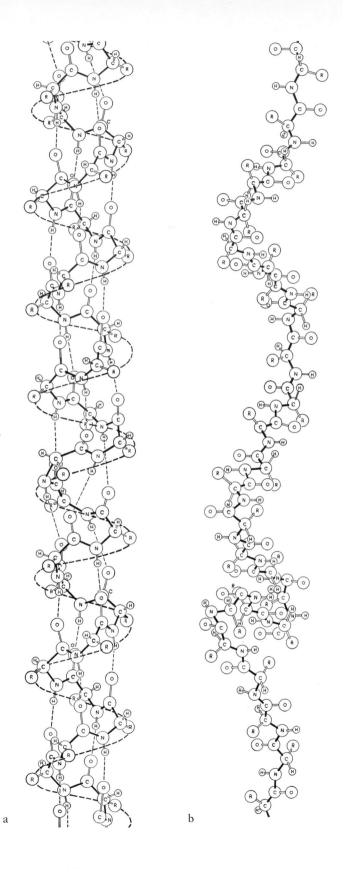

Figure 8-20.
After [8-6]. Copyright (1957)
Scientific American. Used
with permission.
(a) Linear, rod-like, helical
structure of alpha helix.

(b) Random chain of the
polypeptide molecule as the
hydrogen bonds of alpha helix
are disrupted in solution.

a   b

8.5 Two-Dimensional Space Groups

There are altogether 17 symmetry classes of one-sided planar networks (cf. e.g. [8-2]). Fig. 8-21 illustrates them in a way analogous to the seven symmetry classes of the one-sided bands (Fig. 8-9). The most important symmetry elements and the coordinate notations of the symmetry classes are also given. The first letter (p or c) in this notation refers to translation. The next three positions carry information on the presence of various symmetry elements, m – symmetry plane, g – glide-reflection plane, 2, 3, 4, or 6 – rotation axes. The number 1 or blank indicate the absence of a symmetry element. The representations of the symmetry classes in Fig. 8-21 were in some way inspired by the illustrations inside the covers of Buerger's Elementary Crystallography [8-7]. Along with the purely geometrical configurations, Fig. 8-21 presents 17 Hungarian needlework patterns. A brief description of their sources is given in the legend of the figure [8-8].

Figure 8-21.
The 17 symmetry classes of one-sided planar networks with the most important symmetry elements and the notations of the classes indicated. Along with the geometrical configurations, Hungarian needlework patterns are presented with corresponding symmetries. A brief description of the origin of these patterns is given here [8-8]:

p1 and p4 Patterns of indigo dyed decorations on textiles for clothing. Sellye, Baranya county, 1899.

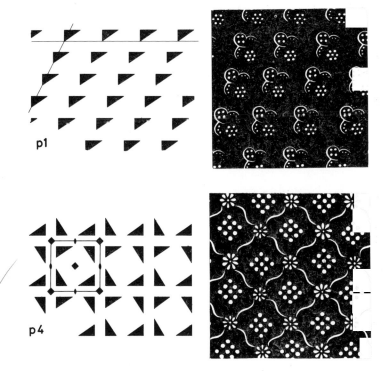

*p*2 Indigo dyed decoration with palmette motif for curtains. Currently very popular pattern.

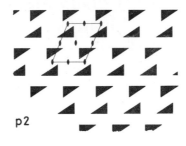

p2

*p*3, *p*6, *p*6*mm*, *p*3*m*1, and *p*31*m* Decorations with characteristic bird motifs from peasant vests. Northern Hungary.

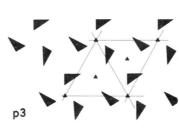

p3

p6

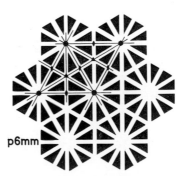

p6mm

p3m1

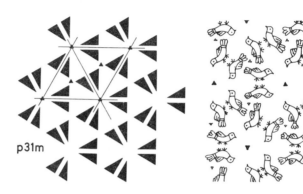

p31m

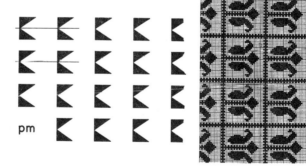

pm Decoration with tulip motif for table-cloth. Cross-stitched needlework. From the turn of the century.

pm

*pmm*2 Bed-sheet border decoration with pomegranate motif. Northwest Hungary, 19th century.

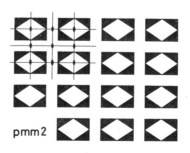

pmm 2

p4mm Pillow-slip decoration with stars. Cross-stitched needlework. Transsylvania, 19th century.

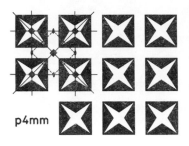

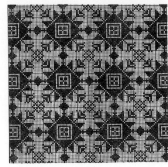

cm Pillow-slip decoration with peacock tail motif. Cross-stitched needlework. Much used throughout Hungary around the turn of the century.

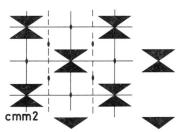

*cmm*2 Bed-sheet border decoration with cockscomb motif. Cross-stitched needlework. Somogy county, 19th century.

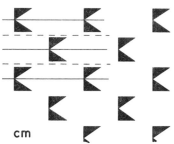

pg From a pattern-book of indigo dyed decoration. Pápa, Veszprém county, 1856.

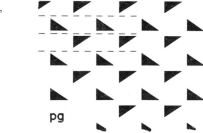

*pgg*2 Children's bag decoration. Transsylvania, turn of the century.

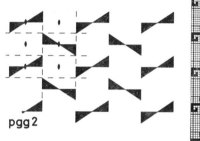

pgg 2

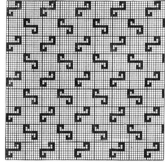

*pmg*2 Pillow-slip decoration with scrolling stem motif. Much used throughout Hungary around the turn of the century.

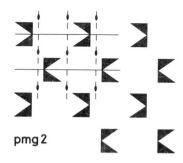

pmg 2

p4gm Blouse-arm embroidery. Bács-Kiskun county, 19th century.

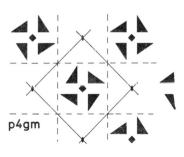

p4gm

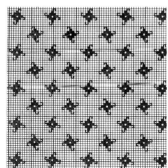

There is a striking limitation on the possible periods of the rotation axes in the space groups. The origin of this limitation will be addressed later in the chapter on crystals (Chapter 9) as it is applicable to the three-dimensional space groups as well.

The lattice of the planar networks with two-dimensional space groups is defined by two non-collinear translations. Such a lattice is shown in Fig. 8-22. Given a particular lattice, the question is which pair of translations should be selected to describe it? An infinite number of choices exists for each translation because a line joining any two lattice points is a trans-

Figure 8-22.
Plane lattice defined by two non-collinear translations.

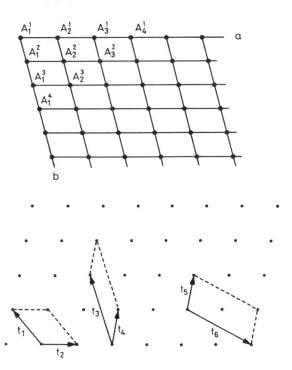

Figure 8-23.
Illustration of primitive and unit cells on a plane lattice, after Azaroff [8-22]. Copyright (1960) McGraw-Hill, Inc. Used with permission.

lation of the lattice. Fig. 8-23 shows a plane lattice and some of the possible choices for translation pairs to describe it. A primitive cell is defined by choices of translation pairs such as t_1 and t_2 or t_3 and t_4. Only one lattice point is associated with each primitive cell. This is understood if each lattice point in Fig. 8-23 is considered to belong to four adjacent cells, or only one fourth of each point to belong to any one cell. As each cell contains four corners, all this adds up to one whole point. Alternatively, by displacing any one primitive cell, each primitive cell will contain only one lattice point. On the other hand, a multiple cell contains one or more lattice points in addition to the one shared at the corners. The translation pair t_5 and t_6, for instance, defines a double cell. A cell is called a unit cell if the entire lattice can be derived from it by translations. Thus a unit cell may be either primitive or multiple. The unit cell is chosen usually to represent best the symmetry of the lattice. The translations selected as the edges of the plane unit cell are a and b, and for a space lattice, a, b, and c. The latter are called the crystallographic axes. The angles between the edges of the three-dimensional unit cell are α, β, and γ, but only γ is needed for the plane lattice.

Fig. 8-24 shows three planar networks based on the same plane lattice. Two and only two lines intersect in each point of all three networks. Accordingly, the parallelograms of all three networks have the same area. All of them are unit cells, in fact primitive cells. Each of these parallelograms is determined by two sides a and b, and the angle γ between them. These are called the cell parameters.

Figure 8-24.
Different networks based on the same plane lattice.

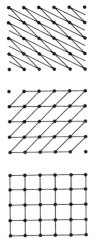

350

The general plane lattice (a) shown in Fig. 8-25 is called a parallelogram lattice. The other four plane lattices of Fig. 8-25 are special cases of the general lattice. The rectangular lattice (b) has a primitive cell with unequal sides. The so-called diamond lattice (c) has a unit cell with equal sides. A special case of the diamond lattice is when the angle between the equal sides of the unit cell is 120° and this lattice (d) is then called rhombic, or triangular since the short cell diagonal divides the unit cell into two equilateral triangles. This lattice may also be considered as having hexagonal symmetry. Finally there is the square lattice (e).

Figure 8-25.
The five unique plane lattices (a through e, see text).

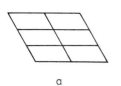 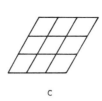  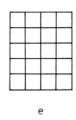

a   b   c   d   e

The five unique plane lattices were described above under the assumption that the lattice points themselves have the highest possible symmetry. In this case these five unique lattices will have the following symmetries (Fig. 8-25):

	Space groups	
	Noncoordinate symbols	Coordinate or international symbols
a) parallelogram lattice	(b/a):2	$p2$
b) rectangular lattice	$(b{:}a)$:2·m	$pmm2$
c) diamond lattice	(a/a):2·m	$cmm2$
d) hexagonal or triangular lattice	(a/a):6·m	$p6mm$
e) square lattice	$(a{:}a)$:4·m	$p4mm$

Combining the point-group symmetries with the plane lattices, altogether 17 two-dimensional space groups may be produced. They are all represented in Fig. 8-21. In fact severe limitations are imposed on the possible point groups that may be combined with lattices to produce space groups. Some symmetry elements, like the five-fold rotation axis, are not compatible with translational symmetry. This will be examined in detail in Chapter 9.

8.5.1 Some Simple Networks

The simplest two-dimensional space group is represented in four variations in Fig. 8-26. This space group does not impose any restrictions on the parameters a, b, and γ. The equal motifs repeated by the translations may be completely separated from one another, they may consist of disconnected parts, they may intersect each other, and, finally, they may fill the entire plane without any gaps. Of course, such variations are possible for any of the more complicated two-dimensional space groups as well.

Figure 8-26.
The simplest two-dimensional space group in four variations.

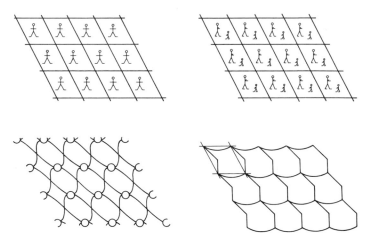

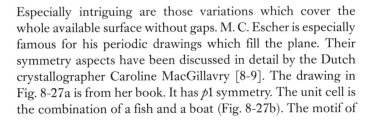

Figure 8-27.
(a) Escher's periodic drawing of fish and boats with space group $p1$ from MacGillavry's book [8-9]. Reproduced with permission from the International Union of Crystallography.

Especially intriguing are those variations which cover the whole available surface without gaps. M. C. Escher is especially famous for his periodic drawings which fill the plane. Their symmetry aspects have been discussed in detail by the Dutch crystallographer Caroline MacGillavry [8-9]. The drawing in Fig. 8-27a is from her book. It has $p1$ symmetry. The unit cell is the combination of a fish and a boat (Fig. 8-27b). The motif of

(b) The unit cell: a fish and a boat as the motif for repetition.

352

Figure 8-28.
(a) Escher's periodic drawing of fish and birds with space group $p2$ from MacGillavry's book [8-9]. Reproduced with permission from the International Union of Crystallography.

(b) The primitive cell and a unit cell displaying two-fold rotation.

another drawing (Fig. 8-28a) from the same book [8-9] is formed from a bird and a fish (Fig. 8-28b). A unit cell may be chosen to contain two birds and two fish, each pair containing a two-fold axis of rotation. The basic motif, the primitive cell, however, contains only one bird and one fish. This is what is also called the asymmetric unit. The unit cell contains two asymmetric units in this network. The whole pattern may be generated by two-fold rotation and translations. Its two-dimensional space group is $p2$. There are altogether four different kinds of two-fold rotation axes in this drawing as was indicated above. They are always at distances of $(1/2)a$, $(1/2)b$, and $(1/2)(a+b)$ from each other regardless of the choice of the unit cell.

Canadian crystallographer François Brisse has designed a series of two-dimensional space-group drawings related to Canada [8-10]. The series was dedicated to the XIIth Congress of the International Union of Crystallography (IUCr), Ottawa, 1981. Drawings have been prepared to represent the Canadian provinces and territories. One of them is shown in Fig. 8-29a. The polar bear is a symbol for Canada's Northwest Territories. Its stylized representation is the asymmetric unit to which first a two-fold rotation is applied (Fig. 8-29b), and then the translations. For unit cell it is convenient to choose two polar bears related by two-fold rotation. White and blue polar bears alternate in the original drawing, but the colors are disregarded in the present discussion.

Figure 8-29.
(a) Brisse's periodic drawing "Northwest Territories". Reproduced with permission from "La symétrie bidimensionelle et le Canada" [8-10].

(b) The primitive cell and a unit cell displaying two-fold rotation.

Figure 8-30.
(a) The maple leaf and its stylized version. The unit cell displaying four-fold rotation was the symbol of the XIIth Congress of the International Union of Crystallography, Ottawa, 1981.

The symbol of the XIIth IUCr Congress was a unit of four stylized maple leaves related by four-fold rotation. It is shown in Fig. 8-30a. The maple leaf is Canada's symbol as is shown in a more natural appearance on a stamp in Fig. 8-30a. The symmetry properties of the unit in Fig. 8-30a have been discussed in detail in the materials of the XIIth Congress [8-11]. A two-dimensional drawing created by repetition from the above unit is shown in Fig. 8-30b. Again the original alternating white/red

(b) Brisse's periodic drawing "Canada". Reproduced with permission from "La symétrie bidimensionelle et le Canada" [8-10].

coloring is disregarded in our discussion. The two-dimensional space group of the pattern is then *p4gm* [8-10]. The pattern shown in Fig. 8-30b has already been used by Pólya [8-12] among his representations of the 17 two-dimensional space groups. Interestingly enough it may also be found as an example of typical decorations in Islamic art as two-dimensional pattern [8-13] and also in its one-dimensional variation as a band ornament [8-14].

The repetition of the flies, butterflies, falcons, and bats in the drawing of Escher in Fig. 8-31a is accomplished by mirror planes. The two-dimensional space group is *pmm* and the mirror planes are indicated separately as the borders of the primitive cell in Fig. 8-31b. The symmetry of yet another of Escher's periodic drawings shown in Fig. 8-32 is sometimes incorrectly described. At first sight the meeting points of the

Figure 8-31.
(a) Escher's periodic drawing of flies, butterflies, falcons and bats from MacGillavry's book [8-9]. Reproduced with permission from the International Union of Crystallography.

(b) The primitive cell framed by the square whose sides are parts of the mirror planes in the periodic drawing.

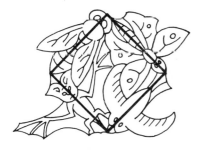

Figure 8-32.
Escher's periodic drawing of clams, starfish and snail shells from MacGillavry's book [8-9]. Reproduced with permission from the International Union of Crystallography.

four clams and the four starfish are considered to have four-fold symmetry. However, this is not the case as the snail shells in between these meeting points do not possess two-fold symmetry. True four-fold axes can be found at the meeting points of the four snail shells and four starfish. The other points have merely two-fold symmetry, and no other symmetry elements are present [8-9].

The following three periodic drawings were created by Khudu Mamedov, an Azerbaidzhani crystallographer, and appeared in a remarkable book on symmetry entitled "Decorations Remember" [8-15]. Fig. 8-33 shows the decoration of a cylindrical mausoleum in Badra, Azerbaidzhan. The basic motif, the word Allah appears 200 times in the original mosaic decoration covering the whole surface of the building. The drawing of Fig. 8-33, if extended to infinity, has the two-dimensional space group *p*4*gm,* the same as the one in Fig. 8-30. Mamedov's drawing entitled "Unity" is shown in Fig. 8-34. The

Figure 8-33.
The decoration of the Badra mausoleum (Azerbaidzhan) [8-15].

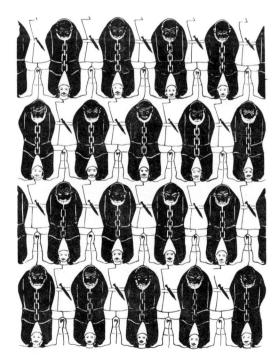

Figure 8-34.
Mamedov's periodic drawing "Unity" [8-15].

356

space group is $p1$ with the basic motif consisting of an old and a young man. The repetition of the uniform shapes truly satisfies the requirement of the two-dimensional space group. A closer look, however, reveals the remarkable individuality of the facial expressions, especially for the old men. The third drawing by Mamedov in Fig. 8-35 is entitled "Sea-gulls" [8-15]. The symmetry of this drawing is the same as that in Fig. 8-29.

Figure 8-35.
Mamedov's periodic drawing "Sea-gulls" [8-15].

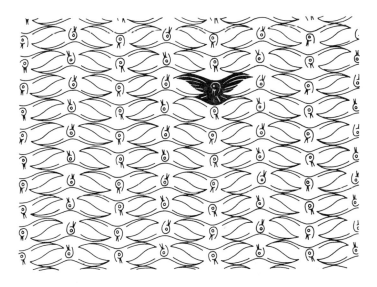

Figure 8-36.
Mackay's jig-saw "puzzles" [8-16], reproduced with permission. The specimens of each piece automatically assemble themselves to give patterns of a specified symmetry. In case of mirror planes small connecting pieces are required to give bonding. The numbering corresponds to the 17 two-dimensional space groups:

1. $p1$ 10. pm
2. $p2$ 11. $pmg2$
3. $p3$ 12. pg
4. $p6$ 13. $p4gm$
5. $p4$ 14. $p4mm$
6. $pgg2$ 15. $p6mm$
7. $cmm2$ 16. $p3m1$
8. $pmm2$ 17. $p31m$
9. cm

Finally an interesting representation for the 17 two-dimensional space groups is shown in Fig. 8-36. It is British crystallographer Alan Mackay's jig-saw "puzzles" [8-16]. Specimens of each piece can automatically assemble themselves to give patterns of a specified symmetry.

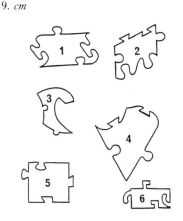

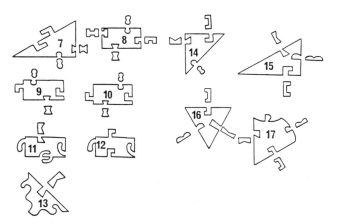

8.5.2 Visual Side-Effects of Decorations

Shubnikov and Koptsik [8-2] have analyzed the influence of the various space groups of bands and networks on people's perception of movement. A one-sided band decoration without a polar axis induces no feeling of movement. The vertical symmetry axes in Fig. 8-37a act as if preventing motion. On the other hand, the bands with polar axes in Fig. 8-37b act as if inducing left-bound (top) and right-bound (bottom) movement. Simple geometrical patterns can achieve similar effects, as seen, e.g., in Fig. 8-7a inducing motion and in Fig. 8-7b preventing it. The two-way traffic analogy may apply to the pattern in Fig. 8-37c.

Figure 8-37.
(a) The vertical symmetry axes seem to prevent motion (Mexican decoration).

(b) The presence of polar axes convey the feeling of motion, left-bound or right-bound, depending on the direction of the axis. The pattern was created after [8-2].

(c) The two-way traffic analogy may apply to the one-dimensional space group with translation upon rotation by 180°.

Two-dimensional space groups with non-orthogonal translation axes and with no symmetry planes emphasize oblique movement, as do, e.g., the patterns in Figs. 8-27a and 8-28a, and the pattern of the brick wall in Fig. 8-38a. The feeling of horizontal movement is transmitted by patterns with horizontal translation axes which have no vertical symmetry plane. An example is cited in Fig. 8-38b. A symmetry plane always conveys the impression that it is preventing motion perpendicular to it. Thus the symmetry of the needlework in Fig. 8-38c seems to present a barrier to horizontal movement, while the up and down movement appears to be unhindered. Each one of the two patterns in Fig. 8-38d conveys the feeling of *either* up *or* down movement. They all have vertical mirror planes. The difference between the patterns in Fig. 8-38c and d is the additional two-fold rotation axes in the former which create the impression of two-way movement in the same pattern.

Figure 8-38.
(a) A brick wall: its two-dimensional pattern with non-orthogonal translation axes and no symmetry planes conveys the impression of oblique movement. Photograph by the authors.

(b) The horizontal translation axis with no vertical symmetry plane conveys the feeling of horizontal movement (Egyptian decoration after [8-2]).

(c) Vertical symmetry planes and two-fold axes together convey the impression of up *and* down motion (Hungarian needlework).

(d) Vertical symmetry planes create the feeling of *either* up *or* down motion (Hungarian pattern).

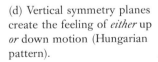

Two-dimensional space groups with only rotation axes in addition to translation convey the feeling of rotation or even dancing. Such patterns are shown in Fig. 8-39.

Figure 8-39.
Two-dimensional patterns with only rotation axes in addition to translation are dynamic, they may induce the feeling of rotation and dancing.

Patterns with glide-mirror planes tend to express chaos and convey the feeling of uncertainty. Examples are cited in Fig. 8-40. On the other hand, symmetry planes in several directions

Figure 8-40.
Patterns with glide-mirror planes after [8-2].

are supposed to radiate authority (Fig. 8-41). Finally, there are patterns in which both mirror planes and glide-mirror planes occur. The details of such patterns convey the feeling of motion, while as a whole they give the impression of calmness,

Figure 8-41.
The decoration of the ceiling under the main cupola of the Capitol in Washington, D.C. Photograph by the authors.

according to Shubnikov and Koptsik [8-2]. Both mirror planes and glide-mirror planes are present in the examples displayed in Fig. 8-42 and Fig. 8-43.

Figure 8-42.
Indian pattern from South America after [8-17].

Figure 8-43.
Pavements with both mirror planes and glide-mirror planes. Photographs by the authors. (a) Main square of Baja, Hungary. The mirror planes are in the horizontal direction, the glide-mirror planes in the vertical direction. The occurrence of symmetry elements is very approximate.

(b) Around the Statehouse in Annapolis, MD.

8.5.3 Moirés

An art exhibition entitled "Moirés" was held a few years ago in Bern, Switzerland [8-18]. The so-called Moiré pattern is created by superimposing infinite planar patterns. The resulting pattern is a new two-dimensional network. The simplest case is illustrated in Fig. 8-44. Two identical systems of lines on transparent paper are superimposed.

Figure 8-44.
Moiré patterns from the superposition of two line systems at increasing angles.

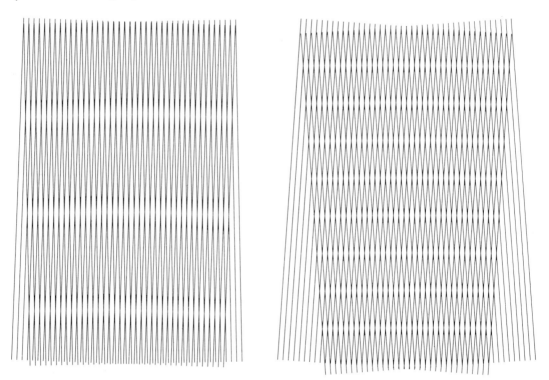

The starting and resulting systems have a period of λ and d, respectively, and they are superimposed at an angle Θ. These parameters then have the following relationship (cf., e.g., [8-2]):

$$\lambda = 2d \sin(\Theta/2).$$

This expression is well known as the Bragg law in X-ray diffraction of crystals, where λ is the wave length of the X-rays, d is the distance between atomic layers in the crystal, and $\Theta/2$ is the angle at which the X-rays hit the atomic layer.

The patterns produced in Fig. 8-44 have two-fold rotation axes perpendicular to their plane. Thus the superposition at angle Θ will produce the same result as at angles $180 + \Theta$.

Fig. 8-45 shows the interference of two identical infinite systems of small circles at a series of angles. The circles are located in the vertices of an orthorhombic network. At small angles Θ the resulting pattern is very much like the starting one. It appears as an enlarged version of the original. With increasing Θ the extent of the enlargement decreases and so does the likeness to the original. However, the centered orthorhombic symmetry is retained.

Figures 8-45.
Moiré patterns from the superposition of two circle systems at increasing angles.

Shubnikov and Koptsik [8-2] describe various examples of Moirés in different fields. A perhaps unexpected application is cited here: Only certain drop sizes cause supernumerary bows in a rainshower. The interference pattern produced by the rain wave front folding over on itself can be simulated with Moiré patterns. This is illustrated in Fig. 8-46 after Fraser [8-19]. The spacing in the bows clearly depends on the drop size. A detailed analysis of Moiré patterns and their applications has been given recently by Giger [8-20]. The potentials of Moirés in artistic expression have been demonstrated by Witschi [8-21].

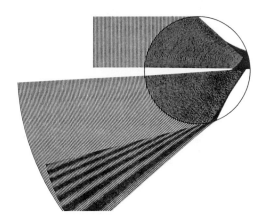

Figure 8-46.
Moiré patterns simulate the formation of supernumerary bows in rain showers. The larger drops produce closely spaced bows at the top, while at the bottom the smaller drops produce more widely spaced bows. After Fraser [8-19], reproduced with permission.

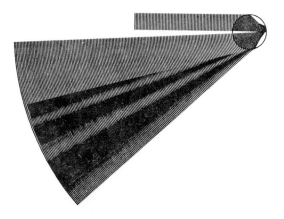

References

[8-1]   F. J. Budden, *The Fascination of Groups,* University Press, Cambridge, 1972.

[8-2]   A. V. Shubnikov and V. A. Koptsik, *Simmetriya v nauke i iskusstve,* Nauka, Moskva, 1972. English translation: *Symmetry in Science and Art,* Plenum Press, New York, London, 1974.

[8-3]   I. Hargittai and Gy. Lengyel, J. Chem. Educ. **61**, 1033 (1984).

[8-4]   K. Peters, W. Ott, and H. G. v. Schnering, Angew. Chem. Int. Ed. Engl. **21**, 697 (1982).

[8-5]   H. Tadokoro, *Structure of Crystalline Polymers,* Wiley-Interscience, New York, Chichester, Brisbane, etc. 1979.

[8-6]   P. Doty, in *The Molecular Basis of Life,* R. H. Haynes and P. C. Hanewalt, eds., W. H. Freeman and Co., San Francisco and London, 1968.

[8-7]   M. J. Buerger, *Elementary Crystallography, An Introduction to the Fundamental Geometrical Features of Crystals* (Fourth printing), Wiley, New York, London, Sydney, 1967.

[8-8]   I. Hargittai and Gy. Lengyel, J. Chem. Educ. **62**, 35 (1985).

[8-9]   C. H. MacGillavry, *Symmetry Aspects of M. C. Escher's Periodic Drawings,* Bohn, Scheltema and Holkema, Utrecht, 1976.

[8-10]   F. Brisse, Can. Mineral. **19**, 217 (1981).

[8-11]   Z. Durski, H. Nowaczek, and H. Boniuk, *Twelfth International Congress of Crystallography, Collected Abstracts,* Paper 20.1-08, p. C-352, Ottawa, 1981.

[8-12]   G. Polya, Z. Krist. **60**, 278 (1924).

[8-13]   I. El-Said and A. Parman, *Geometric Concepts in Islamic Art,* World of Islam Festival Publ. Co., London, 1976.

[8-14]   P. D'Avennes, ed., *Arabic Art in Color,* Dover, New York, 1978.

[8-15]   Kh. S. Mamedov, I. R. Amiraslanov, G. N. Nadzhafov, and A. A. Muzhaliev, *Decorations Remember* (in Azerbaidzhani), Azerneshr, Baku, 1981.

[8-16]   A. L. Mackay, Chimia **23**, 433 (1969).

[8-17]   G. Hartmann, *Textilkunst der Cuna-Indios in Grenzgebiet von Kolumbien und Panama:* eine Ausstellung des Landesmuseums Koblenz, 1980.

[8-18]   W. Witschi, *Moirés,* Kunstmuseum Bern, 1982.

[8-19]   A. B. Fraser, J. Opt. Soc. Am. **73,** 1626 (1983).

[8-20]   H. Giger, *Symmetry, Unifying Human Understanding,*
I. Hargittai, ed., Pergamon Press, New York, Oxford, 1986.

[8-21]   W. Witschi, *Symmetry, Unifying Human Understanding,*
I. Hargittai, ed., Pergamon Press, New York, 1986.

[8-22]   L. V. Azaroff, *Introduction to Solids,* McGraw-Hill, New York,
Toronto, London, 1960.

9 Symmetries in Crystals[1]

"... But I must speak again about crystals, shapes, colors. There are crystals as huge as the colonnade of a cathedral, soft as mould, prickly as thorns; pure, azure, green, like nothing else in the world, fiery, black; mathematically exact, complete, like constructions by crazy, capricious scientists, or reminiscent of the liver, the heart... There are crystal grottos, monstrous bubbles of mineral mass, there is fermentation, fusion, growth of minerals, architecture and engineering art... Even in human life there is a hidden force towards crystallization. Egypt crystallizes in pyramids and obelisks, Greece in columns; the middle ages in vials; London in grinny cubes... Like secret mathematical flashes of lightning the countless laws of construction penetrate the matter. To equal nature it is necessary to be mathematically and geometrically exact. Number and phantasy, law and abundance – these are the living, creative strengths of nature; not to sit under a green tree but to create crystals and to form ideas, that is what it means to be at one with nature!" These are the words of Karel Čapek the Czech writer after his visit to the mineral collection of the British Museum [9-18]. He added a drawing (Fig. 9-1) to his words to express man's humility in front of these miracles of nature.

Figure 9-1.
Čapek's drawing after his visit to the mineral collection of the British Museum [9-18]. Reproduced with permission.

---

1 The fundamental works of Kitaigorodskii [9-1], Wells [9-2] and Azaroff [9-3] were extensively consulted along with other literature also cited in the text.

The word crystal comes from the Greek *krystallos* meaning clear ice. The name originated from the mistaken belief that the beautiful transparent quartz stones found in the Alps were formed from water at extremely low temperatures. Later, by the seventeenth century the name crystal was applied to other solids that were also bounded by many flat faces and had generally beautiful symmetrical shapes. Crystals have also for centuries been considered mystical. A sad angel looks hopelessly at the huge rhombohedric crystal in Dürer's "Melancholia" (Fig. 9-2). The polyhedron in the picture is a truncated rhombohedron, and there has been considerable discussion as to whether Dürer meant a particular mineral by it and if so which one [9-4, 9-5]. The emerging opinion is that this polyhedron "is simply an exercise in accurate draughtsmanship and that the art historians have made rather heavy weather of its explanation... The integral proportions show that no particular mineral was intended" [9-5]. Dürer's drawings have been carefully analyzed by Schröder [9-6] who "has satisfactorily settled the matter with a technological rather than a mystical explanation" [9-5].

Figure 9-2.
Dürer: "Melancholia".

Space-group symmetries have played an outstanding role in Escher's graphic art. So what he wrote about crystals is also of interest [9-7]:

"Long before there were men on this globe, all the crystals grew within the earth's crust. Then came a day when, for the very first time, a human being perceived one of these glittering fragments of regularity; or maybe he struck against it with his stone ax; it broke away and fell at his feet; then he picked it up and gazed at it lying there in his open hand. And he marveled."

"There is something breathtaking about the basic laws of crystals. They are in no sense a discovery of the human mind; they just 'are' – they exist quite independently of us. The most that man can do is to become aware, in a moment of clarity, that they are there, and take cognizance of them."

The symmetry of the shapes of the crystals is their most easily recognizable feature. The great Russian crystallographer E. S. Fedorov remarked that "the crystals glitter with their symmetry". Obviously, this outer symmetry is a consequence of the inner structure. However, with the same inner structure, crystals may grow into different forms. Besides, under natural conditions, crystals seldom grow into their well known regular forms. Under different conditions, in the presence of different impurities, for example, different forms may grow. Fig. 9-3 shows the influence of impurities upon the form of sodium chloride crystals.

Figure 9-3.
The influence of impurities on the habitus of sodium chloride crystals.

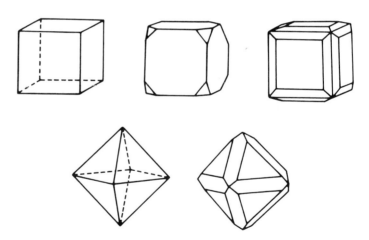

9.1 Basic Laws

It has been recognized already in the earliest stages in the history of crystallography that the most important characteristic of the outer symmetry of the crystals is not really the form itself, but rather two phenomena expressed by two rules. One is the constancy of the angles made by the crystal faces. The other is the law of rational intercepts or the law of rational indices.

Already in 1669 the Danish crystallographer Steno made a detailed study of ideal and distorted quartz crystals (Fig. 9-4). He traced their outlines on paper, and found that the corresponding angles of different sections were always the same regardless of the actual sizes and shapes of the section. Thus all quartz crystals, however much distorted from the ideal, could result from the same fundamental mode of growth, and, accordingly, corresponded to the same inner structure.

Figure 9-4.
Sections of ideal and distorted quartz crystals.

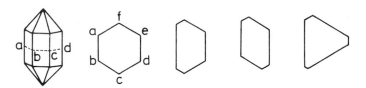

For measuring the angles made by the crystal faces, instruments have been developed. In 1780 the contact goniometer, Fig. 9-5a, was already in usage. Later, for more precise measurement of the interfacial angles, the reflecting goniometer was introduced (Fig. 9-5b). The name 'goniometer' has become so synonymous with crystal structure determination that it is sometimes used colloquially for today's sophisticated X-ray diffractometers.

Another interesting phenomenon early observed in crystals is their cleavage. It is characteristic that they break along well-defined planes. The French crystallographer Haüy noticed that the cleavage rhombs from any calcite crystal always had the same interfacial angles. Thus he suggested that all calcite crystals could be built of these fundamental cleavage rhombs. This is illustrated in Fig. 9-6 which is from Haüy's "Traité de Cristallographie". In fact, it is considered to be so fundamental that few books on crystallography appear without reproducing this drawing. From the units shown in Fig. 9-6, it is possible to build straight edges corresponding to the faces of a cube, as well as inclined edges corresponding to the faces of an octahedron.

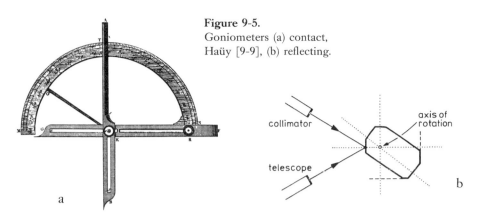

Figure 9-5.
Goniometers (a) contact,
Haüy [9-9], (b) reflecting.

collimator

axis of
rotation

telescope

b

a

Figure 9-6.
Cleavage rhombs and the
stacking of cleavage rhombs,
Haüy [9-9].

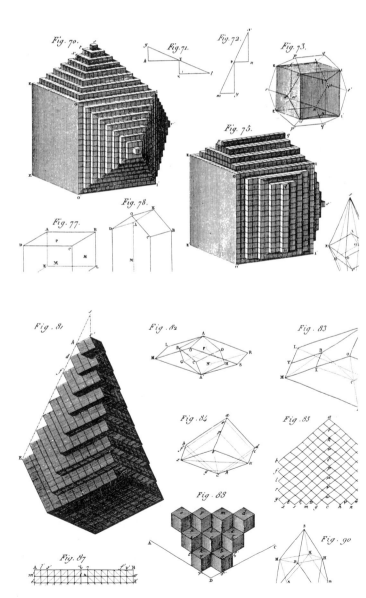

Edges inclined at other edges may also be built. Let the dimensions of the cleavage unit be a and b (Fig. 9-7), then tan $\Theta_1 = b/a$ and tan $\Theta_2 = b/2a$, and generally tan $\Theta = mb/na$, where m and n are rational integers. By extension into the third dimension, we may have a reference face making intercepts a, b, c on three axes. The intercepts made by any other face must be in proportion of rational multiples of these intercepts. This is called the law of rational intercepts.

Figure 9-7.
Inclined edges from cleavage units and illustration for the law of rational intercepts.

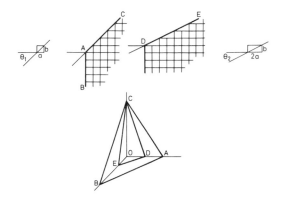

Usually the crystal faces are described by the reciprocals of the multiples of the standard intercepts, hence the name of the law of rational indices. In Fig. 9-7 three lines are adopted as axes which may also be directions of the crystal edges. A reference face ABC makes intercepts a, b, c on these axes. Another face of the crystal, e.g. DEC, can be defined by intercepts a/h, b/k, c/l. Here h, k, l are simple rational numbers or zero. They are called Miller indices. The intercept is infinite if a face is parallel to an axis, and h or k or l will be zero. For orthogonal axes the indices of the faces of a cube are (100), (010), (001). The indices of the face DEC in Fig. 9-7 are (231).

The simple cleavage model of Haüy indeed revealed a lot about the structure of crystals. However, it was not generally applicable since cleavages do not always lead to cleavage forms which can necessarily fill space by repetition. As has already been mentioned in the previous chapter, there is a limited number of space-filling polyhedra.

The characterization of the regularities in the outer form of the crystals has led to the recognition of three-dimensional periodicity in their inner structure. This was long before the possibility of determining the atomic arrangements in crystals by various diffraction techniques had materialized.

Figure 9-8.

(a) Closely packed spheres by Kepler [9-10].

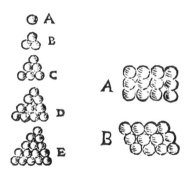

(b) Closely packed cannon balls, Laconia, New Hampshire. Photograph by the authors.

(c) From an open-air sculpture garden near Pécs, Hungary. Photograph by the authors.

It was two hundred years before Dalton and three hundred years before X-ray crystallography, that Kepler discussed the atomic arrangement in crystals. In his "Strena seu de nive sexangula" he presented arrangements of close-packed spheres. These are reproduced in Fig. 9-8a. A close-packed arrangement of cannon balls and a sculpture apparently expressing close packing are shown in Fig. 9-8b and c. The fundamental importance of Kepler's idea is that he correlated for the first time the external forms of solids with their inner structure. Kepler's search for harmonious proportions is the bridge between his epoch-making discoveries in heavenly mechanics and his lesser known but nonetheless seminal ideas in what is called today crystallography. As Schneer describes it [9-11], the renaissance era has provided a stimulating background for the beginnings of the science of crystals.

It is to be noted that even after the discovery of Haüy's model, the attention was focused on the packing in crystals. The aim was to find those arrangements in space which are consistent with the properties of the crystals.

The most important characterization of the crystal structure is the three-dimensional periodicity of the atomic arrangement, for which we find an explanation in the dense packing of the participating species.

The symmetry of the form of the crystal is a consequence of its structure. The same high symmetry of the form, however, may be easily achieved for a piece of glass by artificial mechanical intervention. By acquiring the same outer form as is typical for a piece of diamond, the piece of glass will not acquire all the other properties that the diamond possesses. The difference in value has long ago been recognized. In the India of the sixth century portrayed by Kama Sutra of Vatsayana, one of the arts which a courtesan had to learn was mineralogy (along with chemistry). If she were paid in precious stones, she had to be able to distinguish real crystals from paste [9-12].

Primarily the structure, and accordingly the outer and inner symmetry properties of the crystal determine their many outstanding physical properties. The mechanical, electrical, magnetic, and optical properties of the crystals can all be described in close conjunction with their symmetry properties [9-13].

In an actual crystal the atoms are in permanent motion. However, this motion is much more restricted than that in liquids, let alone gases. As the nuclei of the atoms are much smaller and heavier than the electron clouds, their motion can be very well described by small vibrations about the equilibrium positions. In our discussion of crystal symmetry, as an approximation, the structures will be regarded as completely rigid. However, in modern crystal molecular structure determination atomic motion must be considered. Both the techniques of structure determination and the interpretation of the results must include the consequences of the motion of atoms in the crystal. Let the poet crystallographer be cited here [9-14]:

> My molecule is sick
> And I have caught the illness too.
> Two atoms have temperatures
> Which are negative,
> And two are not resolved at all.
> How can I find a cure –
> The *R*-factor is enormous
> And direct methods fail me?
> Perhaps it is not my métier,
> To be a structure analyst.

9.2 The 32 Crystal Groups

Although the word crystal in its every-day usage is almost synonymous with symmetry, it is important to recognize that there are severe restrictions on crystal symmetry. While there are no restrictions in principle for the number of symmetry classes of molecules, this is not so for the crystals. All crystals, as regards their form, belong to one or other of only 32 symmetry classes. They are also called the 32 *crystal* point groups. Figs. 9-9a, b show them by examples of actual minerals, and by stereographic projections with symmetry elements.

Stereographic projection starts by representing the crystal by a set of lines perpendicular to its faces. The introduction of this method of representation followed soon after the invention of the reflecting goniometer.

Let us place the crystal in the center of a sphere and extend its face-normals to meet the surface of the sphere as seen in Fig. 9-10a. A set of points will occur on the surface of the sphere representing the faces of the crystal. Join now all the points in the northern hemisphere to the South Pole and mark the points on the equatorial plane where these connecting lines intersect this plane. This will create a representation of the faces on the upper half of the crystal within a single circle as seen in Fig. 9-10b. Performing a similar operation for the points of the equator (c) and for the points in the southern hemisphere (d), we arrive at the representation of the whole crystal within the circle (e). The points from the northern hemisphere are marked by dots and those from the southern hemisphere by small circles. Some examples for simple polyhedra are shown in Fig. 9-11.

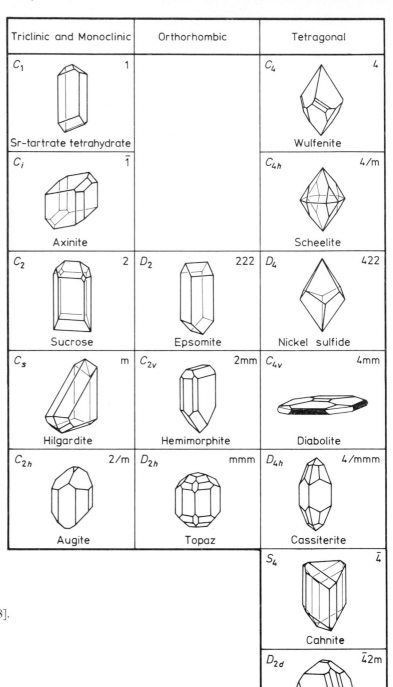

Triclinic and Monoclinic	Orthorhombic	Tetragonal
C_1 1 Sr-tartrate tetrahydrate		C_4 4 Wulfenite
C_i $\bar{1}$ Axinite		C_{4h} 4/m Scheelite
C_2 2 Sucrose	D_2 222 Epsomite	D_4 422 Nickel sulfide
C_s m Hilgardite	C_{2v} 2mm Hemimorphite	C_{4v} 4mm Diabolite
C_{2h} 2/m Augite	D_{2h} mmm Topaz	D_{4h} 4/mmm Cassiterite
		S_4 $\bar{4}$ Cahnite
		D_{2d} $\bar{4}2m$ Chalcopyrite

Figure 9-9.
Representations of the 32
crystal point groups.
(a) Actual minerals after [9-8].

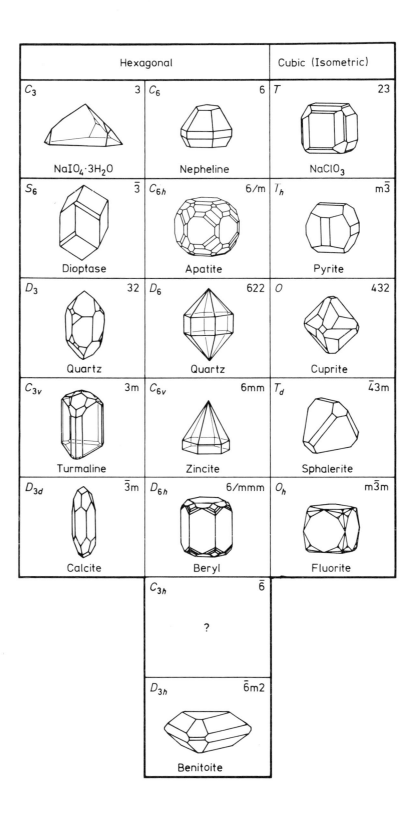

Hexagonal		Cubic (Isometric)
C_3 3 $NaIO_4 \cdot 3H_2O$	C_6 6 Nepheline	T 23 $NaClO_3$
S_6 $\bar{3}$ Dioptase	C_{6h} 6/m Apatite	T_h $m\bar{3}$ Pyrite
D_3 32 Quartz	D_6 622 Quartz	O 432 Cuprite
C_{3v} 3m Turmaline	C_{6v} 6mm Zincite	T_d $\bar{4}3m$ Sphalerite
D_{3d} $\bar{3}m$ Calcite	D_{6h} 6/mmm Beryl	O_h $m\bar{3}m$ Fluorite
	C_{3h} $\bar{6}$?	
	D_{3h} $\bar{6}m2$ Benitoite	

Triclinic and Monoclinic	Orthorhombic	Tetragonal	Hexagonal			Cubic (Isometric)
C_1 1		C_4 4	C_3 3	C_6 6		T 23
C_i $\bar{1}$		C_{4h} 4/m	S_6 $\bar{3}$	C_{6h} 6/m		T_h m$\bar{3}$
C_2 2	D_2 222	D_4 422	D_3 32	D_6 622		O 432
C_S m	C_{2v} 2mm	C_{4v} 4mm	C_{3v} 3m	C_{6v} 6mm		T_d $\bar{4}$3m
C_{2h} 2/m	D_{2h} mmm	D_{4h} 4/mmm	D_{3d} $\bar{3}$m	D_{6h} 6/mmm		O_h m$\bar{3}$m

(b) Stereographic projections.

S_4 $\bar{4}$ C_{3h} $\bar{6}$

D_{2d} $\bar{4}$2m D_{3h} $\bar{6}$m2

Figure 9-10.
The preparation of the stereographic representation.

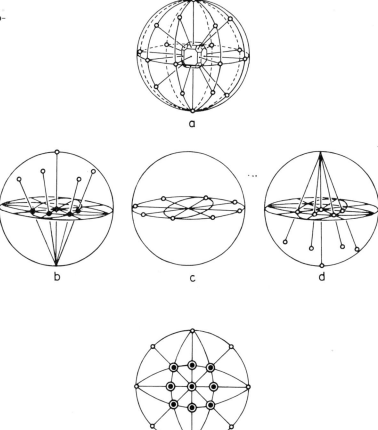

Figure 9-11.
Representation of some
simple highly symmetrical
shapes.
(a) Cube.
(b) Tetrahedron.
(c) Octahedron.
(d) Rhombic dodecahedron.

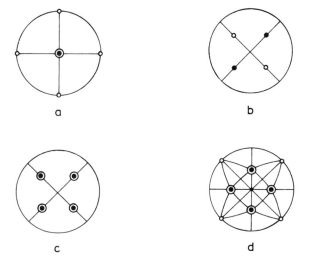

9.3 Restrictions

To have 32 symmetry classes for the external forms of crystals is a definite restriction, and it is obviously the consequence of inner structure. The translation periodicity limits the symmetry elements that may be present in a crystal. The most striking limitation is the absence of five-fold rotation in the world of crystals. Consider, for example, planar networks of polygons possessing two-fold, three-fold, four-fold, five-fold, etc. axes of rotation (Fig. 9-12). Those with two-fold, three-fold, four-fold, and six-fold symmetry cover the available surface without any gaps, while those with five-fold, seven-fold, and eight-fold symmetry leave gaps on the surface.

Figure 9-12.
Planar networks of regular polygons with up to eight-fold symmetry.

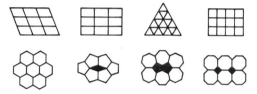

Characteristic examples of completely covering the available area by regular hexagons are shown in Fig. 9-13. The similarity to closely packed circles is remarkable. In fact, when a honeycomb is built from wax, first a network of circles is formed. The bees are of nearly equal size and they are gyrating around themselves. The network of circles will not cover the available surface completely. The hexagons are formed as the liquid wax flows into the cavities between the cylinders.

Figure 9-13.
Networks of regular hexagons covering the surface without gaps or overlaps.
(a) Honeycomb. Photograph and courtesy of Professor Pál Zoltán Örösi, 1982.

(b) Oil platform under construction in the North Sea. Drawn after a lithograph in the 1979 report of the Statoil Company, Norway.

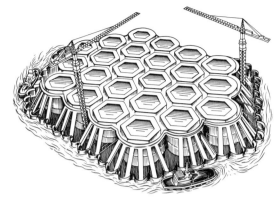

(c) Columnar basalt joints. Drawing by Ferenc Lantos after [9-15].

(d) Moth compound eye (magnified approx. x 2000). Courtesy of Dr. J. Morral, The University of Connecticut, 1984.

(e) Structure of graphite layer.

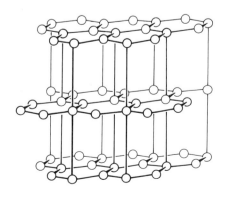

Fig. 9-14 presents two planar networks of octagons. It is evident that the regular octagons cannot cover the surface without gaps. There are smaller squares among the octagons. One of the patterns was drawn after Vasarely, the other is a Hungarian needlework design. It has been suggested [9-16] that Vasarely's octagonal patterns show some influence of folk arts.

Figure 9-14.
Octagonal planar networks.
(a) A pattern after Vasarely [9-16].

(b) Hungarian needlework [9-17].

Let us examine now the possible types of symmetry axes in space groups (cf., e.g., Azaroff [9-3]). Fig. 9-15 shows a lattice row with a period t. An n-fold rotation axis, C_n, is placed in each lattice point. Since n rotations each by an angle φ must lead to superposition, it does not matter in which direction are the rotations performed. Two rotations by φ about two axes but in opposite directions are shown in Fig. 9-15. The two new lattice points produced this way are labeled p and q. These two new

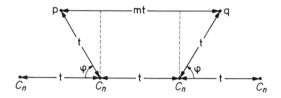

points are equidistant from the original row and hence the line joining them is parallel to the original lattice row. The length of the parallel line joining p and q must be equal to some integer multiple m of the period t. Were it not, then the line joining the two new lattice points p and q would not be a translation of the lattice and the resulting array would not be periodic.

Using Fig. 9-15, it is possible to determine the possible values that the rotation angle φ can have in the lattice,

$$mt = t + 2t \cos\varphi \quad m = 0, \pm 1, \pm 2, \pm 3, \ldots$$

where $+m$ or $-m$ is taken depending on the direction of the rotation,

$$\cos \varphi = \frac{m - 1}{2}.$$

Only the solutions corresponding to the range

$$-1 \leqq \cos \varphi \leqq 1$$

need be considered, and these are shown in Table 9-1. Only five solutions are possible and, accordingly, only five kinds of rotation axes are compatible with a lattice. Thus, not only is five-

Table 9-1. Allowed Rotation Axes n in a Lattice

Possible Values of $m-1$	$\cos \varphi$	$\varphi(°)$	n
-2	-1	180	2
-1	$-\frac{1}{2}$	120	3
0	0	90	4
$+1$	$+\frac{1}{2}$	60	6
$+2$	$+1$	360 or 0	1

fold symmetry not allowed in crystal structures, but all periods larger than six are impossible. Naturally, this applies to the planar networks as well.

The permissible periods of mirror-rotation axes have the same limitations as those of the proper rotation axes.

Let us examine now the limitations on the screw axes. In a lattice the screw axes must be parallel to a translation direction. After n rotations by an angle φ and n translations by the distance T, that is after n translations along the screw axis, the total amount of translation distance in the direction of this axis must be equal to some multiple of the lattice translation mt,

$$nT = mt$$

where n and m are integers. Rearranging this equation,

$$T = \frac{mt}{n},$$

Table 9-2. Possible Values of the Pitch T of an n-fold Screw Axis

n	T						
1	$0t$,	$1t$,	$2t$, ...				
2	$0t$,	$(1/2)t$,	$(2/2)t$,	$(3/2)t$, ...			
3	$0t$,	$(1/3)t$,	$(2/3)t$,	$(3/3)t$,	$(4/3)t$, ...		
4	$0t$,	$(1/4)t$,	$(2/4)t$,	$(3/4)t$,	$(4/4)t$,	$(5/4)t$, ...	
6	$0t$,	$(1/6)t$,	$(2/6)t$,	$(3/6)t$,	$(4/6)t$,	$(5/6)t$,	$(6/6)t$,	$(7/6)t$, ...

n	T(redundancies eliminated)				
1					
2	$(1/2)t$,				
3	$(1/3)t$,	$(2/3)t$			
4	$(1/4)t$,	$(2/4)t$,	$(3/4)t$		
6	$(1/6)t$,	$(2/6)t$,	$(3/6)t$,	$(4/6)t$,	$(5/6)t$

n	Notation of screw axes allowed in a lattice				
2	2_1				
3	3_1	3_2			
4	4_1	4_2	4_3		
6	6_1	6_2	6_3	6_4	6_5

where m, of course, may be 0, 1, 2, 3, etc., but n may only be 1, 2, 3, 4, or 6. It is then possible to determine the permissible values of the pitch of the screw axes in lattices. They are summarized in Table 9-2, taking also into consideration that $(3/2)t = t + (1/2)t$, $(5/4)t = t + (1/4)t$, etc. There are only eleven screw axes that are allowed in a lattice, n_m according to Table 9-2. The subscript in the notation is the m of the expression $T = (mt)/n$. The proper rotation axes may be considered to be special cases of the screw axes, with $m = 0$ and $m = n$. The eleven screw axes are shown in perspective in Fig. 9-16. It is seen there that some pairs are identical except for the direction of the screw motion. Such screw axes are enantiomorphous. The enantiomorphous screw axis pairs are the following:

$$3_1 \text{ and } 3_2$$
$$4_1 \text{ and } 4_3$$
$$6_1 \text{ and } 6_5$$
$$6_2 \text{ and } 6_4.$$

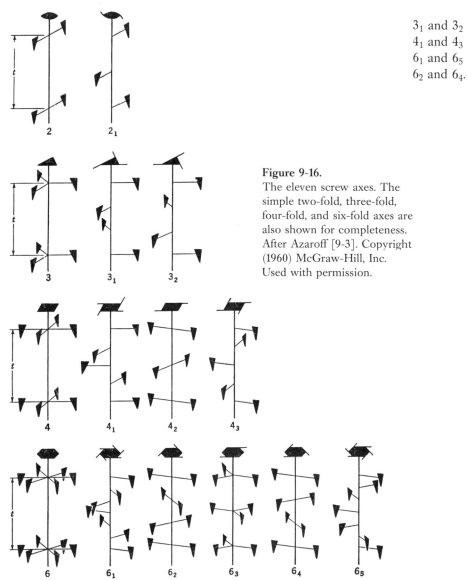

Figure 9-16.
The eleven screw axes. The simple two-fold, three-fold, four-fold, and six-fold axes are also shown for completeness. After Azaroff [9-3]. Copyright (1960) McGraw-Hill, Inc. Used with permission.

Finally the only remaining symmetry element is considered, the glide-symmetry plane. It causes glide reflection as a result of reflection *and* translation. The translation component T of a glide plane is one-half of the normal translation of the lattice in the direction of the glide. A glide along the a axis is $T = (1/2)a$ and this is called an a glide. Similarly, a diagonal glide can have $T = (1/2)a + (1/2)c$. The different possible glides are summarized in Table 9-3.

Table 9-3. Possible Glide Planes

Glide Type	Symbol	Translation Component
Axial	a	$a/2$
Axial	b	$b/2$
Axial	c	$c/2$
Diagonal	n	$a/2 + b/2$; $b/2 + c/2$; or $c/2 + a/2$
Diamond[a]	d	$a/4 + b/4$; $b/4 + c/4$; or $c/4 + a/4$

[a] Translation component is one half of the true translation along the face diagonal of a centered plane lattice.

The fact that the crystal has a lattice structure imposes strict limitations on the symmetry of its outer form. On the other hand, the question arises as to whether it is possible to derive any information about the crystal lattice from the knowledge of the symmetry of its outer form?

The 32 crystal point groups can be classified by symmetry criteria. They are usually grouped according to the highest-ranking rotation axis that they contain. The resulting groups are called crystal systems. There are altogether seven of them and they are listed in Table 9-4. The crystal point groups have to be combined with all possible space lattices in order to produce the space groups.

Table 9-4. Characterization of Crystal Systems

System	Minimal Symmetry (diagnostic symmetry elements)	Relations Between Edges and Angles of Unit Cell	Lattice Type	Numbering in Fig. 9-17
Triclinic	1 (or $\bar{1}$)	$a \neq b \neq c$ $\alpha \neq \beta \neq \gamma \neq 90°$	P	1
Monoclinic	2 (or $\bar{2}$)	$a \neq b \neq c$ $\alpha = \gamma = 90° \neq \beta$	P C (or A)	2 3
Orthorhombic	222 (or $\overline{222}$)	$a \neq b \neq c$ $\alpha = \beta = \gamma = 90°$	P C (or B or A) I F	4 5 6 7
Trigonal (Rhombohedral)	$3(\bar{3})$	$a = b = c$ $\alpha = \beta = \gamma \neq 90°$	R	8
Hexagonal	$6(\bar{6})$	$a = b \neq c$ $\alpha = \beta = 90°$ $\gamma = 120°$	P	9
Tetragonal	4 (or $\bar{4}$)	$a = b \neq c$ $\alpha = \beta = \gamma = 90°$	P I	10 11
Cubic	four 3 (or $\bar{3}$)	$a = b = c$ $\alpha = \beta = \gamma = 90°$	P I F	12 13 14

9.4 The 230 Space Groups

There are 14 infinite lattices, called Bravais lattices, in three-dimensional space. They are shown in Fig. 9-17. These lattices are the analogs of the five infinite lattices in two-dimensional space. The Bravais lattices are presented as points at vertices of parallelepipeds. The corresponding parallelepipeds are capable of filling space without any gaps or overlap. The representation of the lattices by systems of points is especially useful as it makes it possible to join the lattice points in any desired way conforming with the symmetry requirements. In this way, not only the original parallelepipedal forms, but any other possible figures may be used as building units for the space lattice.

The 14 Bravais lattices are enumerated in Table 9-4. The lattices are characterized as the following types: primitive (P, R), side-centered (C), face-centered (F), body-centered (I). The numbering of the Bravais lattices in Table 9-4 corresponds to that in Fig. 9-17. The lattice parameters are also enumerated in the table. In addition, the distribution of lattice types among the crystal systems is shown.

Figure 9-17.
The fourteen Bravais lattices.

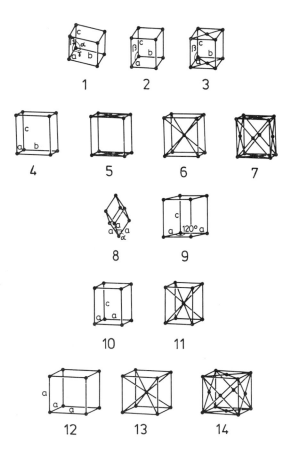

The actual infinite lattices are obtained by parallel translations of the Bravais lattices as unit cells. Some Bravais cells are also primitive cells, others are not. For example, the body-centered cube is a unit cell but not a primitive cell. The primitive cell in this case is an oblique parallelepiped constructed by using as edges the three directed segments connecting the body center with three nonadjacent vertices of the cube.

The three-dimensional space groups are produced by combining the 32 crystallographic point groups with the Bravais lattices. Since the symmetry elements in a space lattice can have translation components, indeed not only the 32 groups but also the analogous groups, which have screw axes and glide planes, have to be considered. There are altogether 230 three-dimensional space groups! Their complete description can be found in the International Tables for X-ray Crystallography [9-19]. Only a few examples are discussed here.

There are only two combinations possible for the triclinic system, they are named $P1$ and $P\bar{1}$. For the monoclinic system three point groups are to be considered and two lattice types. Combining P and I lattices on one hand, and point group 2 and symmetry 2_1 on the other hand, the four possible combinations are $P2$, $P2_1$, $I2$, and $I2_1$. The latter two, however, are equivalent, only their origins differ.

The description of the symmetry elements of the space groups is similar to that of the point groups [9-20]. The main difference is that the order by which the symmetry elements of the space groups are listed may be of great importance, except for the triclinic system. The order of the symmetry elements expresses their relative orientation in space, that is with respect to the three crystallographic axes. For the monoclinic system the unique axis may be the c or the b axis. For the $P2$ space group, the complete symbol may be $P112$ or $P121$, depending on this choice and using the sequence abc. The two variations are called first setting and second setting, respectively. The ordering of symbols for the orthorhombic system is especially important. The symmetry elements are usually listed in the order abc. The space groups which belong to the crystal class $2mm$, are properly presented as $Pmm2$, c being the unique axis.

In the tetragonal system, the c axis is the four-fold axis. The sequence for listing the symmetry elements is c, a, [110], since the two crystallographic axes orthogonal to c are equivalent. For example, the three-dimensional space group notation $P\bar{4}m2$ has the following meaning: the unique axis in a primitive tetragonal lattice is a $\bar{4}$ axis, the two a axes are parallel to m, and the [110] direction has two-fold symmetry. A similar sequence is used for listing the symmetry elements of the hexagonal system for which the c axis again is the unique axis and the other two are equivalent. P denotes the primitive hexagonal lattice while

R denotes the centered hexagonal lattice in which the primitive rhombohedral cell is chosen as the unit cell.

All three crystallographic axes are equivalent in the cubic system. The order of listing the symmetry elements is a, [111], [110]. When the number 3 appears in the second position, it merely serves to distinguish the cubic system from the hexagonal one.

It may be of interest to add some new symmetry to a group or to decrease its symmetry, and examine the consequences. If the addition produces a new group, it is called a supergroup of the original group. If eliminating symmetry leads to a new group, it is usually a subgroup of the original one. For example, the point group 1 is obviously a subgroup of all the other 31 groups as it has the lowest possible symmetry. On the other hand, the highest symmetry cubic group can have no supergroups.

Two of Escher's drawings are of special value in representing periodicity in three dimensions (cf. [9-21]). Besides, their comparison reveals the important difference between lattice and structure. The drawing in Fig. 9-18 is entitled "Cubic Space

Figure 9-18.
Escher: "Cubic space division". Collection Haags Gemeente-museum – The Hague. Reproduced with permission.
© M. C. Escher Heirs c/o Cordon Art – Baarn – Holland.

Division" [9-22] and emphasizes clearly the uniformity of the environment about each lattice point located at the centers of the cubes. The drawing in Fig. 9-19 was prepared about three years later than the previous one. It is called "Depth" [9-22]. Its three-dimensional pattern may have the same translational properties as the other drawing, but as a whole it certainly has a much lower symmetry. It is also a good example for what is called pseudo-symmetry, implying a higher symmetry in the lattice associated with the actual structure. Brock and Linga-felter [9-23] pointed out commonly existing misconceptions of the difference between a crystal and a lattice. A crystal is an array of units (atoms, or ions, or molecules) in which a structural

Figure 9-19.
Escher: "Depth". Collection
Haags Gemeentemuseum –
The Hague. Reproduced with
permission. © M. C. Escher
Heirs c/o Cordon Art –
Baarn – Holland.

motif is repeated in three dimensions. A lattice is an array of points, and every point has the same environment of points in the same orientation. Each crystal has an associated lattice, whose origin and basis vectors can be chosen in various ways. From the above it is clear, for example, that it would be improper to speak about "interpenetrating lattices", while it is correct to talk about interpenetrating arrays of atoms [9-23].

As we have reached in our discussion the system of the 230 three-dimensional space groups, it appears, as it indeed is, a perfect system. It was established a long time ago, in fact well before X-ray diffraction could have been applied to the determination of crystal structure. That these 230 three-dimensional space groups were derived in their entirety by Fedorov, Schoenflies, and Barlow working independently, will always be considered a great scientific feat. No crystal can ever be produced, either in nature or artificially, whose structure would not fall into one or other of these 230 groups.

An interesting statistical test was performed concerning the total number of three-dimensional space groups some time in the mid-sixties [9-24]. It was a uniquely appropriate point in the history of crystallography for such a test: Already a large number of crystal structures have been determined, but not all the space groups have yet been found examples for among the actual crystals. The total number of three-dimensional space groups had long before been firmly established. Thus the test was considered as much to be a check of the applied statistical method as to be a source for crystallographic information.

Although there are 230 space groups, not all of them are in practice distinguishable. So 11 enantiomorphous groups were excluded from the count as were two more groups for other reasons. Thus the number of space groups to be considered was 217. The reviewed 3782 crystal structures showed a wide variation in frequencies among the different space groups. One group occurred 355 times, while 33 groups occurred only once each. It was also interesting that only 178 groups out of the 217 total occurred. We are not concerned here with the details of the statistical test applied. Based on the available data of the distribution of the space groups among the determined structures, the findings were extrapolated to an indefinitely large sample. The statistical test led to an extrapolated value of 216. The estimated accuracy of the procedure was 2%. Thus the estimate agreed with the accepted value of the total number of the practically distinguishable space groups of 217.

The statistical analysis has also been applied separately to the data on inorganic and organic crystals. In both cases the extrapolated estimate for the total number of three-dimensional space groups was smaller than when all data had been considered together. The total numbers estimated for the inorganic and organic structures were 209 and 185, respectively. Thus the conclusion could be made that the inorganic and organic crystals belong to space groups with different population distributions.

9.5 Rock Salt and Diamond

Let us look in some more detail at the symmetry systems of two crystals following a discussion by Shubnikov and Koptsik [9-20]. The unit cell of the rock salt structure and the projection of this structure along the edges of the unit cell onto a horizontal plane are shown in Fig. 9-20a and b.

The equivalent ions are related by translations $a=b=c$ along the edges of the cube, or $(a+b)/2$, $(a+c)/2$, $(b+c)/2$ along the face diagonals. All this corresponds to the face-centered cubic group (F). The structure coincides with itself not only after the above enumerated translations, but also after the operations of the point group $m\bar{3}m$ (or by other notation $\tilde{6}/4$). The point-group symmetry elements are shown also in Fig. 9-20c. The symmetry elements of this group intersect at the centers of all atoms and thus they become symmetry elements for the whole unit cell, and, accordingly, for the whole crystal.

Figure 9-20.
The crystal structure of the
rock salt after Shubnikov and
Koptsik [9-20].
(a) A unit cell.

(b) Projection of the structure
along the edges of the unit
cell onto a horizontal plane.

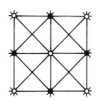

(c) Projection of some sym-
metry elements of the $Fm\overline{3}m$
space group onto the same
plane.

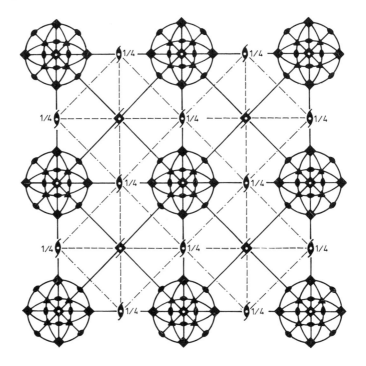

Among the projected symmetry elements in Fig. 9-20c, there
are some which are derived from the generating elements. This
is the case, for example, for the vertical glide-reflection planes
with elementary translations $a/2$ and $b/2$ (represented by
broken lines), translations (dot-dash lines), vertical screw axes
2_1 and 4_2

and symmetry centers (small hollow circles, some of which lie
above the plane by 1/4 of the elementary translation).

Shubnikov and Koptsik [9-20] suggest also two very simple
descriptions of the rock salt crystal structure. According to one,
the sodium and chloride ions occupy positions with point-
group symmetry $m\overline{3}m$ forming a checkered pattern in the $Fm\overline{3}m$
space group. According to the other description, the structure
consists of two cubic sublattices in parallel orientation, one of
sodium ions and the other of chloride ions.

Figure 9-21.
The diamond structure after Shubnikov and Koptsik [9-20]. (a) A unit cell, the edges of the cube are the *a, b,* and *c* axes.

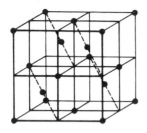

(b) Two face-centered cubic sublattices displaced along the body diagonal of the cube.

Fig. 9-21 illustrates the diamond structure after Shubnikov and Koptsik [9-20]. It can be regarded as a set of two face-centered cubic sublattices displaced relative to each other by 1/4 of the body diagonal of the cube. Each of the two sublattices has the $F\bar{4}3m$ space group and in addition there are some opera-

(c) Projection of some symmetry elements of the $Fd\bar{3}m$ space group onto a horizontal plane.

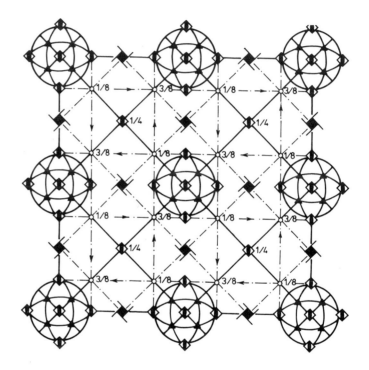

tions transforming one to the other. The complete diamond structure has the space group $Fd\bar{3}m$ where "*d*" stands for a "diamond" plane.

Among the projected symmetry elements in Fig. 9-21c, there are again some which are produced by the generating elements. Special for the diamond structure are the symmetry elements which connect the two subgroups $F\bar{4}3m$. They include vertical left-handed and right-handed screw axes, 4_1 and 4_3, respectively,

4_1

4_3

symmetry centers (small hollow circles, 1/8 and 3/8 of the elementary translation c above the plane), vertical "diamond" glide-reflection planes d represented by dot-dash lines with arrows, and similar systems of connecting elements in the horizontal directions.

The subgroup $F\bar{4}3m$ is common both to the rock salt space group $Fm\bar{3}m$ and to the diamond space group $Fd\bar{3}m$. The space group $Fd\bar{3}m$ is obtained from $Fm\bar{3}m$ by replacing the symmetry planes m by glide-reflection planes d with the latter displaced 1/8 along the cube edges.

The above descriptions are far from complete. They are intended to give some feeling for the characteristics of these two highly symmetrical structures rather than to be a rigorous treatment.

9.6 Beyond the Perfect System

The 230 space groups exhaustively characterize all the symmetries possible for infinite lattice structures. So "exhaustively" that according to some views this perfect system is a little too perfect and a little too rigid. These views may well point towards the further development of our ideas on structures and symmetries [9-12, 9-25].

There is an inherent deficiency in crystal symmetry in that crystals are not really infinite. Mackay [9-25] argues that the crystal formation is not the insertion of components into a three-dimensional framework of symmetry elements. On the contrary, the symmetry elements are the consequence. The crystal arises from the local interactions between individual atoms. He furthermore says that a regular structure should mean a structure generated by simple rules, and the list of rules considered to be simple and "permissible", should be extended. These rules would not necessarily form groups.

Mackay finds [9-25] the formalism of the International Tables for X-Ray Crystallography [9-19] to be too restrictive and quotes Bell the historian of mathematics on the rigidity of the Euclidean geometry formalism:

"The cowboys have a way of trussing up a steer or a pugnacious bronco which fixes the brute so that it can neither move nor think. This is the hog-tie and it is what Euclid did to geometry."

Mackay has a long list [9-26] covering a whole range of transitions from classical crystallographic concepts to what is termed the modern science of structure at the atomic level. This list is reproduced in Table 9-5. It is impossible not to notice

Table 9-5. Mackay's List of Transition from the Classical Concepts of Crystallography to the Modern Concepts of a Science of Structure [9-26]

Classical Concepts	Modern Concepts
Absolute identity of components	Substitution and non-stoichiometry
Absolute identity of the environment of each unit	Quasi-identity and quasi-equivalence
Operations of infinite range	Local elements of symmetry of finite range
"Euclidean" space elements (Plane sheets, straight lines)	Curved space elements. Membranes, micelles, helices. Higher structures by curvature of lower structures
Unique dominant minimum in free energy configuration space	One of many quasi-equivalent states; metastability recording arbitrary information (pathway); progressive segregation and specialization of information structure
Infinite number of units. Crystals	Finite numbers of units. Clusters; "crystalloids"
Assembly by incremental growth (one unit at a time)	Assembly by intervention of other components ("crystalase" enzyme). Information-controlled assembly. Hierarchic assembly
Single level of organization (with large span of level)	Hierarchy of levels of organization. Small span of each level
Repetition according to symmetry operations	Repetition according to program. Cellular automata
Crystallographic symmetry operations	General symmetry operations (equal "program statements")
Assembly by a single pathway in configuration space	Assembly by branched lines in configuration space. Bifurcations guided by "information", i.e., low-energy events of the hierarchy below

some resonance of many of Mackay's ideas with other directions in modern chemistry, and structural chemistry in particular, where the non-classical, the non-stoichiometrical, the non-stable, the non-regular, the non-usual, the non-expected are gaining importance every day. For crystallography it seems to be a long way yet to perform all the suggested transitions but the initial break-throughs are fascinating and promising. Impressive progress has been reported in the studies of liquids, amorphous materials, metallic alloys as regards the description of their structural regularities [9-27].

Liquid structures, for example, cannot be characterized by any of the 230 three-dimensional space groups and yet it is unacceptable to consider them as possessing no symmetry whatsoever. Bernal noted [9-28] that the major structural distinction between liquids and crystalline solids is the absence of long-range order in the former. The description of non-crystallographic symmetry properties has yet to appear. It will characterize liquid structures and colloids, as well as the structures of amorphous substances. It will also account for the greater variations in their physical properties as compared with those of the crystalline solids. Bernal's ideas [9-29] have greatly encouraged further studies in this field which may be called generalized crystallography. Referring to Bernal's geometrical theory of liquids, Belov [9-30] noted in Bernal's obituary: "... his last enthusiasm was for the laws of lawlessness."

The paradoxical incompleteness and inadequacy of perfect symmetry compared with less-than-perfect symmetry are well expressed in two short poems by the English poet Anna Wickham [9-31]:

GIFT TO A JADE

For love he offered me his perfect world.
This world was so constricted and so small
It had no loveliness at all,
And I flung back the little silly ball.
At that cold moralist I hotly hurled
His perfect, pure, symmetrical, small world.

THE WOMAN AND HER INITIATIVE

Give me a deed, and I will give a quality.
Compel this colloid with your crystalline.
Show clear the difference between you and me
By some plain symmetry, some clear stated line.
These bubblings, these half-actions, my revolt from unity.
Give me a deed, and I will show my quality.

The structures intermediate between the perfect order of crystals and the complete disorder of gases are not merely rare exceptions. On the contrary, they are often found in substances which are very common in our environment or are widely used in various technologies. They include plastics, textiles and rubber. Many of the constituents of the living organisms are of this kind. Guinier [9-32] envisaged a continuous passage from the exact scheme of neighboring atoms in a crystal to the very flexible arrangement in an amorphous body. The term paracrystal was coined [9-32] for domains with approximate long-distance order in the range of a few tens to a few hundreds of atomic diameters. Fig. 9-22 is a schematic representation of a paracrystal lattice with one atom per unit cell. The shaded areas indicate the regions where an atom is likely to be found around the atom fixed at the origin. At greater distances the neighboring sites first overlap, then merge, and thus eventually the long-range order completely vanishes.

Figure 9-22.
Paracrystal lattice with one atom per unit cell after Guinier [9-32]. Used with permission.

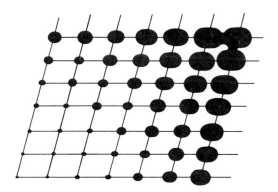

One of the most fascinating examples of non-periodic regular arrangements is described in Mackay's paper [9-25] "De nive quinquangula" – on the pentagonal snowflake. A regular, but "non-crystalline" structure is built from regular pentagons in a plane. It starts with a regular pentagon of given size (zeroth order pentagon). Six of these pentagons are combined to make a larger one (first order pentagon). As is seen in Fig. 9-23, the resulting triangular gaps are covered by pieces from cutting up a seventh zeroth order pentagon. This indeed yields five triangles

Figure 9-23.
Tiling with regular pentagons after Mackay [9-12].

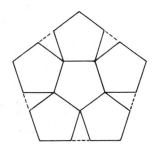

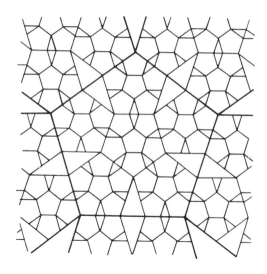

plus yet another regular pentagon of the order −1. This construction is then repeated on an ever increasing scale as indicated in Fig. 9-23. The hierarchic packing of pentagons builds up like a pentagonal snowflake. It is shown in Fig. 9-24 from a

Figure 9-24.
"Pentagonal snowflake". Computer drawing and courtesy of Robert H. Mackay, London, 1982.

computer drawing. Shubnikov illustrated his New Year's greetings with tiles in a pentagonal tessellation (Fig. 9-25) in accord with Kepler's hexagonal snowflake which was a New Year's gift [9-10].

Figure 9-25.
Illustration of pentagonal symmetry:
Shubnikov's New Year's greeting with pentagonal tessellation.

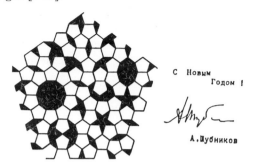

Mackay [9-24] calls attention to yet another limitation of the 230 space-group system. It covers only those helices that are compatible with the three-dimensional lattices. All other helices which are finite in one or two dimensions are excluded. Some important virus structures with icosahedral symmetry are among them. Also, there are very small particles of gold that do not have the usual face-centered cubic lattice of gold. They are actually icosahedral shells. The most stable configurations contain 55 or 147 atoms of gold. But icosahedral symmetry is not treated in the International Tables and crystals are only defined for infinite repetition.

Crystals are really advantageous for the determination of the structure of molecules. A crystal provides an amplification which multiplies the scattering of the X-rays from a single molecule by the number of molecules in the array, perhaps by 10^{15}. They also minimize the damage to individual molecules by the viewing radiation. The spots are emphasized in the diffraction pattern and the background neglected. The damaged molecules transfer their scattering contribution to the background as do those which are not repeated with regular lattice periodicity. But defects and irregularities may be important and may well be lost in present day sophisticated structure analyses [9-24, 9-32].

It is perhaps worth to point out that every crystal is in fact defective, even if its only defect is that it has surfaces. However, if a crystal is only a ten unit cell cube, about half of the unit cells lie in the surface and thus have environments very different from those of the other half. The physical observation is that very small aggregates need not be crystalline, although they may nevertheless be perfectly structured. Mackay's proposal is to apply the name crystalloid to them. His proposed definitions [9-24]:

Crystal: the unit cell consisting of one or more atoms, or other identical components, is repeated a large number of times by three noncoplanar translations. Corresponding atoms in each unit cell have almost identical surroundings. The fraction of atoms near the surface is small and the effects of the surface can be neglected.

Crystallite: a small crystal where the only defect is the existence of the external surface. The lattice may be deemed to be distorted but it is not dislocated. Crystallites may further be associated into a mosaic block.

Crystalloid: a configuration of atoms, or other identical components, finite in one or more dimensions, in a true free energy minimum, where the units are not related to each other by three lattice operations.

The above ideas are expected to further develop in the future mainly by translating them into more quantitative descriptions as being applied to various structural problems. They by no means belittle the great importance of the 230 three-dimensional space groups and their wide applicability. What is really expected is that the less easily handled systems with varying degrees of regularity in their structures will eventually be given helpful systematization and characterization. Let us illustrate this need with the words of the poet crystallographer [9-14] about flying through clouds:

> We cruise through the hydrosphere
> Our world is of water, like the sea,
> But the molecules more sparsely spread,
> Not independent, not touching
> But somewhere in between,
> Clustering, crystallizing, dispersing
> In the delicate balance of radiation
> And the adiabatic lapse rate.

Symmetry considerations will undoubtedly play an important role in the envisaged development above. Things are already happening "beyond the perfect system" [9-33]. Recently there was a report of an electron diffraction study on a metastable solid with long-range orientational order but with icosahedral point-group symmetry [9-33a], paralleled by a theoretical paper [9-33b] on the symmetries of a state between crystal and liquid, called quasicrystals with "quasiperiodic" lattice. Thus a "forbidden" symmetry was not only constructed (cf. Figs. 9-23, 9-24) [9-26] but was also found in real experiment. A sample of titles of reports and commentaries that immediately followed, conveys the flavor of impact this discovery was having [9-33c]: "Theory of New Matter Proposed" (The New York Times); "Towards fivefold symmetry?" (Nature); "Forbidden fivefold symmetry may indicate quasicrystal phase" (Physics Today); "The rules of crystallography fall apart" (New Scientist); "The Fivefold Way of Crystals" (Science News); "Some answers but more questions" (Nature). Now, however, let us return to a basic problem of crystal symmetry, and that of dense packing.

9.7 Dense Packing

Dalton [9-34] envisaged the structural difference between water and ice in packing properties. Fig. 9-26 reproduces a drawing from his **1808** book "A new system of chemical philosophy". According to Dalton, the "atoms" of ice arrange themselves in a hexagonal scheme, while the "atoms" of water do not. In any case it is remarkable that the principal difference between the water and ice structures is expressed in the packing density. Fig. 9-27 originates from a different age [9-35]. It shows the atomic and molecular arrangements in the crystals of 2Zn-insulin from the work of Dorothy Hodgkin and her associates. The molecular structure of insulin is extremely complicated but the molecular packing, especially the arrangement of the insulin hexamers, certainly reminds us of Dalton's hexagonal ice.

The symmetry of the crystal structure is a direct consequence of dense (close) packing. The densest (closest) packing is when each item makes the maximum number of contacts in the structure. First the packing of equal spheres in atomic and ionic systems will be discussed. Then molecular packing will be considered. Only characteristic features and examples will be dealt with here since systematic treatises mentioned at the beginning of our discussion of crystal symmetries, are available for consultations [9-1–9-3].

Figure 9-26.
Dalton's models for water (1,3) and ice (2,4-6) in his book [9-34]. We appreciate the kindness of Professor D. Hodgkin who sent us photographs of these drawings.

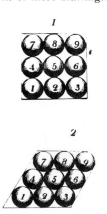

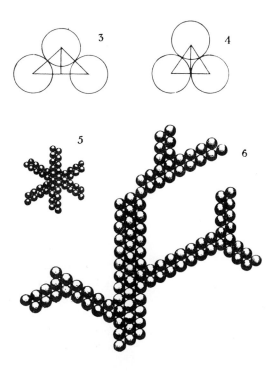

Figure 9-27.
Atomic arrangement in the
2Zn-insulin crystal. The
smaller projection drawing
shows the molecular packing
in the insulin hexamers. We
appreciate the kindness of
Professor D. Hodgkin who
sent us photographs of these
drawings.

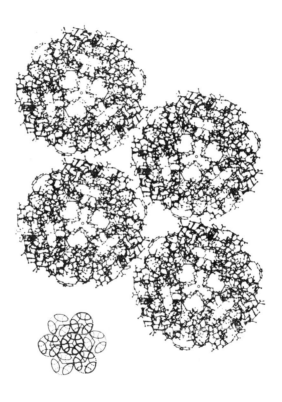

9.7.1 Sphere Packing

The most efficient packing results in the largest possible
density. The density is the fraction of the total space occupied
by the packing units. Only those packings will be considered
in which each sphere is in contact with at least six neighbors.
The densities of some packings are given in Table 9-6. There
are stable arrangements with smaller numbers of neighbors,
meaning lower coordination numbers. This occurs only when
directed bonds are present. In our discussion, however, the
existence of chemical bonds is not a prerequisite at all.

Table 9-6. Densities of Sphere Packing

Coordination Number	Name	Density
6	Simple cubic	0.5236
8	Simple hexagonal	0.6046
8	Body-centered cubic	0.6802
10	Body-centered tetragonal	0.6981
12	Closest packing	0.7405

For three-dimensional six-coordination the most symmetrical packing is when the spheres are at the points of a simple cubic lattice (Fig. 9-28a). Each sphere is in contact with six others situated at the vertices of an octahedron. In order to increase clarity the atoms are shown separated in the figure as is usual in such figures. The packing is more realistically represented when the spheres touch each other. Already Kepler (Fig. 9-8) and Dalton (Fig. 9-26) also employed such representations.

The structure of crystalline arsenic provides an example for somewhat distorted simple cubic packing. It is illustrated in Fig. 9-28b. The atoms are in the positions of the cubic structure. Each has three nearest and three more distant neighbors. The layers formed by the nearest bonded atoms may also be derived from a plane of hexagons. These layers buckle as the bond angle decreases from 120°.

Figure 9-28.
Various types of sphere packing after Wells [9-2]. Reproduced with permission.

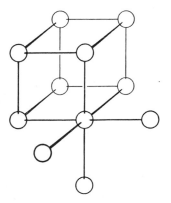

(a) Simple cubic.

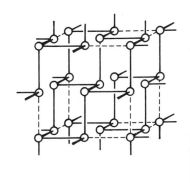

(b) The somewhat distorted cubic packing of arsenic.

(c) Simple hexagonal.

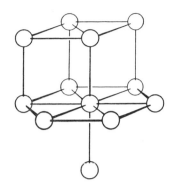

(d) Body-centered cubic.

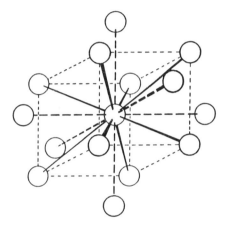

The simple hexagonal sphere packing is shown in Fig. 9-28c. The coordination number is eight. It is not very important for crystal structures.

Fig. 9-28d shows the body-centered packing with 8-coordination. For the central atom the six next nearest neighbors are at the centers of neighboring unit cells. In terms of polyhedral domains, a truncated octahedron is adopted here. The central atom, in fact, has 14-coordination.

It may often be convenient to describe the crystal structure in terms of the domains of the atoms [9-2]. The domain is the polyhedron enclosed by planes drawn midway between the atom and each neighbor, these planes being perpendicular to the lines connecting the atoms. The number of faces of the polyhedral domain is the coordination number of the atom and the whole structure is a space-filling arrangement of such polyhedra.

The closest packing of equal circles on a plane surface has already been considered. The closest packing of spheres on a plane surface is a similar problem. Again, the densest arrange-

Figure 9-29.
Closest packing of ABC layers after Wells [9-2]. Reproduced with permission.

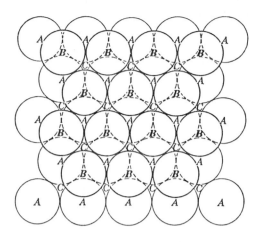

Figure 9-30.
Close packing of spheres, Shubnikov and Koptsik [9-20]. (a) Hexagonal closest packing. (b) Cubic closest packing.

ment is when a sphere is in contact with six others. Layers of spheres may then be superimposed in various ways. The closest packing is when each sphere touches three others in each adjacent layer, the total number of contacts then being 12. Closest packing is thus based on closest packed layers. Fig. 9-29 illustrates this after Wells [9-2]. The spheres in one layer are labeled A. A similar layer can be placed above the first so that the centers of the spheres in the upper layer are vertically above the positions B (or C). The third layer can be placed in two ways. The centers of the spheres may lie above either the C or A positions. The two simplest sequences of layers are then ABABAB... and ABCABC... They will have the same density (0.7405).

The packing based on the sequence ABAB... is called hexagonal closest packing and is illustrated by Fig. 9-30a. Each sphere has twelve neighbors situated at the vertices of a coordination polyhedron.

The packing based on the sequence ABCABC... is called cubic closest packing. It is illustrated in Fig. 9-30b, and is characterized by cubic symmetry.

The closest packing of equal spheres is achieved in an arrangement in which each sphere touches three others in each adjacent layer. The total number of neighbors is then 12. Although the packing in any layer is evidently the densest possible packing, it may not be assumed that this is necessarily true of the space-filling arrangements resulting from stacking such layers. Thus consider the addition of a fourth sphere to the most closely packed triangular arrangement [9-2]. The maximum number of contacts is three in the emerging tetrahedral group. The space-filling arrangement would require each tetrahedron to have faces common with four other tetrahedra. However, regular tetrahedra are not suitable to fill space without gaps or overlaps because the angle of the tetrahedron, 70°32', is not an exact submultiple of 360°.

Alternatively, continue placing spheres around a central one, all spheres having the same radius. The maximum number that can be placed in contact with the first sphere is 12. However, there is a little more room around the central sphere than just for twelve, but not enough for a thirteenth sphere. Because of the extra room there is an infinite number of ways of arranging the twelve [9-2].

9.7.1.1 Icosahedral Packing

The most symmetrical arrangement is to place the twelve spheres at the vertices of a regular icosahedron, which is the only regular polyhedron with twelve vertices. Thus the icosahedral packing is the most symmetrical. However, it is not the densest packing. The spheres in the shell are further from each other than in any other arrangement. Also, it is not a crystallographic packing. When icosahedra are packed together, they will not form a plane, but will gradually curve up and will eventually form a closed system as is illustrated in Fig. 9-31 [9-36].

The length of an edge of a regular icosahedron is some five percent greater than the distance from the center to vertex. Thus the sphere of the outer shell of twelve makes contact only with the central sphere. Conversely, if each sphere of an icosahedral group of twelve, all touching the central sphere, is in contact with its five neighbors, then the central sphere must have a radius of some 10 % smaller than the radius of the outer spheres. The relative size considerations are important in the structures of free molecules as well if the central atom or group of atoms is surrounded by 12 ligands [9-37].

An interesting case, and a step forward from the isolated molecule towards more extended systems is when an icosahedron of 12 spheres about a central sphere is surrounded by a second icosahedral shell exactly twice the size of the first [9-38]. This shell will contain 42 spheres and will lie over the first so that spheres will be in contact along the five-fold axes. Further layers can be added in the same fashion. The third layer is shown in Fig. 9-32 as an example for icosahedral packing of equal spheres. On each triangular face the layers of spheres succeed each other in cubic close packing sequence. Each sphere which is not on an edge or vertex, touches only 6 neighbors, 3 above and 3 below. Each such sphere is separated by a distance of 5 % of its radius from its neighbors in the plane of the face of the icosahedron. The whole assembly can be distorted to cubic close packing in the form of a cuboctahedron. This distortion may be envisaged as a reversible process by the kind of transformation discussed earlier, see Fig. 3-88b.

Figure 9-31.
Icosahedral Polyoma virus drawn after [9-36].

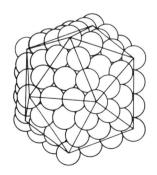

Figure 9-32.
Illustration to icosahedral packing after Mackay [9-38]: Icosahedral packing of spheres showing the third shell.

407

Figure 9-33.
Cuboctahedron and "twinned"
cuboctahedron.

While the most symmetrical arrangement of 12 neighbors, viz. the icosahedral coordination does not lead to the densest possible packing, others do. The cuboctahedron and its "twinned" version, alone or in combination, lead to infinite sphere packing with the same high density (0.7405). Both coordination polyhedra are shown in Fig. 9-33. The "twinned" polyhedron is obtained by reflecting one half of a cuboctahedron cut parallel to a triangular face across the plane of section.

9.7.1.2 Connected Polyhedra

There are, of course, more complex forms of closest packing than those considered so far. Besides, the species to be packed need not be identical. Thus close packing of atoms of two kinds could be considered. Close-packed structures with atoms in the interstices are also important. The interstice arrays may have very different arrangements in various structures. A shorthand notation of some configurations has been worked out to facilitate the description of more complicated systems. Such a notation is illustrated in Fig. 9-34. Suppose, for example, that in a compound with composition AX_2, each atom A is bonded to

Figure 9-34.
Shorthand notation for some common structural units after Wells [9-2].
(a) Notations for tetrahedron.

(b) Notations for octahedron.

four X atoms and that all four X atoms are equivalent. Each X atom must then be bonded to two A atoms. The lines of the squares in Fig. 9-34 do not represent chemical bonds, rather these squares stand for polyhedral arrangements. Among the AX_n polyhedral groups the most common are the AX_4 tetrahedra and AX_6 octahedra. They may appear in various orientations in the crystal structures. Similar structural features have already been discussed for the polyhedral molecular geometries. Whereas in molecules only two or at most a few polyhedra were joined, here we deal with their infinite networks.

Many crystal structures may be built from the two most important coordination polyhedra, the tetrahedron and octahedron. As was shown earlier, they may share vertices, edges, or faces. The ways how the polyhedra are connected introduce certain geometrical limitations with important consequences as to the variations of the interatomic distances and bond angles [9-2].

If two tetrahedra or octahedra share a *face*, the resulting systems as well as the angles X-A-X are unambiguously defined. The angles arc marked f in Fig. 9-35. For *edge*-sharing or *vertex*-sharing, in each case there is a maximum value of the angles A-X-A, cf. the points e and v in Fig. 9-35. These maximum

Figure 9-35.
Variations of bond angles X-A-X in the systems of joined tetrahedra and octahedra after Wells [9-2]. Used with permission. The notation for bond angles X-A-X is f when the polyhedra join by faces, e – by edges, and v – by vertices.

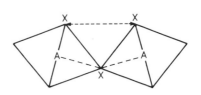

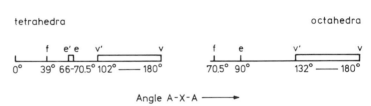

values correspond to the fully extended systems with maximum separation of the A atoms – i.e. the centers of the polyhedra. From the A-X-A angles the distance between the centers of the polyhedra may be calculated. These distances may be expressed either in terms of the edge-length X...X or the distance from the center, A-X. Face-sharing does not occur for tetrahedral polyhedra and is rare for the octahedral ones. Significant stereochemical information may be drawn from the comparison of these ideal values with those measured in actual crystals and in those free molecules which are dimers, or higher associates of tetrahedral or octahedral units.

The maximum number of regular tetrahedra which can meet at one point is eight and the analogous number is six for the regular octahedra [9-2]. Of course, the number of polyhedra which can meet will much depend on the system of closest packing realized in the crystal.

Examples are shown in Figs. 9-36 to 9-41 for a variety of ways to connect tetrahedral and octahedral units. Tetrahedra share two vertices or/and three vertices in Fig. 9-36. For one of these an Islamic decoration analogous to its projection is shown in Fig. 9-37. Octahedra share adjacent vertices and form a tetramer in two representations in Fig. 9-38a and b. Two more exam-

Figure 9-36.
Connected tetrahedra.
(a) All tetrahedra share two vertices.

409

(b) and (c) All tetrahedra share three vertices. After [9-39].

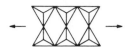

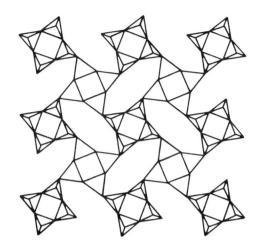

(d) and (e) Some tetrahedra share two, others share three vertices.

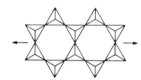

Figure 9-37.
Islamic decoration, drawn after [9-40], analogous to the one-dimensional pattern of Fig. 9-36d but extending in two dimensions.

Figure 9-38.
Connected octahedra.
(a) and (b) Two representations of four octahedra sharing adjacent vertices and forming a tetramer.

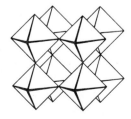
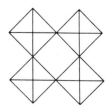

410

(c) Infinite chain of octahedra connected at adjacent vertices.

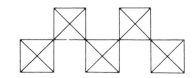

(d) Infinite chain of octahedra connected at nonadjacent vertices.

ples show infinite chains of octahedra sharing adjacent (c) and nonadjacent (d) vertices. Octahedra sharing 2, 4, or 6 edges are presented in Fig. 9-39. An example of octahedra sharing faces and edges is seen in Fig. 9-40 together with an analogous pattern from a Formosan basket weaving [9-41]. Finally, a composite structure from tetrahedra and octahedra is shown in Fig. 9-41.

Figure 9-39.
Octahedra sharing edges.
(a) Two edges.
(b) Four edges.
(c) Six edges.

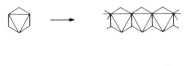

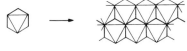

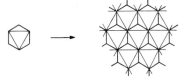

Figure 9-40.
Joined octahedra.
(a) Sharing faces and edges: Nb_3S_4 crystal after [9-2].

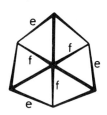

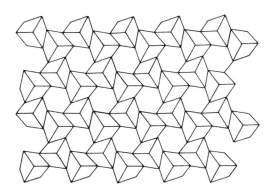

411

(b) Analogy from Formosan basket weaving pattern, after [9-41].

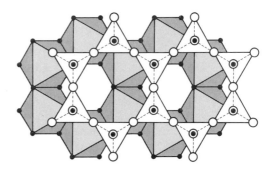

Figure 9-41.
Joined tetrahedra and octahedra:
A composite structure (kaolin) built from tetrahedra and octahedra [9-2]. Reproduced with permission.

The tetrahedra and octahedra are important building blocks of the crystal structures. The great variety of structures combining these building blocks, on one hand, and the conspicuous absence of some of the simplest structures, on the other hand, together suggest that the immediate environment of the atoms is not the only factor which determines these structures. Indeed the relative sizes of the participating atoms and ions are of great importance.

9.7.1.3 Atomic Sizes

The interatomic distances are primarily determined by the position of the minimum in the potential energy function describing the interactions between the atoms in the crystal. The question is then, what are the sizes of the atoms and ions? Since the electron distribution function for an atom or an ion extends indefinitely, no single size can be rigorously assigned to it. They change relatively little when forming a strong chemical bond, and even less for weak bonds. But these smaller changes in the sizes of atoms and ions depend upon the physical properties that are actually being considered. Thus they will in fact be somewhat different for different physical properties. For the present discussion of the crystal structures, the atomic and ionic radii should, when added appropriately, yield the interatomic and interionic distances characterizing these structures.

Coverlent and metallic bondings suppose a strong overlap of the outermost atomic orbitals and so the atomic radii will be approximately the radii of the outermost orbitals. The atomic radii [9-42] are empirically obtained from interatomic distances. For example C-C is 1.54 Å in diamond, Si-Si is 2.34 Å in disilane, and so on. The consistency of this approach is shown by the agreement between the Si-C bond lengths determined experimentally and calculated from the corresponding atomic radii. The interatomic distances appreciably depend on the coordination. With decreasing coordination number the bonds usually get shorter. For coordinations 8, 6, and 4, the bonds get shorter by about 2, 4, and 12 %, respectively, as compared with the coordination of 12.

The covalent bond is directional and multiple covalent bonds are much shorter than the corresponding single ones. For carbon bonds as well as for nitrogen, oxygen, or sulfur bonds, the decrease for double and triple bonds amounts to 10-12 and 20-22 %, respectively.

Establishing the system of ionic radii is even less unambiguous than that for atomic radii. The starting point is a system of analogous crystal structures. Such is, for example, the structure of sodium chloride and the analogous series of other alkali halide face-centered crystals. In any case the ionic radii represent relative sizes, and if the alkali and halogen ions are chosen for starting point, then they represent the relative sizes of the outer electron shells of the ions as compared with those of the alkali and halogen ions.

Consider now the sodium chloride crystal structure shown in Fig. 9-42a. It is built from sodium ions and chloride ions, and it is kept together by electrostatic forces. The chloride ions are much larger than the sodium ions (ionic radii 1.80 Å versus 0.95 Å). As equal numbers of cations and anions build up this structure, the maximum number of neighbors will be the number of the larger chloride ions that can be accommodated around the smaller sodium ion. The opposite would not work: although more sodium ions could surround a chloride ion, the same coordination could not be achieved around the sodium. Thus the coordination number will obviously depend on the relative sizes of the ions. In the simple ionic structures, however, only such coordination numbers may be accomplished that make a highly symmetrical arrangement possible. The relative sizes of the sodium and chloride ions allow six chloride ions to surround each sodium ion in six vertices of an octahedron. Fig. 9-42b shows the arrangement of ions in cube-face layers of alkali halide crystals with the sodium chloride structure. As the relative size of the metal ion increases with respect to the size of the halogen anion, greater coordination may be possible. Thus,

Figure 9-42.
Sodium chloride structures. (a) The sodium chloride crystal structure in two representations. The space-filling model is from W. Barlow, Z. Krist. **29**, 433 (1898) reproduced here after Pauling [9-42].

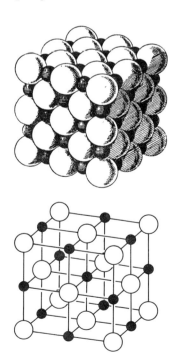

(b) The arrangement of ions in cube-face layers of alkali halide crystals with the sodium chloride structure. Adaptation of Figure 13-6 from Pauling [9-42]. Copyright © 1960 by Cornell University. Used by permission of the publisher, Cornell University Press.

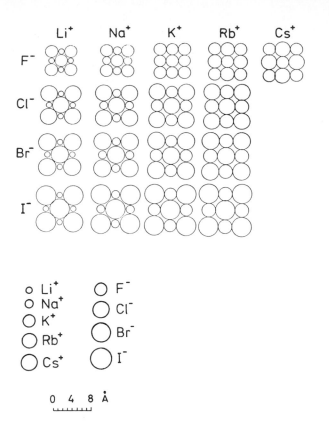

Figure 9-43.
Cesium chloride crystal structure.

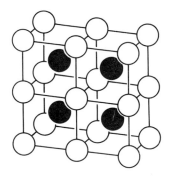

for example, the radius of the cesium ion is 1.69 Å, and this ion may be surrounded by eight chloride ions in eight vertices of a cube in the cesium chloride crystal as shown in Fig. 9-43.

9.7.2 Molecular Crystals

A molecular crystal is built from molecules. It is easily distinguished from an ionic/atomic crystal on a purely geometrical basis. At least one of the intramolecular distances of an atom in the molecule is significantly smaller than its distances to the adjacent molecules. Every molecule in the molecular crystal may be assigned a certain well-defined space in the crystal. In terms of interactions, there are the much stronger intramolecular interactions and the much weaker intermolecular interactions. Of course, even among the intramolecular interactions, there is a range of interactions of various energies. Bond stretching, for instance, requires a proportionately higher energy than angular deformation, and the weakest are those interactions that determine the conformational behavior of the molecule. On the other hand, there are differences among the intermolecular interactions as well. For example, intermolecular hydrogen bond energies are equal to or greater than the conformational energy differences. Thus, there may be some overlap in the energy ranges of the intramolecular and intermolecular interactions.

The majority of molecular crystals are organic compounds. There is usually very little electronic interaction between the molecules in these crystals, although as will be discussed later even small interactions may have appreciable structural consequences. The physical properties of the molecular crystals are primarily determined by the packing of the molecules.

9.7.2.1 Geometrical Model

As structural information for large numbers of molecular crystals has become available, general observations and conclusions have appeared [9-1]. An interesting observation was that there are characteristic shortest distances between the molecules in molecular crystals. The intermolecular distances of a given type of interaction are fairly constant. From this observation a geometrical model has been developed for describing the molecular crystals. First, the shortest intermolecular distances were found, then the so-called "intermolecular atomic radii" were postulated. Using these quantities, spatial models of the molecules were built. Fitting together these models, the densest packing could be found empirically. A simple but ingenious equipment was even constructed for fitting the molecules. A packing example is shown in Fig. 9-44a. The mole-

Figure 9-44.
Dove-tail molecular packing.
(a) Dense packing of 1,3,5-tri-phenylbenzene molecules after [9-1].

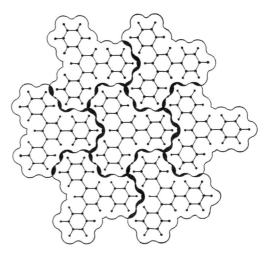

415

cules are packed together in such a way as to minimize the empty space among them. The concave part of one molecule accommodates the convex part of the other molecule. The example is the packing of 1,3,5-triphenylbenzene molecules in their crystal structure. If the areas of the molecules are blackened, a characteristic pattern of Eastern decorations [9-44] is obtained. This is shown in Fig. 9-44b. The complementary character of molecular packing is well expressed by the term of dove-tail packing [9-43]. The arrangement of the molecules in Fig. 9-45a can be called head-to-tail. On the other hand,

(b) Characteristic Eastern decoration from the middle ages [9-44].

the molecules of a similar compound are arranged head-to-head as seen in Fig. 9-45b after Wudl and Zellers [9-45]. The head-to-head arrangement seems to be less advantageous for packing. Thus in this particular case some other intermolecular

Figure 9-45.
Molecular packings after [9-45].
(a) Head-to-tail packing.

416

(b) Head-to-head packing.

interactions may be responsible for stabilizing this structure. Many of Escher's periodic drawings with interlocking motifs are also excellent illustrations for the dove-tail packing principle. Fig. 9-46 reproduces one of them [9-46]. Note how the toes of the black dogs are the teeth of the white dogs and *vice versa* in this figure.

Figure 9-46.
Escher's periodic drawing of dogs from MacGillavry's book [9-46]. Reproduced with permission from the International Union of Crystallography.

Because of this interlocking character, the packing in organic molecular crystals is usually characterized by large coordination numbers, i.e. by a relatively large number of adjacent or touching molecules. Experience shows that the most often occurring coordination number in the organic structures is 12, so it is the same as for the densest packing of equal spheres. The coordinations 10 or 14 occur also but less often.

417

The geometrical model allowed Kitaigorodskii [9-1, 9-43] to predict the structure of organic crystals in numerous cases, knowing only the cell parameters and, obviously, the size of the molecule itself. In the age of fully automated, computerized diffractometers this may not seem to be so important but it has indeed enormous significance for our understanding the packing principles in the molecular crystals.

The packing as established by the geometrical model is what is expected to be the ideal arrangement. Usually it does not differ from the real packing as determined by X-ray diffraction measurements. When there are differences between the ideal and experimentally determined packings, it is of interest to examine the reasons of their occurrence. The geometrical model has some simplifying features. One of them is that it considers uniformly the intermolecular atom... atom distances. Another is that it considers interactions only between adjacent atoms.

The development of the experimental techniques and the appearance of more sophisticated models have recently pushed the application of the geometrical model into the background. However, its simplicity and the facility of visualization ensure this model a lasting place in molecular crystallography. It has also exceptional didactic value.

The so-called coefficient of molecular packing (k) has proved useful in characterizing molecular packing. It is expressed the following way:

$$k = \frac{\text{molecular volume}}{\text{crystal volume/molecule}} .$$

The molecular volume is calculated from the molecular geometry and the atomic radii. The quantity crystal volume/molecule is determined from the X-ray diffraction experiment. For most crystals k is between 0.65 and 0.77. This is remarkably close to the coefficient of the dense packing of equal spheres. The density of closest packing of equal spheres is 0.7405 [9-2].

If the form of the molecule does not allow the coefficient of molecular packing to be greater than 0.6, then the substance is predicted to transform into a glassy state with decreasing temperature. It has also been observed that morphotropic changes associated with loss of symmetry led to an increase in the packing density. Comparison of analogous molecular crystals shows that sometimes the decrease in crystal symmetry is accompanied by an increase in the density of packing.

Another interesting comparison involves benzene, naphthalene, and anthracene. When their coefficient of packing is greater than 0.68, they are in the solid state. There is a drop in this coefficient to 0.58 when they go into liquid phase. Then with increasing temperature their k is decreasing gradually down to the point where they start to boil.

9.7.2.2 Densest Molecular Packing

Kitaigorodskii [9-1, 9-43] examined the relationship between densest packing and crystal symmetry by means of the geometrical model. He determined that real structures will always be among those that have the densest packing. First of all he established the symmetry of those two-dimensional layers that allow a coordination number six in the plane at an arbitrary tilt angle of the molecules with respect to the axes of the layer unit cell. In the general case for molecules with *arbitrary* form, there are only two kinds of such layers. One has inversion centers and is associated with a non-orthogonal lattice. The other has a rectangular net, from which the associated lattice is formed by translations, plus a second-order screw axis parallel to a translation. The next task was to select the space groups for which such layers are possible. This is an approach of great interest since the result will answer the question as to why there is a high occurrence of a few space groups among the crystals while most of the 230 groups almost never occur.

We present here some of the highlights of Kitaigorodskii's considerations [9-1]. First, the problem of dense packing is examined for the plane groups of symmetry. The distinction between dense-packed, densest-packed and maximum density was introduced for the plane layer of molecules. The plane was called dense-packed when coordination of six was achieved for the molecules. Densest-packed meant six-coordination with any orientation of the molecules with respect to the unit cell axes. The term maximum density was used for the packing if coordination six was possible at any orientation of the molecules with respect to the unit cell axes while the molecules retained their symmetry.

For the plane group $p1$ it is possible to achieve densest packing with any molecular form if the translation periods t_1 and t_2 and the angle between them are chosen appropriately as illustrated in Fig. 9-47. The same is true for the plane group $p2$ shown also in Fig. 9-47. On the other hand, the plane groups pm

Figure 9-47.
Densest packing with space groups $p1$ and $p2$ after [9-1].

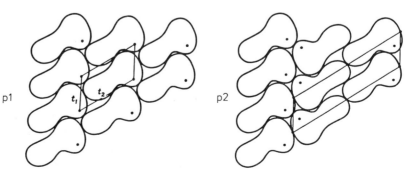

p1

t_1 t_2

p2

419

and *pmm* are not suitable for densest packing. As is seen in Fig. 9-48, the molecules are oriented in such a way that their convex parts face the convex parts of other molecules. This arrangement, of course, counteracts dense packing. The plane groups *pg* and *pgg* may be suitable for 6-coordination as an example

Figure 9-48.
The symmetry planes in the space groups *pm* and *pmm* prevent dense packing [9-1].

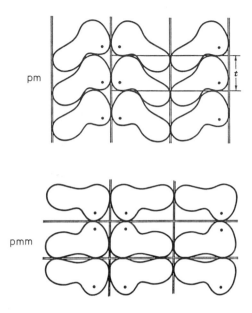

shows it in Fig. 9-49a. This layer is not of maximum density and in a different orientation of the molecules only 4-coordination is achieved as seen in Fig. 9-49b. For the plane groups *cm, cmm,* and *pmg* 6-coordination cannot be achieved for a molecule with arbitrary shape. For higher symmetry groups, e.g. tetragonal *p*4

Figure 9-49.
Varieties in the packing with *pgg* space groups [9-1].
(a) densest packing of molecules with arbitrary shape.

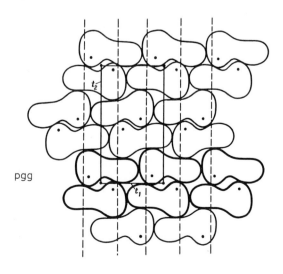

420

(b) Another orientation of the molecules which reduces the coordination number to four.

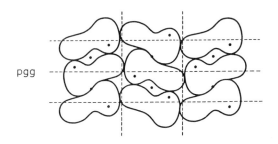

pgg

Figure 9-50.
Molecules of arbitrary shape cannot be packed in space group $p4$ without overlaps. After [9-1].

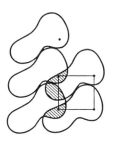

or hexagonal $p6$, the axes of the unit cell are equivalent, and the packing of the molecules is not possible without overlaps. This is illustrated for group $p4$ in Fig. 9-50.

If the molecule, however, retains a symmetry plane, then it may be packed with 6-coordination in at least one of the plane groups, pm, pmg, or cm. The form shown in Fig. 9-51 is suitable for such packing in pmg and cm, though not in pm. Thus, depending on the molecular shape, various plane groups may be applicable in different cases.

The criteria and examples for the suitability as well as incompatibility of plane groups for achieving molecular 6-coordination have been considered. The next step is to apply the geometrical model to the examination of the suitability of three-dimensional space groups for densest packing. The task in this case is to select those space groups in which layers can be packed allowing the greatest possible coordination number. Obviously, for instance, mirror planes would not be applicable for repeating the layers.

Figure 9-51.
Molecules with a symmetry plane achieve 6-coordination in the space groups cm and pmg. After [9-1].

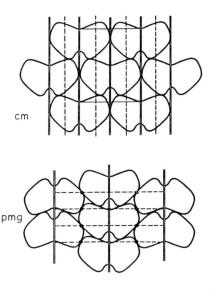

cm

pmg

421

Low-symmetry crystal classes are typical for the organic compounds. Densest packing of the layers may be achieved either by translation at an arbitrary angle formed with the layer plane, or by inversion, glide plane, or by screw-axis rotation. In rare cases closest packing may also be achieved by two-fold rotation.

Kitaigorodskii [9-1] has analyzed all 230 three-dimensional space groups from the point of view of densest packing. Only the following space groups were found to be available for the densest packing of molecules of arbitrary form:

$$P\bar{1}, P2_1, P2_1/c, Pca, Pna, P2_12_12_1.$$

For molecules with symmetry centers, there are even fewer suitable three-dimensional space groups, namely:

$$P\bar{1}, P2_1/c, C2/c, Pbca.$$

In these cases all mutual orientations of the molecules are possible without losing the six coordination.

Table 9-7 compiles all possibilities for densest packing, taking into consideration the molecular symmetry in the crystal. The above six space groups that are suitable for maximum density packing of molecules in general positions in the crystal, are indeed the ones that have occured to date most often in experimental structure determinations.

Table 9-7. Densest and Maximum Density Packings in Molecular Crystals and Their Distribution among Molecular Symmetries after Kitaigorodskii [9-1]

	Molecular Symmetries							
	1	2	m	$\bar{1}$	mm	$2/m$	222	mmm
Densest packing	$P\bar{1}$ $P2_1$ $P2_1/c$ Pca Pna $P2_12_12_1$	none	none	$P\bar{1}$ $P2_1/c$ $C2/c$ $Pbca$	none	none	none	none
Maximum density packing	none	$C2/c$ $P2_12_12_1$ $Pbcn$	Pmc Cmc $Pnma$	none	Fmm $Pmma$ $Pmmn$	$C2/m$ $Pbaa$ $Cmca$	$C222$ $F222$ $I222$ $Ccca$	$Cmmm$ $Fmmm$ $Immm$

The space group $P2_1/c$ occupies a strikingly special position among the organic crystals. It is the unique feature of this space group that it allows the formation of layers of densest packing in all three coordinate planes of the unit cell.

The space groups $P2_1$ and $P2_12_12_1$ are also among those providing densest packing. However, their possibilities are more limited than those of the space group $P2_1/c$ and occur only for molecules which take either left-handed or right-handed forms.

According to statistical examinations performed some time ago these three groups are the first three in frequency of occurrence.

An interesting and really fundamental question is about the conservation of molecular symmetry in the crystal structure. This question will be looked at here subsequently from different viewpoints. Densest packing may often be well facilitated by partial or complete loss of molecular symmetry in the crystal structure. There are, however, space groups in which the molecular symmetry may "survive" densest packing when building the crystal. Among those elements of symmetry that are not convenient for the purposes of densest packing there is usually just one symmetry plane or one two-fold rotation axis which it seems relatively easy for the molecules to preserve. Preserving higher symmetry usually costs too great a sacrifice of packing density.

It is quite common on the other hand, for molecules originally belonging to the point group C_{2v}, to preserve a symmetry plane or a two-fold axis. This may indicate the energetic advantage of some well-defined symmetrical arrangements. The alternative to the geometrical model for discussing and establishing the molecular packing in organic crystals has been the calculations of energy, based on carefully constructed potential energy functions.

9.7.2.3 Energy Calculations

It is important to be able to determine a priori the arrangement of molecules in crystals. The correctness of such predictions is a test for our understanding of how crystals are built. A further benefit is the possibility of calculating even those structures whose determination is not amenable to experimental analysis. But even as part of an experimental study, it is instructive to build good models which can then be refined.

There are two main non-experimental approaches for finding the densest packing in molecular crystals. One is the utilization of the geometrical model. Its main advantages have been seen above. Its main limitations are the following: It cannot account for the structural variations in a series of analogous compounds. It is very restricted in correlating structural features with various other physical properties. Finally, it is unable to make detailed predictions for unknown structures.

Currently, increasing attention is being paid to another approach, viz. considerations of energy. The spatial arrangement of molecules in a crystal corresponds to the minimum of free energy. If the system is considered completely rigid, the molecular packing may be determined by minimizing the potential energy of intermolecular interactions [9-47].

Considering the molecules to be rigid, i.e. ignoring the vibrational contribution, the energy of the crystal structure is expressed as a function of geometrical parameters including the cell parameters, the coordinates of the centers of gravity of the symmetrically independent molecules, and parameters characterizing the orientation of these molecules. In particular cases the number of independent parameters can be reduced. On the other hand, considerations for the nonrigidity of the molecules necessitates additional parameters. Minimizing the crystal structure energy leads to structural parameters corresponding to optimal molecular packing. Then it is of great interest to compare these findings with those from the real experiment.

To determine the deepest minimum on the multidimensional energy surface as a function of many structural parameters is a formidable mathematical task. Usually simplifications and assumptions are introduced concerning, for example, the space-group symmetry. Accordingly, the conclusions from these theoretical calculations cannot be considered to be entirely a priori.

The considerations on the intermolecular interactions can be conveniently reduced to considerations of atom-atom nonbonded interactions. Although these interactions can be treated quantum mechanically, empirical and semi-empirical approaches have also proved successful in dealing with them. In the description of the atom-atom non-bonded interactions it is supposed that the van der Waals forces originate from a variety of sources.

In addition to the intermolecular interactions, the intramolecular interactions may also be taken into account in a similar way. This rather limited approach may nevertheless be useful for calculating the molecular conformation and even the molecular symmetry. Deviations from the ideal conformations and symmetries may also be estimated this way provided they are due to steric effects. For further details we may refer to more specialized literature, for example Dashevskii's monograph [9-48].

By summation over the interaction energies of the molecular pairs, the total potential energy of the molecular crystal may be obtained in an atom-atom potential approximation. The result is expected to be approximately the same as the heat of sublimation extrapolated to 0 K provided that no changes take place in the molecular conformation and vibrational interactions during evaporation.

Combinations of quantitative molecular packing analyses in different space groups along with *ab initio* molecular orbital calculations have been described [9-49]. Several different potential energy functions have been tested, and the results proved invariant to the choice of the potential functions employed. As these calculations were aimed at studying conformational polymorphism, we shall return to them later in the relevant section.

In the majority of the molecular packing studies, the crystal classes are taken from the experimental X-ray diffraction determinations. The optimal packing is then determined for the thus assumed crystal class. In other cases the crystal classes have also been established in the optimization calculations.

The crystal structures of $Sb(C_6H_5)_3$, $P(C_6H_5)_3$, and $As(C_6H_5)_3$ have been compared for example. The antimony coordination was found to be tetragonal pyramidal in the space group $P\bar{1}$, while the coordination of phosphorus and arsenic is trigonal pyramidal in space group Cc. The calculations of the potential energy minimization were in agreement with the experimental results but only when electrostatic contributions were also taken into account [9-50].

The packing peculiarities of a series of other compounds have also been investigated by similar atom-atom potential calculations in addition to the experimental determinations of their structures. A deeper insight was achieved this way into the understanding of the choices of space groups and the densest packing for various molecular crystals.

9.8 Hypersymmetry

There are some crystal structures in which extra symmetries are present in addition to those prescribed by their three-dimensional space group. The phenomenon is called hypersymmetry and has been discussed recently in detail by Zorkii and Koptsik [9-51]. Thus hypersymmetry refers to symmetry features not included in the system of the 230 three-dimensional space groups.

For example, phenol molecules, connected by hydrogen bonds, form spirals with three-fold screw axes as indicated in Fig. 9-52. This screw axis does not extend, however, to the whole crystal, and it does not occur in the three-dimensional space group characterizing the phenol crystal [9-51]. Tolane

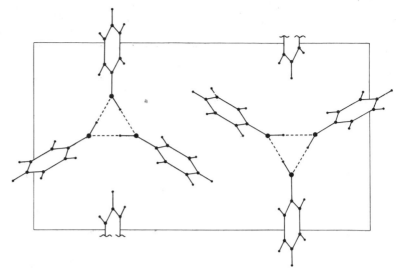

(diphenylacetylene), $C_6H_5-C\equiv C-C_6H_5$, is another example. As shown in Fig. 9-53, the tolane molecules A and B are related by two-fold rotation plus translation and these symmetry elements are not part of the crystal space group [9-51].

Figure 9-53.
The molecules A and B in the tolane (diphenylacetylene) crystal are related by two-fold rotation plus translation. These symmetry elements are not part of the three-dimensional space group of the tolane crystal. After [9-51].

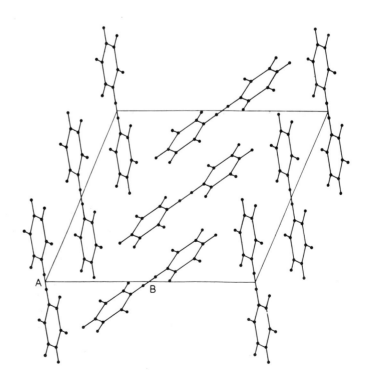

426

A typical characteristic of hypersymmetry operations is that they exercise their influence in well-defined discrete domains. These domains do not overlap, they do not even touch each other. The examples mentioned in the introduction to the present section were characterized by such hypersymmetry elements which led to point-group properties. This means that no infinite molecular chains could be selected, for example, to which these hypersymmetry operations would apply. They effected, instead, pairs of molecules, or very small groups of molecules. Thus, they can really be considered as local point-group operations. These hypersymmetry elements, accordingly, divide the whole crystalline system into numerous small groups of molecules, or transform the crystal space into a layered structure.

A prerequisite for hypersymmetry is that there should be chemically identical (having the same structural formula), but symmetrically independent, molecules in the crystal structure – symmetrically independent, that is, in the sense of the three-dimensional space group to which the crystal belongs. The question then arises, as to whether these symmetrically independent but chemically identical molecules will have the same structure or not? Only if they do have the same structure, conformation as well as bond configuration, can we talk about the validity of the hypersymmetry operations. Here, preferably, quantitative criteria should be introduced, which is the more difficult since, for example, with increasing accuracy, structures that could be considered identical before, may no longer be considered so later when more accurate data become available. In any case it is emphasized that for the validity of the hypersymmetry operations, the symmetrically independent molecules are supposed to have identical structures.

On the other hand, since even a slightly different environment will have some influence on the molecular structure, the hypersymmetry operations will not be absolute. In this the hypersymmetry operations are somewhat different from the usual symmetry operations. For their mathematical description, however, this approximate character has been ignored [9-51]. The ultimate goal is to find such a generalized formulation of the space-group system that would allow the simultaneous consideration of the usual symmetry as well as the hypersymmetry. When such a generalized formulation of space groups encompassing usual and hypersymmetry operations becomes available, the task of discovering crystals with hypersymmetry will be greatly facilitated.

A special case of hypersymmetry is when the otherwise symmetrically independent molecules in the crystal are related by hypersymmetry operations making them enantiomorphous pairs.

Hypersymmetry is a rather widely observed and sometimes ignored phenomenon which is not restricted to a special class of compounds. It may be supposed, however, that certain types of molecules are more apt to have this kind of additional symmetry in their crystal structures, than others.

According to recent statistics [9-52], about 8 % of all homomolecular crystals contain symmetrically independent molecules. Homomolecular is the term applied to crystals which are built of chemically identical molecules [9-53]. Recognizing the importance of structural differences in symmetrically independent molecules, a special name, to wit quasi-heteromolecular has been proposed for their crystals [9-52].

According to some quantitative criteria for defining the extent of differences in considering two structures different indeed, some 20 % of all quasi-heteromolecular organic crystals were found to have appreciable structural differences in their symmetrically independent molecules. This ratio was even higher among quasi-heteromolecular coordination compounds of transition metals [9-54]. The symmetrically independent molecules in fact showed a great variety of structural differences [9-55].

There are hypersymmetry phenomena in some crystal structures which are characterized by extra symmetry operations applicable to infinite chains of molecules. This kind of hypersymmetry has proved to be more easily detectable and has been reported often in the literature.

Hypersymmetry may be interpreted on the basis of the symmetry of the potential functions describing the conditions of the formation of the molecular crystal. The molecules around a certain starting molecule will be related by the symmetry of the potential function itself, or the symmetry of certain combinations of the potential energy functions. The occurrence of some screw axes of rotation by hypersymmetry elements has been successfully interpreted this way. In some instances energy calculations as well as geometrical reasons have shown the physical importance of hypersymmetry. The consequence of some of the kinds of hypersymmetry considered (e.g., for tolane), is stronger chemical bonding among molecules. Hypersymmetry may be expressed in the layered structure of a molecular crystal. This again may have advantages for geometrical and energy considerations. Thus the phenomenon of hypersymmetry is another good example of how symmetry properties and other properties are related to each other.

All in all, however, hypersymmetry is a relatively recent discovery. The investigation of its occurrence and nature as well as the examination of its consequences are expected to be vigorously pursued in the future.

9.9 Crystal Field Effects

Elucidating the effects of intermolecular interactions may greatly facilitate our understanding of the structure and energetics of crystals. The geometrical changes of molecular structures cover a wide range in energy. Molecular shape, symmetry, and conformation change more easily upon the molecule becoming part of a crystal than do bond angles and especially bond lengths. The consequences of molecular packing are succinctly expressed in the following metaphor attributed to Kitaigorodskii "The molecule also has a body. When you strike it, the molecule feels hurt all over".

Kitaigorodskii [9-43] suggested four approaches to investigating the effect of the crystalline field on molecular structure: (1) Comparison of gaseous (i.e. free) and crystalline molecules. (2) Comparison of symmetrically (i.e. crystallographically) independent molecules in the crystal. (3) Analysis of the structure of molecules whose symmetry in the crystal is lower than their free molecular symmetry. (4) Comparison of the molecular structure in different polymorphic modifications. Bernstein and Hagler [9-49] called attention to the fact that the energetics of intramolecular deformations may be such that some molecules will be able to achieve a more symmetrical conformation under crystal field effects than when they are isolated in the vapor. Some of their results will be mentioned later. Similarly, biphenyl has a higher molecular symmetry (a coplanar structure) in the crystal [9-56] than in the vapor [9-57] (where the two benzene rings are rotated by about 45° relative to each other (Fig. 9-54)). Thus a more general formulation of the above point 3 is the analysis of the structure of molecules whose symmetry in the crystal is different from their free molecular symmetry.

Figure 9-54.
The molecular structure of biphenyl is coplanar in the crystal [9-56] while its two benzene rings are rotated by about 45° relative to each other in the vapor [9-57].

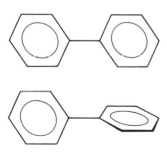

9.9.1 Structure Differences in Free and Crystalline Molecules

The points 1 and 3 above both refer to the comparison of the structures of free and crystalline molecules. They provide, perhaps, the most straightforward information, since the structure of the free molecule is determined exclusively by intramolecular interactions. Thus any difference that is reliably detected will carry information as to the effects of the crystal field on the molecular structure. However, before discussing

Figure 9-55.
Part of the sodium chloride crystal structure with NaCl, $(NaCl)_2$, $(NaCl)_4$ units indicated. The species of 3x3x3 ions itself has a high relative abundance in cluster formation.

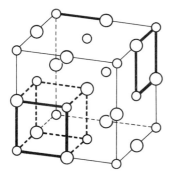

more subtle structural differences in molecular crystals as compared with free molecules, it is appropriate to point out some more striking differences between ionic crystals and the corresponding vapor-phase molecules.

Although molecules cannot be identified as the building blocks of ionic crystals, the free molecules of *some* compounds may be considered as if they were taken out of the crystal. A nice example is sodium chloride whose main vapor components are monomeric and dimeric molecules. They are indicated in the crystal structure in Fig. 9-55 as is a tetrameric species. Recent mass spectrometric studies of cluster formation determined a great relative abundance of a species with 27 atoms in the cluster. The corresponding 3x3x3 cube may again be considered as a small crystal [9-58].

Another series of simple molecules whose structure may easily be traced back to the crystal structure is shown in Fig. 9-56. It is evident, for example, that various MX_2 and MX_3 molecules may take different shapes and symmetries from the same kind of crystal structure. The crystal structure is represented by the octahedral arrangement of six "ligands" around the "central atom".

Figure 9-56.
Different shapes of MX_2 and MX_3 molecules derived from the crystal structure in which the central atom has an octahedral environment.

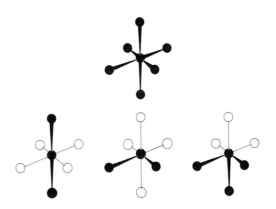

The unit corresponding to the copper chloride trimeric molecule seems to be already present in the crystal. The molecule has a ring structure with alternating copper and chloride atoms. The low-temperature phase of copper chloride has a zinc-blende structure which transforms on heating into a wurtzite structure [9-2]. In both arrangements the six-membered ring is the smallest unit which can be isolated from the structure upon evaporation [9-59].

Even though the shapes of the free molecules mentioned above may be considered as originating from the crystal structure, the interatomic distances change considerably. They are collected for some lithium halides in Table 9-8 [9-60–9-62]. As

Table 9-8. Variation of Interatomic Distances of Lithium Halides upon Sublimation

	Lithium Chloride	Lithium Bromide	Lithium Iodide
Crystal [9-60]	2.57 Å	2.75 Å	3.00 Å
Vapor			
dimer [9-60, 9-61]	2.23 ± 0.03 Å	2.35 ± 0.02 Å	2.54 ± 0.02 Å
monomer [9-62]	2.02067 ± 0.00006 Å	2.17042 ± 0.00004 Å	2.39191 ± 0.00004 Å

we go from the crystal to the free dimers and then to the monomers, the interatomic distances drastically decrease.

There is even less structural similarity for many other metal halides as the crystalline systems are compared with the molecules in the vapor phase. Aluminium trichloride as well as iron trichloride crystallizes in a hexagonal layer structure. Upon melting and then upon evaporation at relatively low temperatures, dimeric molecules are formed. At higher temperatures they dissociate into monomers. This is illustrated in Fig. 9-57 [9-63].

Figure 9-57.
Structural changes upon evaporation of aluminium trichloride and iron trichloride.

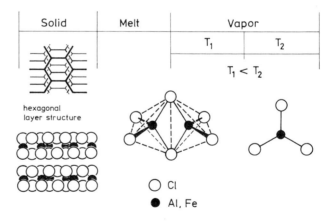

In some cases the vapor of molecular crystals contains more rotational isomers than the crystal. Thus, for example, the vapor [9-64] of ethane-1,2-dithiol consists of *anti* and *gauche* forms with respect to rotation about the central bond (Fig. 9-58) while only the *anti* form was found in the crystal [9-65].

aaa   gag   aga

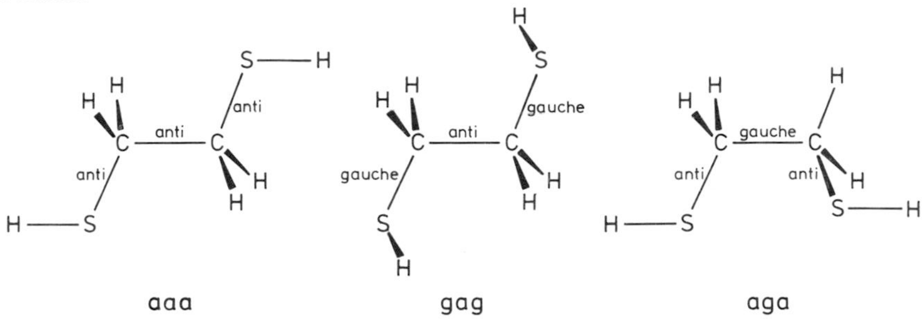

Bis(cyclopentadienyl)titanium borohydride has also been studied both in the crystal [9-66] and vapor [9-67]. The rings seem to take an orientation of high symmetry relative to each other and the Ti... B axis in the vapor as seen in Fig. 9-59. The structure is less symmetrical in the crystal.

Figure 9-59.
The projection models of the vapor-phase bis(cyclopentadienyl)titanium borohydride molecule in two different planes [9-67]. The angles P and Q are 0° when one of the apexes of the five-membered rings coincides with the Ti... B axis.

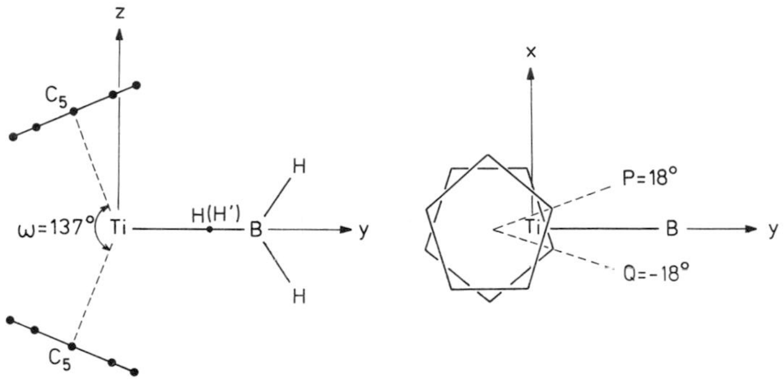

The comparison of the structures of free and crystalline molecules is obviously based on the application of various experimental techniques although theoretical calculations play an increasing role. Thus it is important to comment upon the inherent differences in the physical meaning of the structural information originating from such different sources. The consequences of intramolecular vibrations on the geometry of free molecules have already been mentioned. The effects of molecular vibrations and librational motion in the crystal are not less important. To minimize their effects it is desirable to examine the crystal molecular structure at the lowest possible temperatures. Also the corrections for thermal motion are of great importance. Especially when employing older data in comparisons and discussing subtle effects, these problems have to be considered. There is another important source for differences in structural information, which may have no real structural implications. Apparent differences may originate from the difference

432

in the physical meaning of the physical phenomena utilized in the experimental techniques. When all sources of apparent differences have been eliminated, and the molecular structure still differs in gas and crystal, the intermolecular interactions in the crystal may indeed be responsible [9-68].

Variations in the ring angular deformations of gaseous and crystalline substituted benzene derivatives may be a sensitive indicator of intermolecular interactions. Recent parallel gas electron diffraction and X-ray crystallographic studies of *para*-dicyanobenzene [9-69] and *para*-diisocyanobenzene [9-70] have led to interesting results in this respect. Table 9-9 presents

Table 9-9. The Benzene Ring Ipso Angles in *para*-Dicyano-benzene [9-69] and *para*-Diisocyanobenzene [9-70]

	Free Molecule	Crystal Molecule	Difference
NCC_6H_4CN	122.1(2)°	121.6(1)°	−0.5°
CNC_6H_4NC	121.7(2)°	122.2(3)°	+0.5°

Figure 9-60.
Schematic representation of the possible intermolecular interaction between anti-parallel cyano groups of *para*-dicyanobenzene in the crystal.

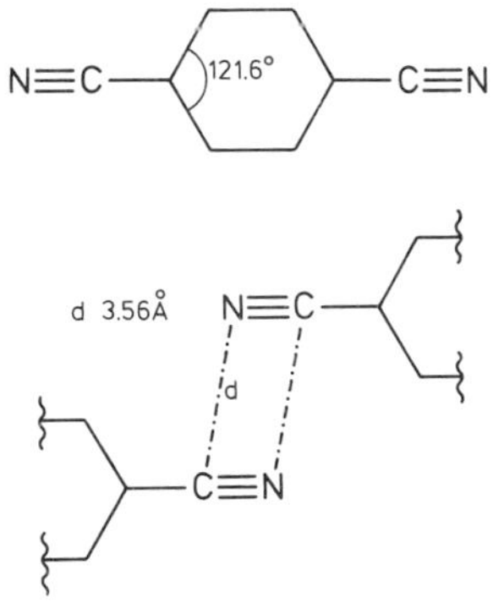

the variations of the so-called ipso angle, the ring angle at the substituent, in the two derivatives in the two phases. The changes are marginal but their trend can be interpreted in a reasonable way invoking the marked difference in the molecular packing in the two crystals. For *para*-dicyanobenzene the smaller deformation in the crystal versus the free molecule indicates an attenuation of the polar character of the cyano substituent under the impact of intermolecular interactions. The *para*-dicyanobenzene molecules are parallel to each other in the crystal and they form layers [9-71]. Each layer is displaced with respect to the next, leading to antiparallel contacts between neighboring cyano groups in addition to parallel contacts. The shortest distance between two antiparallel cyano groups is 3.56 Å as seen in Fig. 9-60. These interactions are also the origin of the relatively high melting point of this compound, viz. 223-226°C.

The gas/crystal difference in the ring angular deformation is just the opposite for *para*-diisocyanobenzene [9-70]. Here the ring is more deformed in the crystal than in the free molecule. The crystal structure of *para*-diisocyanobenzene is strikingly different from that of *para*-dicyanobenzene. The intermolecular interactions occur between the isocyano groups and the benzene rings, as shown schematically in Fig. 9-61. They apparently lead to some increase in the polar character of the isocyano substituent, and accordingly to somewhat enhanced ring

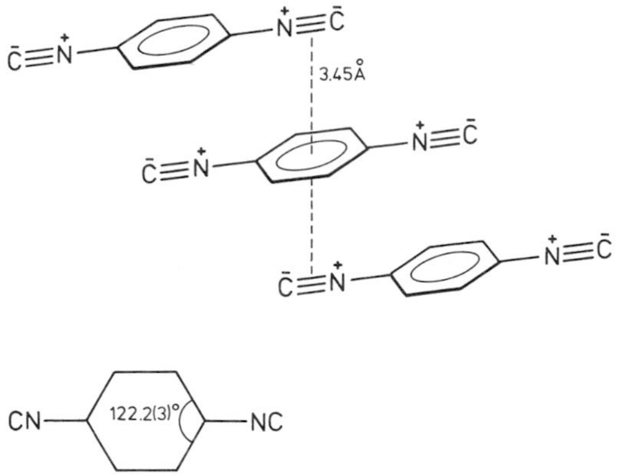

deformation in the crystal as compared with the free *para*-diiso-
cyanobenzene molecule or, for that matter, with the crystalline
para-dicyanobenzene molecule. Marginal as these findings may
be, they could serve as a stimulus for the structural chemist in
search of gas/crystal differences in accurately determined
systems, and for the theoretical chemist who may build models
and perform calculations on them in which both the intramolecular and intermolecular interactions are adequately represented.

The gas/crystal structural changes obviously depend on the
relative strengths of the intramolecular and intermolecular
interactions. More pronounced changes are expected, for
example, in relatively weak coordination linkages under the
influence of the crystal field than in stronger bonds.

The N-B bond of trimethylamine – boron trichloride,
$(CH_3)_3N \cdot BCl_3$, donor-acceptor complex is considerably longer
in the gas [9-72] than in the crystal [9-73]. The two sets of
structural parameters are compared in Fig. 9-62. The N-Si
dative bond of 1-methylsilatrane is much shorter in crystal

Figure 9-62.
Selected geometrical parameters of the trimethylamine-
boron trichloride molecule in
gas [9-72] and crystal [9-73].

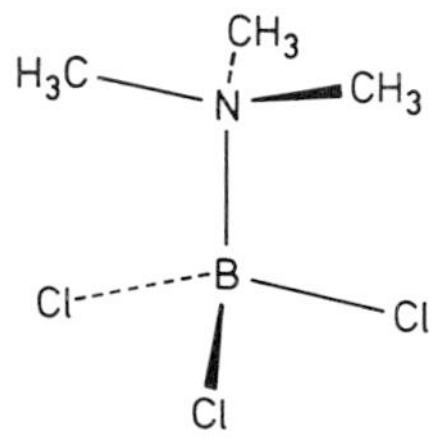

gas (ED)		crystal (XD)
1.659(6) Å	B–N	1.609(6) Å
108.7(5)°	C–N–C	108.6(14)°
110.8(3)°	Cl–B–Cl	109.5(2)°

[9-74] than in gas [9-75] so much so that in the latter according to the geometrical data cited in Fig. 9-63 there is hardly any such bond. It may be supposed that the intermolecular forces compress the molecules somewhat along the coordination linkage in the crystal. Part of the gas/crystal bond length differences in these cases originates, of course, from the differing physical meaning of the gas electron diffraction and crystal X-ray diffraction results, since these coordination linkages are related to considerable charge transfer. The other part of the differences, however, is supposed to be really structural, and the geometrical changes in the rest of the molecule are fully consistent with this notion.

Figure 9-63.
Selected geometrical parameters of the 1-methylsilatrane molecule in gas [9-75] and crystal [9-74].

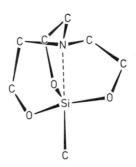

gas (ED)		crystal (XD)
2.45(5) Å	N···Si	2.175(4) Å
79.0(15)°	N–Si–O	82.7(2)°
116.5(9)°	O–Si–O	118.4(4)°
101.0(15)°	C–Si–O	97.2(5)°

The bond angles of the donor part of $(CH_3)_3 N \cdot BCl_3$ are the same within experimental error in the gas as in the crystal. It is generally observed [9-76] that the donor geometry is insensitive to complex formation. On the other hand, the acceptor configuration usually changes considerably upon complex formation. The free BCl_3 molecule has D_{3h} symmetry with 120° bond angles. The greater angular deformation of the BCl_3 moiety of the complex in the crystal, than in the gas, is consistent with a stronger coordination linkage in the crystal.

Consider now the bond angles in 1-methylsilatrane (Fig. 9-63). Supposing the pentacoordination of silicon, the bond angles determined in the crystal are closer to the ideal trigonal bipyramidal configuration than those in the gas. However, if one ignores the N-Si linkage and supposes four-coordination of silicon, then the structure in gas is nearer to the ideal tetrahedral configuration than that in crystal. Thus the bond angle differences are consistent with the notion of the N-Si bond being stronger in the crystal than in the gas. Part of the bond length difference, however, originates from the difference

435

in the meaning of the two bond lengths, as the N-Si linkage involves considerable charge transfer.

Gas/crystal comparisons are as of yet mainly confined to registering structural differences. The interpretation of these results is at a qualitative initial stage. Further investigation of such differences will enhance our understanding of the inter-molecular interactions in crystals.

9.9.2 Conformational Polymorphism

The investigation of different rotational isomers of the same compound in different crystal forms (polymorphs) is also an efficient tool in elucidating intermolecular interactions. The phenomenon is called conformational polymorphism. The energy differences between the polymorphs of organic crystals are similar to the free energy differences of rotational isomers of many free molecules, viz. a few kcal/mol. When the molecules adopt different conformations in the different polymorphs, the change in rotational isomerism is attributed to the influence of the crystal field since the difference in the intermolecular forces is the single variable in the polymorphic systems. Bernstein and co-workers [9-49, 9-77] have extensively studied conformational polymorphism of various organic compounds with a variety of techniques in addition to X-ray crystallography. Among the molecules investigated were N-*(para*-chloroben-zylidine)-*para*-chloroaniline (I)

(I)

which exists in at least two forms, and *para*-methyl-N-*(para*-methylbenzylidine)aniline (II)

(II)

which exists in at least three forms. For (I) a high-energy planar conformation was shown to occur with a triclinic lattice. A lower-energy form with normal exocyclic angles was found in the orthorhombic form. It was an intriguing question as to why molecule (I) would not always pack with its lowest-energy conformation.

The X-ray diffraction work has been augmented by lattice energy calculations employing different potential functions. The results did not depend on the choice of the potential function, and they showed that the crystal packing and the (intra)molecular structure together adopt an optimal compromise. The minimized lattice energies were analyzed in terms of partial atomic contributions to the total energy. Even for the trimorphic molecule (II) the relative energy contributions of various groups in all polymorphs were similar. However, this obviously could only be achieved in some lattices by adopting a conformation, different from the most favorable, with respect to the structure of the isolated molecule. The investigation of conformational polymorphism proved to be a suitable and promising tool for investigating the nature of those crystal forces influencing molecular conformation, or possibly molecular structure in a broader sense.

Obviously, possible variations in bond angles and bond lengths have been ignored in the considerations described above. The energy requirements for changing bond angles and bond lengths are certainly higher than those for conformational changes, and, accordingly, higher than what may be available in polymorphic transitions. However, some relaxation of the bond configuration may take place, especially if considering the (intra)molecular structure also adopted as a compromise by the bond configurations and the rotational forms.

Bond configuration relaxation during internal rotation has in fact been investigated by quantum chemical calculations for a series of 1,2-dihaloethanes [9-78]. The bond angle C-C-X may change as much as 4° during internal rotation according to these calculations. If there is then a mixture of, say, *anti* and *gauche* forms, as is often the case, and the relaxation of the bond configuration is ignored, this may lead to considerable errors in the determination of the *gauche* angle of rotation.

9.10 Forget Me Not

Our discussion of crystals has become rather down to earth and, we admit, rather molecule oriented towards the end. As a token of compensation, the words of the nineteenth-century English writer John Ruskin are cited here after Azaroff [9-3], together with a fitting drawing of "crystal characters" in Fig. 9-64 from Bunn's book [9-79].

Figure 9-64.
"Crystal characters" from Bunn's book [9-79]. Reproduced with permission.

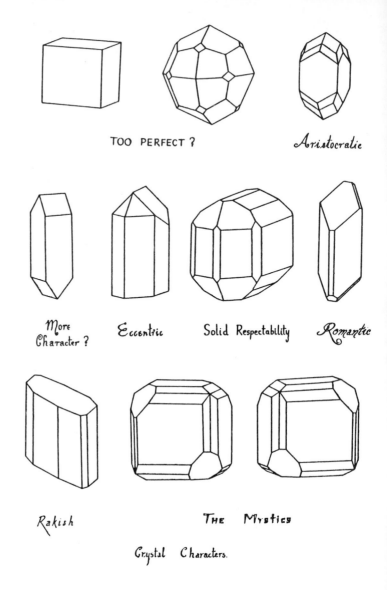

TOO PERFECT ?

Aristocratic

More Character ?

Eccentric

Solid Respectability

Romantic

Rakish

THE MYSTICS

Crystal Characters.

"And remember, the poor little crystals have to live their lives, and mind their own affairs, in the midst of all this, as best they may. They are wonderfully like humane creatures – forget all that is going on if they don't see it, however dreadful; and never think what is to happen tomorrow. They are spiteful or loving, and indolent or painstaking, with no thought whatever of the lava or the flood which may break over them any day; and evaporate them into air-bubbles, or wash them into a solution of salts. And you may look at them, once understanding the surrounding conditions of their fate, with an endless interest. You will see crowds of unfortunate little crystals, who have been forced to constitute themselves in a hurry, their dissolving element being fiercely scorched away; you will see them doing their best, bright and numberless, but tiny. Then you will find indulged crystals, who have had centuries to form themselves in, and have changed their mind and ways continually; and have been tired, and taken heart again; and have been sick, and got well again; and thought they would try a different diet, and then thought better of it; and made but a poor use of their advantages, after all.

And sometimes you may see hypocritical crystals taking the shape of others, though they are nothing like in their minds; and vampire crystals eating out the hearts of others; and hermit-crab crystals living on the shells of others; and parasite crystals living on the means of others; and courtier crystals glittering in the attendance upon others; and all these, besides the two great companies of war and peace, who ally themselves, resolutely to attack, or resolutely to defend. And for the close, you see the broad shadow and deadly force of inevitable fate, above all this: you see the multitudes of crystals whose time has come; not a set time, as with us, but yet a time, sooner or later, when they all must give up their crystal ghost – when the strength by which they grew, and the breath given them to breathe, pass away from them; and they fail, and are consumed, and vanish away; and another generation is brought to life, framed out of their ashes."

References

[9-1]   A. I. Kitaigorodskii, *Molekulyarnie Kristalli*, Nauka, Moskva, 1971.

[9-2]   A. F. Wells, *Structural Inorganic Chemistry*, Fifth Edition, Clarendon Press, Oxford, 1984.

[9-3]   L. V. Azaroff, *Introduction to Solids*, McGraw-Hill, New York, Toronto, London, 1960.

[9-4]   P. C. Ritterbush, Nature **301**, 197 (1983).

[9-5]   A. L. Mackay, Nature **301**, 652 (1983).

[9-6]   E. Schröder, *Dürer. Kunst und Geometrie*, Akademie Verlag, Berlin, 1980.

[9-7]   B. Ernst, *The Magic Mirror of M. C. Escher*, Ballantine Books, New York, 1976.

[9-8]   M. J. Buerger, *Elementary Crystallography, An Introduction to the Fundamental Geometrical Features of Crystals* (Fourth Printing), Wiley, New York, London, Sydney, 1967; E. S. Dana, *A Textbook of Mineralogy*, Fourth Edition, revised and enlarged by W. E. Ford, Wiley, New York, London, Sydney, 1932; P. M. Zorkii, *Arkhitektura Kristallov*, Nauka, Moskva, 1968.

[9-9]   R. J. Haüy, *Traité de Cristallographie*, 1822. Reprinted by Culture et Civilisation, Bruxelles, 1968.

[9-10]   J. Kepler, *Strena seu de nive sexangula*, Francofurti ad Moenum: Godefridum Tampach, 1611. English translation, *The Six-Cornered Snowflake*, Clarendon Press, Oxford, 1966.

[9-11]   C. J. Schneer, Am. Sci. **71**, 254 (1983).

[9-12]   A. L. Mackay, Izv. Jugosl. Centra Kristallogr. **10**, 15 (1975).

[9-13]   L. A. Shuvalov, A. A. Urosovskaya, I. S. Zheludev, A. V. Zaleskii, S. A. Semiletov, B. N. Grechushnikov, I. G. Chistyakov, and S. A. Pikin, *Fizicheskie Svoistva Kristallov* (Physical properties of crystals), *Sovremennaya Kristallografiya* (Modern Crystallography), Vol. 4, Nauka, Moskva, 1981.

[9-14]   A. L. Mackay, *The Floating World of Science*, Poems, The RAM Press, London, 1980.

[9-15]   A. Holmes, *Principles of Physical Geology*, The Ronald Press Co., New York, 1965.

[9-16]   M. Joray, *Vasarely*, Griffon, Neuchatel, 1976.

[9-17]   Gy. Lengyel, *Kézimunkák,* Kossuth, Budapest, 1978.

[9-18]   K. Čapek, *Anglické Listy,* Československý Spisovatel, Praha, 1970. The English version cited in our text was kindly provided by Dr. Alan L. Mackay, Birkbeck College, 1982.

[9-19]   *International Tables for X-Ray Crystallography,* 3rd Edition, The Kynoch Press, Birmingham, 1969.

[9-20]   A. V. Shubnikov and V. A. Koptsik, *Simmetriya v nauke i iskusstve,* Nauka, Moskva, 1972. English translation: *Symmetry in Science and Art,* Plenum Press, New York, London, 1974.

[9-21]   L. Glasser, J. Chem. Educ. **44,** 502 (1967).

[9-22]   M. C. Escher, *His Life and Complete Graphic Work,* with fully Illustrated Catalogue, J. L. Locher, ed., Harry N. Abrams, Inc. Publ., New York, 1982.

[9-23]   C. P. Brock and E. C. Lingafelter, J. Chem. Educ. **57,** 552 (1980).

[9-24]   A. L. Mackay, Acta Crystallogr. **22,** 329 (1967).

[9-25]   A. L. Mackay, Physics Bulletin 495 (1976).

[9-26]   A. L. Mackay, Kristallografiya **26**(5), 910 (1981); A. L. Mackay, Physica **114A,** 609 (1982).

[9-27]   I. Hargittai and W. J. Orville-Thomas, eds., *Diffraction Studies on Non-Crystalline Substances.* Elsevier-Akadémiai Kiadó, Amsterdam-Budapest, 1981.

[9-28]   J. D. Bernal, Acta Phys. Acad. Sci. Hung. **8,** 269 (1958).

[9-29]   J. D. Bernal and C. H. Carlisle, Kristallografiya, **13**(5), 927, (1969).

[9-30]   N. V. Belov, Kristallografiya **17,** 208 (1972).

[9-31]   A. Wickham, *Selected Poems,* Chatto and Windus, London, 1971.

[9-32]   A. Guinier, in *Diffraction Studies on Non-Crystalline Substances,* I. Hargittai and W. J. Orville-Thomas, eds., p. 411, Elsevier-Akadémiai Kiadó, Amsterdam-Budapest, 1981; R. Hosemann and B. Steffen, in *Diffraction Studies on Non-Crystalline Substances,* I. Hargittai and W. J. Orville-Thomas, eds., p. 491, Elsevier-Akadémiai Kiadó, Amsterdam-Budapest, 1981.

[9-33]   (a) D. Shechtman, I. Blech, D. Gratias, and J. W. Chan, Phys. Rev. Lett. **53,** 1951 (1984). (b) D. Levine and P. J. Steinhardt, Phys. Rev. Lett. **53,** 2477 (1984). (c) The New York Times January 8, 1985, p. C2; J. Maddox, Nature **313,** 263 (1985); Physics Today February 1985, p. 17; L. Milgrom, New Scientist 24 January, 1985, p. 34; I. Peterson, Science News **127,** 188 (1985); A. L. Mackay and P. Kramer, Nature **316,** 17 (1985).

[9-34]   J. Dalton, *A New System of Chemical Philosophy*, p. 128, plate III, Manchester, 1808.

[9-35]   D. Hodgkin, Kristallografiya, **26**(5), 1029 (1981).

[9-36]   K. W. Adolph, D. L. D. Caspar, C. J. Hollingshed, E. E. Lattman, W. C. Phillips, and W. T. Murakami, Science **203**, 1117 (1979).

[9-37]   R. E. Benfield and B. F. G. Johnson, J. Chem. Soc. Dalton 1743 (1980).

[9-38]   A. L. Mackay, Acta Crystallogr. **15**, 916 (1962).

[9-39]   B. C. Chakoumakos, R. J. Hill, and G. V. Gibbs, Am. Mineral. **66**, 1237 (1981).

[9-40]   I. El-Said and A. Parman, *Geometric Concepts in Islamic Art*, World of Islam Festival Publ. Co., London, 1976.

[9-41]   Chen Chi-Lu, *Material Culture of the Formosan Aborigines*, The Taiwan Museum, Taipei, 1968.

[9-42]   L. Pauling, *The Nature of the Chemical Bond*, Third Edition, Cornell University Press, Ithaca, NY, 1973.

[9-43]   A. I. Kitaigorodsky, in *Advances in Structure Research by Diffraction Methods*, R. Brill and R. Mason, eds., Vol. 3, p. 173, Pergamon Press, Oxford, etc., Friedr. Vieweg and Sohn, Braunschweig, 1970.

[9-44]   Kh. S. Mamedov, I. R. Amiraslanov, G. N. Nadzhafov, and A. A. Muzhaliev, *Decorations Remember* (in Azerbaidzhani), Azerneshr, Baku, 1981.

[9-45]   F. Wudl and E. T. Zellers, J. Am. Chem. Soc. **102**, 4283; 5430 (1980).

[9-46]   C. H. MacGillavry, *Symmetry Aspects of M. C. Escher's Periodic Drawings*, Bohn, Scheltema & Holkema, Utrecht, 1976.

[9-47]   T. V. Timofeeva, N. Yu. Chernikova, and P. M. Zorkii, Uspekhi Khimii **49**, 966 (1980).

[9-48]   V. G. Dashevskii, *Konformatsiya Organicheskikh Molekul*, Khimiya, Moskva, 1974.

[9-49]   J. Bernstein and A. T. Hagler, J. Am. Chem. Soc. **100**, 673 (1978).

[9-50]   C. P. Brock, Acta Crystallogr. **A33**, 193 (1977); C. P. Brock and J. A. Ibers, Acta Crystallogr. **A32**, 38 (1976).

[9-51]   P. M. Zorkii and V. A. Koptsik, in *Sovremennie Problemi Fizicheskoi Khimii*, Ya. I. Gerasimov and P. A. Akishin, eds., Izd. Moskov. Univ., Moskva, 1979.

[9-52]   P. M. Zorkii and A. E. Razumaeva, Zh. Strukt. Khim. **20**, 463 (1979).

[9-53]   V. K. Belsky and P. M. Zorkii, Acta Crystallogr. **A33**, 1004 (1977).

[9-54]   N. Yu. Chernikova, E. E. Lavut, and P. M. Zorkii, Koord. Khim. **5**, 1266 (1979).

[9-55]   E. E. Lavut, P. M. Zorkii, and N. Yu. Chernikova, Zh. Strukt. Khim. **22**, 89 (1981).

[9-56]   G.-P. Charbonneau and Y. Delugeard, Acta Crystallogr. **B33**, 1586 (1977); H. Cailleau, J. L. Bauduor, and C. M. E. Zeyen, Acta Crystallogr. **B35**, 426 (1979); H. Takeuchi, S. Suzuki, A. J. Dianoux, and G. Allen, Chem. Phys. **55**, 153 (1981).

[9-57]   O. Bastiansen, Kristallografiya **26**(5), 935 (1981). First investigation: O. Bastiansen, Acta Chem. Scand. **6**, 205 (1952). Latest reinvestigation: A. Almenningen, O. Bastiansen, L. Fernholt, B. N. Cyvin, S. J. Cyvin, and S. Samdal, J. Mol. Struct. **128**, 59 (1985).

[9-58]   T. P. Martin, Phys. Rev. **95**, 167 (1983).

[9-59]   R. A. J. Shelton, Trans. Faraday Soc. **57**, 2113 (1961).

[9-60]   S. H. Bauer, T. Ino, and R. F. Porter, J. Chem. Phys. **33**, 685 (1960).

[9-61]   P. A. Akishin and N. G. Rambidi, Zh. Strukt. Khim. **5**, 23 (1960).

[9-62]   D. R. Lide, P. Cahill, and L. P. Gold, J. Chem. Phys. **40**, 156 (1964).

[9-63]   M. Hargittai, Kém. Közlem. **50**, 371 (1978).

[9-64]   Gy. Schultz and I. Hargittai, Acta Chim. Hung. **75**, 381 (1973).

[9-65]   M. Hayashi, Y. Shiro, T. Oshima, and H. Murata, Bull. Chem. Soc. Japan **38**, 1734 (1975).

[9-66]   K. M. Melmed, D. Coucouvanis, and S. J. Lippard, Inorg. Chem. **12**, 232 (1973).

[9-67]   G. I. Mamaeva, I. Hargittai, and V. P. Spiridonov, Inorg. Chim. Acta **25**, L123 (1977).

[9-68]   I. Hargittai and M. Hargittai, in *Molecular Structure and Energetics*, Vol. 2, Chapter 1, J. F. Liebman and A. Greenberg, eds., VCH Publishers, Deerfield Beach, FL, 1986.

[9-69]   M. Colapietro, A. Domenicano, G. Portalone, Gy. Schultz, and I. Hargittai, J. Mol. Struct. **112**, 141 (1984).

[9-70]   M. Colapietro, A. Domenicano, G. Portalone, I. Torrini, I. Hargittai, and Gy. Schultz, J. Mol. Struct. **125**, 19 (1984).

[9-71]   C. Van Rij and D. Britton, Acta Crystallogr. **B33**, 1301 (1977); U. Drück and W. Littke, Acta Crystallogr. **B34**, 3095 (1978).

[9-72]   M. Hargittai and I. Hargittai, J. Mol. Struct. **39**, 79 (1977).

[9-73]   P. H. Clippard, J. C. Hanson, and R. C. Taylor, J. Cryst. Mol. Struct. **1**, 363 (1971).

[9-74]   L. Párkányi, L. Bihácsi, and P. Hencsei, Cryst. Struct. Comm. **7**, 435 (1978).

[9-75]   Q. Shen and R. L. Hilderbrandt, J. Mol. Struct. **64**, 257 (1980).

[9-76]   M. Hargittai and I. Hargittai, *The Molecular Geometries of Coordination Compounds in the Vapour Phase,* Elsevier-Akadémiai Kiadó, Amsterdam-Budapest, 1977.

[9-77]   J. Bernstein, T. E. Anderson, and C. J. Eckhardt, J. Am. Chem. Soc. **101**, 541 (1979); J. Bernstein, Acta Crystallogr. **B35**, 360 (1979); J. Bernstein, Y. M. Engel, and A. T. Hagler, J. Chem. Phys. **75**, 2346 (1981); I. Bar and J. Bernstein, Acta Crystallogr. **B37**, 569 (1981); I. Bar and J. Bernstein, J. Phys. Chem. **86**, 3223 (1982).

[9-78]   P. Scharfenberg and I. Hargittai, J. Mol. Struct. **112**, 65 (1984).

[9-79]   C. Bunn, *Crystals: Their Role in Nature and Science,* Academic Press, New York, London, 1964.

Subject Index[*]

---

[*] Numerals in bold refer to pages where definitions can be found; f refers to following page and ff refers to following pages.

Formula Index

Author Index*

Abbott, E. A.   48, *73*
Acton, N.   106f, *149*
Adams, W. J.   129, *151,* 145, *155*
Adolph, K. W.   407, *442*
Akishin, P. A.   425ff, *442,* 430, *443*
Allen, G.   429, *443*
Allen, L. C.   113, *150*
Almenningen, A.   429, *443*
Almlof, J.   278, *281*
Amiraslanov, I. R.   356f, *365,* 416, *442*
Ammeter, J. H.   278, *281*
Anderson, T. E.   436, *444*
Anet, F. A. L.   62ff, *74*
Apagyi, M.   18, *72*
Appleton, L. R. H.   23, *72*
Azaroff, L. V.   350, *366,* 367, 382f, 385, 402, 437, *440*

Bach, J. S.   54f, 329
Bader, R. F.   140, *154*
Baer Capitman, B.   85, *148*
Balogh, D. W.   98, 105, *148*
Bar, I.   436, *444*
Baranyi, A.   131, *152*
Barcsay, J.   14
Barlow, W.   391, 413
Barshad, Y. Z.   140, *154*
Bartell, L. S.   114ff, *150,* 120, 129, 140, *151,* 131, *152,* 140, *153f,* 144f, *155*
Bartók, B.   1, 17, 18
Bastiansen, O.   429, *443*
Bauduor, J. L.   429, *443*
Bauer, S. H.   430, *443*
Beach, A. L.   129, *152*
Beer, A.   9, *72*
Beer, P.   9, *72*
Beier, B. F.   131, *152*
Belov, N. V.   58f, 67, *74,* 397, *441*
Belsky, V. K.   428, *443*

Benbow, R. M.   57, *74*
Benfield, R. E.   146f, *155,* 407, *442*
Benson, S. W.   302, *325*
Bentley, R. B.   121, *151*
Bentley, W. A.   33, 35, 37, *73*
Bernal, J. D.   34, 60, *74,* 397, *441*
Bernardi, F.   285, *324*
Bernstein, J.   425, 429, 436, *442,* 436, *444*
Berry, R. S.   142, *154,* 143f, *155*
Bersuker, I. B.   275f, *281*
Berthelot, M.   1
Bertolucci, M. D.   201, 208, *218,* 219, 221ff, 241, 249f, *280*
Bethe, H.   268, *280*
Beyens, D.   273, *281*
Bickart, P.   3, *7*
Bihácsi, L.   435, *444*
Blake, W.   13
Blech, I.   401, *441*
Bock, H.   285, *324*
Boelema, E.   110, *150*
Boggs, J. E.   124f, 132ff, 140, *151,* 137, *153*
Bohm, H.   113, *150*
Bohn, M. D.   146, *155*
Bohn, R. K.   146, *155*
Boniuk, H.   354, *365*
Bracewell, R. N.   56, *74*
Branca, S. J.   106f, 143, *149*
Brener, L.   143, *155*
Brill, R.   416, 418f, 429, *442*
Brisse, F.   353ff, *365*
Britton, D.   433, *444*
Brock, C. P.   390f, *441,* 425, *442*
Brunvoll, J.   124, *151*
Buday, G.   14
Budden, F. J.   327f, 330, *365*
Buenker, R. J.   311, *325*

Buerger, M. J.   345, *365,* 376, *440*
Bunn, C.   437f, *444*
Bunnett, J. F.   283, *324*
Burfoot, J. C.   54, *73*
Burns, G.   157, *200*
Burns, W.   110, *150*

Cailleau, H.   429, *443*
Čapek, K.   367, *441*
Carlisle, C. H.   397, *441*
Carlson, K. D.   273, *281*
Carr, R. W. Jr.   302, *325*
Caspar, D. L. D.   407, *442*
Cahill, P.   430, *443*
Cavell, R. G.   130, *152*
Ceulemans, A.   273, *281*
Chakounakos, B. C.   410, *442*
Chan, J. W.   401, *441*
Charbonneau, G.-P.   429, *443*
Chen, Chi-Lu   411f, *442*
Chernikova, N. Yu.   424, *442,* 428, *443*
Chikaraishi, T.   125, 134f, *151*
Chistyakov, I. G.   374, *440*
Christe, K. O.   137f, *153*
Clark, T.   111, *150*
Clippard, P. H.   434, *444*
Colapietro, M.   96, *148,* 433, *443f*
Cole, T. J.   105, 107, *149*
Copernicus, N.   70, *74*
Cotton, F. A.   83, *148,* 112, *150,* 157, 168, 171, 193, 196, *200,* 201, 209f, *218,* 219f, 241, 268, 271, 273, *280,* 298, 318, *325*
Cotton, J. D.   322, *326*
Coucouvanis, D.   432, *443*
Coulson, C. A.   2, 4, *7,* 229, *280*
Coxeter, H. S. M.   3, *7,* 66, 68, 70, *74,* 98, *148*
Cross, P. C.   201, 208, *218*

---

*The numerals in Roman refer to pages on which the author is cited, while those in italics refer to pages where the references can be found. This Author Index was compiled by Balázs and Eszter Hargittai.

454